ALGEBRA
An Intermediate Course

ALGEBRA
An Intermediate Course

MERVIN L. KEEDY
PURDUE UNIVERSITY

MARVIN L. BITTINGER
INDIANA UNIVERSITY
PURDUE UNIVERSITY AT INDIANAPOLIS

ADDISON-WESLEY PUBLISHING COMPANY

Reading, Massachusetts ∎ Menlo Park, California
London ∎ Amsterdam ∎ Don Mills, Ontario ∎ Sydney

Sponsoring Editor:	*Patricia Mallion*
Production Editor:	*Robert Hartwell Fiske*
Text Designer:	*Vanessa Piñeiro*
Illustrator:	*VAP Group Limited*
Cover Designer:	*Richard Hannus*
Art Coordinator:	*Susanah H. Michener*
Production Manager:	*Herbert Nolan*

The text of this book was composed in Century Schoolbook by Syntax International.

Library of Congress Cataloging in Publication Data

Keedy, Mervin Laverne.
  Algebra, an intermediate course.

  Includes index.
  1. Algebra.   I. Bittinger, Marvin L.   II. Title.
QA154.2.K425   1983   512.9   82–18178
ISBN 0–201–14799–8

ISBN 0–201–14799–8
BCDEFGHIJ–DO–89876543

PREFACE

This book is intended for use by students who have finished a first course in algebra. It starts "at the beginning," but proceeds faster than an introductory book does, covering the usual topics of intermediate algebra. Students who have never studied algebra or whose background is otherwise weak are advised to use the companion volume, *Algebra: An Introductory Course*. A placement test is available to help place students in the proper book. (See the list of supplements that follows this preface.)

There is strong emphasis on algebraic skills, but the "whys" of algebra also receive significant attention. The style and format are designed to make the text easy to read and comprehend. Examples to illustrate each skill and concept have been chosen carefully to make sure they are appropriate and precisely to the point. Many applications of algebra have been included in the examples and exercises. The exercises in the exercise sets are very much like the examples. Exercises are paired and graded, with answers to the odd-numbered exercises in the back of the book. Answers to the other exercises are to be found in the *Answer Booklet*. (See the list of supplements.) Some (optional) exercises are designed to be done with a calculator and are marked ▦. There are also exercises of a somewhat challenging nature.

ACKNOWLEDGMENTS

We wish to thank Don Campbell of Cochise College for his suggestions regarding the alternate factoring method in Chapter 4. For their assistance, we also wish to thank Jack A. Golden, of the University of Montana; Barbara A. Poole, of North Seattle Community College; and Donald R. Johnson, of Scottsdale Community College. To the following participants in the Focus Group, we are likewise grateful: Laura Cameron, of the University of New Mexico; Marianna McClymonds, of Phoenix College; Jacky Peterson, of Mesa Community College; Dennis H. Sorge, of Purdue University; James W. Stark, Jr., of Lansing Community College; and John D. Whitesitt, of Southern Oregon State College.

M.L.K.
M.L.B.
January 1983

SUPPLEMENTS

This book is one of a series. The series is accompanied by the following kinds of supplements.

- *Computerized Testbank.* Test items can be randomly selected and a test constructed by a computer, using the testbank. The bank can be used with almost all computer hardware currently available. The bank contains approximately ten test items for each objective considered to be a terminal objective. The test question format is multiple choice, but the questions are worded in such a way that an ordinary constructed-response test can also be constructed, merely by leaving off the answer choices. The computerized testbank is available to users in three ways:

1. *On computer tape.* In this mode the bank can be used with most large computing systems. Users may have to specify the kind of tape their system accepts.

2. *On mini-floppy discs for TRS-80 microcomputers.*

3. *On mini-floppy discs for Apple II microcomputers.*

Final copies of tests can be produced from our programs where a letter quality printer is available (such as NEC spinwriter, DIABLO, or CUME) by using the printout and writing in only some special characters and certain graphs. If a printer is not available, the computer will still generate tests. Notation such as fractions and exponents, requiring spacing other than the usual one-line-at-a-time, will be printed in coded form. That notation is then to be added and the tests can then be typed and duplicated.

The authors owe very special thanks to John Whitesitt, of Southern Oregon State College, for his work in designing and producing the computerized testbank.

■ *Printed Testbank.*   The same test items appear in the printed form of the testbank as in the computerized testbank. The printed bank is usable by those who do not wish to use a computer or who do not have access to a computer.

■ *Videotape Cassettes.*   Color TV is used to help students over the rough spots. A lecturer (John Jobe, of Oklahoma State University) speaks to students and works out examples, with lucid explanations. These video cassettes can be used to supplement lectures or in individualized learning situations, such as learning laboratories.

■ *Audiotape Cassettes.*   In the tape recorder, the student has the voice of a friendly tutor, who helps him or her by talking through examples and problems, using the book itself as a visual device, to make a truly audiovisual presentation. These audiotapes are especially useful in individualized learning situations, such as learning laboratories, and are an excellent help for students who have missed a number of classes.

■ *Instructor's Manual with Tests.*   In this booklet there are five alternative forms of each chapter test and five alternative forms of the final examination. They are classroom ready, and all answers are provided.

■ *Answer Booklet.*   This booklet contains answers to *all* exercises. Answers to the odd-numbered exercises are in the text. The booklet is a great time-saver for the instructor, and it can also be made available to students, if desired, at low cost.

■ *Student's Guide to Exercises.*   Complete worked-out solutions are provided in this booklet. There are solutions to about 70 percent of the odd-numbered exercises in the exercise sets. That represents a change from the third edition, where most of the worked-out exercises were margin exercises, as found in the paperback books of the series. The change was made because it was found that help with exercises from the exercise sets themselves was preferred by students.

■ *Placement Test.* To help place a student properly in one of the algebra books of the series, a test has been constructed. The test is handy because it is short, although its brevity may make it appear less useful than it actually is. It was thoroughly evaluated statistically and is known to give a good screening with respect to placement into one of four categories into which the books of the series fall.

I ARITHMETIC, Fourth Edition

II INTRODUCTORY ALGEBRA, Fourth Edition
ALGEBRA: AN INTRODUCTORY COURSE*

III INTERMEDIATE ALGEBRA, Fourth Edition
ALGEBRA: AN INTERMEDIATE COURSE*

IV COLLEGE ALGEBRA, Third Edition
FUNDAMENTAL COLLEGE ALGEBRA, Second Edition*
ALGEBRA AND TRIGONOMETRY, Third Edition
FUNDAMENTAL ALGEBRA AND TRIGONOMETRY,
Second Edition*
TRIGONOMETRY: TRIANGLES AND FUNCTIONS,
Third Edition

*These texts are hardcover versions of the paperbound texts.

More precise placement within a specific book can be made by using an alternative form of the final examination as a pretest.

CONTENTS

6 EXPONENTS, POWERS, AND ROOTS

7 QUADRATIC EQUATIONS AND FUNCTIONS

ALGEBRA
An Intermediate Course

1

REAL NUMBERS AND THEIR PROPERTIES

1.1 NUMBERS AND ALGEBRA

Numbers are important in algebra as well as arithmetic. The most important set of numbers in ordinary algebra is the set of real numbers.

**SOME SUBSETS OF
THE REAL NUMBERS**

There is just one real number for each point of a line.

The *positive* numbers are shown to the right of 0. The *negative* numbers are shown to the left. Zero is neither positive nor negative. There are several kinds of real numbers.

Natural numbers: **Those used for counting: 1, 2, 3, and so on.**

When we include 0 with this set we have the system of whole numbers.

Whole numbers: **The set of natural numbers, with 0 included: 0, 1, 2, 3, and so on.**

We extend the set of whole numbers to the integers by including some negative numbers.

Integers: **0, 1, −1, 2, −2, 3, −3, and so on.**

Rational numbers: **This set consists of the integers and all quotients of integers (excluding division by zero).**

Thus the rational numbers can be named by fractional symbols with integers for numerator and denominator. The following are rational numbers:

$$\frac{4}{5}, \quad \frac{-4}{7}, \quad \frac{9}{1}, \quad 6, \quad -4, \quad 0,$$

$$\frac{78}{-5}, \quad -\frac{2}{3}, \quad 1.5, \quad -0.12 \quad \left(-\frac{2}{3} \text{ can be named } \frac{-2}{3} \text{ or } \frac{2}{-3}\right),$$

$$\left(1.5 \text{ can be named } \frac{3}{2}\right).$$

Real numbers that are not rational are called *irrational*. They will be considered in a later chapter.

> In arithmetic you usually write $2\frac{1}{2}$, rather than $\frac{5}{2}$. In algebra you will find that the so-called *improper* symbols such as $\frac{5}{2}$ are more useful.

DECIMAL AND FRACTIONAL NOTATION

Rational numbers can also be named using *decimal* notation.

Example 1 Find decimal notation for $\frac{5}{8}$.

$\frac{5}{8}$ means $5 \div 8$, so we divide.

```
     0.625
8) 5.000
   4 8
   ‾‾‾
     20
     16
     ‾‾
     40
     40
```

Decimal notation for $\frac{5}{8}$ is thus 0.625. We also call 0.625 a *decimal*.

Decimal notation for some numbers repeats.

Example 2   Find decimal notation for $\frac{6}{11}$.

We divide:

$$
\begin{array}{r}
0.5454\ldots \\
11\overline{)6.0000} \\
\underline{5\,5} \\
50 \\
\underline{44} \\
60 \\
\underline{55} \\
50 \\
\underline{44} \\
6
\end{array}
$$

Repeating decimal notation can be abbreviated by writing a bar over the repeating part, in this case, $0.\overline{54}$.

Example 3   Find fractional notation for 3.6.

$3.6 = \dfrac{36}{10}$ The last decimal place is *tenths* so the denominator is 10.

Example 4   Find fractional notation for 0.75.

$0.75 = \dfrac{75}{100}$ The denominator is 100 because the last decimal place is *hundredths*.

PERCENT NOTATION

We can also name rational numbers using percent notation. Let us recall what percent means.

> **The symbol % means "$\times 0.01$" or "$\times \dfrac{1}{100}$."**

We use this definition to change to or from percent notation.

Example 5   Find decimal notation and fractional notation for 67.9%.

$$
\begin{aligned}
67.9\% &= 67.9 \times 0.01 &&\text{Replacing \% by } \times 0.01 \\
&= 0.679 &&\text{Multiplying. This is decimal notation.} \\
&= \frac{679}{1000} &&\text{Changing to fractional notation}
\end{aligned}
$$

Example 6   Find decimal notation and percent notation for $\frac{5}{8}$.

$$\begin{array}{r} .625 \\ 8\overline{)5.000} \end{array}$$

Decimal notation is 0.625.

$$\frac{5}{8} = 0.625 \times 100 \times 0.01 \qquad \text{Multiplying by } 100 \times 0.01, \text{ or } 1$$

$$= (0.625 \times 100) \times 0.01$$

$$= 62.5 \times 0.01$$

$$= 62.5\% \qquad \text{Replacing } \times 0.01 \text{ by } \%$$

Percents can be less than 1% or greater than 100%.

Examples   Find percent notation.

7. $0.005 = 0.005 \times 100 \times 0.01$

$$= (0.005 \times 100) \times 0.01$$

$$= 0.5 \times 0.01$$

$$= 0.5\%$$

8. $1.35 = 1.35 \times 100 \times 0.01$

$$= 135 \times 0.01$$

$$= 135\%$$

Examples   Find decimal notation.

9. $120\% = 120 \times 0.01$

$$= 1.2$$

10. $0.25\% = 0.25 \times 0.01$

$$= 0.0025$$

> There is a table of decimal, fractional, and percent equivalents inside the back cover. If you do not already know those facts, you should memorize them.

ORDER

The number line shows the real numbers lined up, in order. We say that one number is *greater* than another if it is to the right of the other on the number line. A number is *less* than another if it occurs to the left of the other. The symbol < means "is less than." The symbol > means "is greater than."

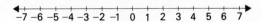

Examples

11. $-5 < 2$

12. $-7 < 0$

13. $2.314 < 2.315$

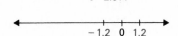

14. $1.2 > -1.2$

15. $-6 > -8$

16. $-\dfrac{4}{5} < -\dfrac{1}{3}$

ABSOLUTE VALUE

The absolute value of a number is its distance from 0 on a number line. The absolute value of a number a can be named $|a|$. Note on the number line above that 3 and -3 are both a distance of three units from 0.

Examples Simplify.

17. $|-7|$ The distance of -7 from 0 is 7, so $|-7| = 7$.

18. $|5|$ The distance of 5 from 0 is 5, so $|5| = 5$.

19. $|0|$ The distance of 0 from 0 is 0, so $|0| = 0$.

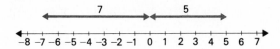

Absolute value is simple, so try not to make it difficult. If a number is positive or zero, then when you take its absolute value, just leave it alone. If a number is negative, to find its absolute value, just make it positive.

EXERCISE SET 1.1

Given the numbers $-5, 0, 2, -\dfrac{1}{2}, -3, \dfrac{7}{8}, 14, -\dfrac{8}{3}, 2.43, 7\dfrac{1}{2}$:

1. Name the natural numbers.

2. Name the whole numbers.

3. Name the rational numbers.

4. Name the integers.

Find decimal notation.

5. $\dfrac{3}{8}$

6. $\dfrac{1}{8}$

7. $\dfrac{5}{3}$

8. $\dfrac{7}{6}$

Find fractional notation.

9. 2.7

10. 13.91

11. 0.145

12. 0.0213

Find percent notation.

13. $\dfrac{4}{5}$

14. $\dfrac{7}{25}$

15. $\dfrac{3}{8}$

16. $\dfrac{1}{8}$

17. $\dfrac{12}{5}$

18. $\dfrac{14}{5}$

19. $\dfrac{3}{80}$

20. $\dfrac{3}{160}$

21. 0.412

22. 0.325

23. 0.001

24. 0.0025

25. 1.25

26. 2.34

Find decimal and fractional notation.

27. 56.7%

28. 47.5%

29. 6.85%

30. 8.23%

31. 120%

32. 180%

33. 0.5%

34. 0.1%

Insert $>$ or $<$ to make true sentences.

35. 55   40

36. 31   94

37. 2.3   3.2

38. 4.7   47

39. -10 -4

40. -6 -20

41. -2.3 -3.2

42. -16 -1.6

Simplify.

43. $|16|$

44. $\left|\dfrac{7}{8}\right|$

45. $|-32|$

46. $\left|-\dfrac{2}{3}\right|$

47. $|0|$

48. $|3 - 3|$

☆
Find decimal notation.

49. $\dfrac{3}{14}$ **50.** $\dfrac{17}{128}$ **51.** $\dfrac{1403}{3300}$

52. $\dfrac{49}{200}$ **53.** $\dfrac{2{,}635{,}269}{12{,}345{,}679}$ **54.** $\dfrac{41}{333{,}000}$

55. All rational numbers can be named using decimal notation. Without dividing, is it possible to know if the decimal notation for the rational number $\frac{p}{q}$ will be repeating? If yes, explain.

56. Given that the decimal notation for a rational number $\frac{p}{q}$ is repeating, what is the greatest number of digits that can be in the repeating part?

57. Place the following rational numbers in increasing order.

$$1.1\,\%,\ \frac{1}{11},\ \frac{2}{7},\ 0.3\,\%,\ 0.11,\ \frac{1}{8}\,\%,\ 0.009,\ \frac{99}{1000},\ 0.286,\ \frac{1}{8},\ \frac{9}{100},\ 1\,\%$$

58. a) What is the largest integer that is less than $-\frac{29}{5}$?
     b) What is the smallest integer that is greater than $-\frac{29}{5}$?
     c) What integer is nearest to $-\frac{29}{5}$?

1.2 ADDITION AND SUBTRACTION

ADDITION

To explain addition of real numbers we can use a number line. To perform the addition $a + b$, we start at a and then move according to b. If b is positive, we move to the right. If b is negative, we move to the left.

Examples

1. $6 + (-4) = 2$

Start at 6. Move 4 units to the left. The answer is 2.

2. $-3 + 2 = -1$

Start at -3. Move 2 units to the right. We get -1.

3. A halfback made a gain of 7 yards, followed by a gain of -5 yards. What was the net gain?

$7 + (-5) = 2$, so the net gain was 2 yards.

What happens when we add two negative numbers?

Examples   Add.

4. $-5 + (-9) = -14$

5. $-10 + (-21) = -31$

6. $-9.4 + (-3.2) = -12.6$

7. $-\dfrac{9}{6} + \left(-\dfrac{11}{6}\right) = -\dfrac{20}{6},$ or $-\dfrac{10}{3}$

8. $-\dfrac{1}{2} + \left(-\dfrac{1}{4}\right) = -\dfrac{1}{2} \cdot \dfrac{2}{2} + \left(-\dfrac{1}{4}\right) = -\dfrac{2}{4} + \left(-\dfrac{1}{4}\right) = -\dfrac{3}{4}$

| Here we have multiplied by 1 to get a common denominator. |

9. $-\dfrac{1}{5} + \left(-\dfrac{1}{15}\right) = -\dfrac{1}{5} \cdot \dfrac{3}{3} + \left(-\dfrac{1}{15}\right) = -\dfrac{3}{15} + \left(-\dfrac{1}{15}\right) = -\dfrac{4}{15}$

The sum of two negative numbers is negative. Add their absolute values and then make the answer negative.

What happens when we add a positive number and a negative number? There are several different cases.

Examples   Add.

10. $-6 + 6 = 0$

11. $-10 + 10 = 0$

12. $7 + (-7) = 0$

13. $4.2 + (-4.2) = 0$

> **If a positive number and a negative number have the same absolute value, their sum is 0.**

Now suppose we add a positive number and a negative number with different absolute values. The answer may be positive or negative.

Examples   Add.

14. $-5 + 7 = 2$

15. $-5 + 1 = -4$

16. $6.2 + (-9.8) = -3.6$

17. $-5.89 + 2.11 = -3.78$

18. $\dfrac{1}{4} + \left(-\dfrac{1}{2}\right) = \dfrac{1}{4} + \left(-\dfrac{2}{4}\right) = -\dfrac{1}{4}$

19. $-\dfrac{3}{8} + \dfrac{1}{2} = -\dfrac{3}{8} + \dfrac{4}{8} = \dfrac{1}{8}$

> **To add a positive and a negative number with different absolute values, find the difference of the absolute values. If the negative number has the greater absolute value, the answer is negative. If the positive number has the greater absolute value, the answer is positive.**

ADDITIVE INVERSES

The additive inverse of a number is the number opposite to it on the number line. The additive inverse (or simply *inverse*, or *opposite*) of a number x is named $-x$.

Examples   Find the additive inverse.

20. 5

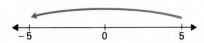

We reflect to the opposite side of 0. The additive inverse of 5 is -5. (-5 is read as "the inverse of 5." It is also read *negative 5*.)

21. -2

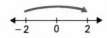

We reflect to the opposite side of -2. The additive inverse of -2 is 2. In symbols, $-(-2) = 2$.

> This symbol should be read as "the inverse of negative 2," or "the inverse of the inverse of 2." You should not say "negative negative 2" or "minus minus 2."

Examples

22. Find $-a$ when a is 8.

$-8 = -8$

23. Find $-a$ when a is -2.

$-(-2) = 2$

24. Find $-a$ when a is 0.

$-0 = 0$

25. Find $-(-a)$ when a is -3.

$-(-(-3)) = -3$

> **Every real number has an additive inverse.**
> 1. **The additive inverse of a number x can be named $-x$.**
> 2. **The additive inverse of a positive number is negative. The inverse of a negative number is positive. The inverse of 0 is 0.**
> 3. **A number and its additive inverse have the same absolute value. A number and its inverse add to 0.**
> 4. **Replacing a number by its additive inverse is sometimes called** *changing the sign.*

SUBTRACTION

Subtraction is defined as follows.

> **The difference $a - b$ is the number that when added to b gives a.**

> To see that the definition is right, think of $5 - 3$. When you subtract, you get 2. To check, you add 2 to 3, to see if 2 is the number which when added to 3 gives 5.

Using the definition we can explain subtraction on a number line.

Example 26   Subtract: $-3 - 5$.

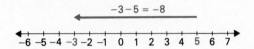

To subtract 5 from -3 we find the number which when added to 5 gives -3. We start at 5 and see that a move of 8 units to the left gives us -3. The answer is -8.

Compare the following.

a) $3 - 8 \quad\;\; = -5$
$\quad\; 3 + (-8) = -5$

b) $-2 - (-9) = 7$
$\quad\; -2 + 9 \quad\;\; = 7$

The pattern above shows us how to subtract without a number line. We first change the subtraction to an addition.

> **For any real numbers a and b, $a - b = a + (-b)$. (We can subtract by adding the additive inverse of the subtrahend.)**

Examples   Subtract.

27. $5 - (-3) = 5 + 3 = 8$ Changing the sign of -3 and changing the subtraction to addition

28. $-7 - 3 = -7 + (-3) = -10$

29. $-19.4 - 5.6 = -19.4 + (-5.6) = -25$

30. $-\dfrac{4}{3} - \left(-\dfrac{2}{5}\right) = -\dfrac{4}{3} + \dfrac{2}{5} = -\dfrac{20}{15} + \dfrac{6}{15} = -\dfrac{14}{15}$

THE COMMUTATIVE AND ASSOCIATIVE LAWS

An important property of real numbers is that the order in which they are added makes no difference. For example, $4 + 7 = 7 + 4$.

> For any numbers a and b, $a + b = b + a$. This is the *Commutative Law of Addition*. **(It says that numbers can be added in any order.)**

Numbers can also be grouped in various ways for adding. Before we state the law let's consider the use of parentheses to show grouping. Calculations inside parentheses are to be done first.

Example 31

$$(3 \times 7) + 10 \quad \text{means} \quad 21 + 10, \text{ or } 31, \quad \text{but}$$

$$3 \times (7 + 10) \quad \text{means} \quad 3 \times 17, \text{ or } 51.$$

The way we group, or place parentheses, can affect an answer, unless there are only additions to do. In that case we can group, or place parentheses, as we please.

> For any real numbers a, b, and c, $a + (b + c) = (a + b) + c$. This is the *Associative Law of Addition*.

Examples Tell which property of real numbers is illustrated by each sentence (look for regrouping or reordering).

32. $4.5 + \left(9 + \dfrac{1}{2}\right) = (4.5 + 9) + \dfrac{1}{2}$ Associative law of addition (grouping has been changed)

33. $7.8 + \dfrac{1}{4} = \dfrac{1}{4} + 7.8$ Commutative law of addition (order has been changed)

EXERCISE SET 1.2

Add.

1. $-12 + (-16)$ **2.** $-11 + (-18)$ **3.** $-8 + (-8)$ **4.** $-6 + (-6)$

5. $8 + (-3)$ **6.** $9 + (-4)$ **7.** $11 + (-7)$ **8.** $12 + (-8)$

9. $-16 + 9$ **10.** $-23 + 8$ **11.** $-24 + 0$ **12.** $-34 + 0$

13. $-8.4 + 9.6$ **14.** $-6.3 + 8.2$

15. $-2.62 + (-6.24)$ **16.** $-5.83 + (-7.43)$

17. $-\dfrac{2}{7} + \dfrac{3}{7}$ **18.** $-\dfrac{5}{6} + \dfrac{1}{6}$

19. $-\dfrac{11}{12} + \left(-\dfrac{5}{12}\right)$

20. $-\dfrac{3}{8} + \left(-\dfrac{7}{8}\right)$

21. $\dfrac{2}{5} + \left(-\dfrac{3}{10}\right)$

22. $-\dfrac{3}{4} + \dfrac{1}{8}$

Find $-a$ when a is

23. 7

24. -8

25. -2.8

26. 0

Change the sign.

27. 10

28. -8

29. 0

30. $-2x$

Subtract.

31. $5 - 7$

32. $9 - 12$

33. $-5 - 7$

34. $-9 - 12$

35. $-6 - (-11)$

36. $-7 - (-12)$

37. $10 - (-5)$

38. $28 - (-16)$

39. $15.8 - 27.4$

40. $17.2 - 34.9$

41. $-18.01 - 11.24$

42. $-19.04 - 15.76$

43. $-\dfrac{21}{4} - \left(-\dfrac{7}{4}\right)$

44. $-\dfrac{16}{5} - \left(-\dfrac{3}{5}\right)$

45. $-\dfrac{1}{2} - \left(-\dfrac{1}{12}\right)$

46. $-\dfrac{3}{4} - \left(-\dfrac{3}{2}\right)$

Tell which property of real numbers is illustrated by each sentence.

47. $1.23 + 3.21 = 3.21 + 1.23$

48. $8 + (-2.3) = -2.3 + 8$

49. $9 + (2 + 5) = (9 + 2) + 5$

50. $(2.6 + 5.2) + 3.4 = 2.6 + (5.2 + 3.4)$

☆ _____

Any exercises marked are to be worked with a calculator.

51. What laws or properties of real numbers are illustrated by this sentence?

$(3 + 2) + (-5) = (-5 + 3) + 2$

52. Add.

$$\begin{array}{r} 475.362 \\ -123.814 \\ 30.105 \\ -649.914 \\ \hline \end{array}$$

Compute.

53. $(8 - 10) - (-6 + 2)$

54. $|8 - 10| - |-6 + 2|$

55. $-10 + 18 - 6 - 14 + 31$

56. $-10 + 18 - (6 - 14) + 31$

57. $-10 + 18 - 6 - (14 + 31)$

58. $|(8 - |-9|)| + |-6 + 12|$

59. $|-6 - (-3)| - |3 - 6|$

60. Find a number that is 27 less than -6.

61. What number can be added to 11.7 to obtain $-7\frac{3}{4}$?

1.3 MULTIPLICATION AND DIVISION

MULTIPLICATION OF REAL NUMBERS

We know how to multiply positive numbers. What happens when we multiply a positive number by a negative number?

> **When we multiply a positive number and a negative number, we multiply their absolute values. Then we make the answer negative.**

Examples Multiply.

1. $-3 \cdot 5 = -15$

2. $6 \cdot (-7) = -42$

3. $-1.2 \times 4.5 = -5.4$

4. $3 \cdot \left(-\frac{1}{2}\right) = \frac{3}{1} \cdot \left(-\frac{1}{2}\right) = -\frac{3}{2}$

What happens when we multiply two negative numbers?

> **To multiply two negative numbers, we multiply their absolute values. The answer is positive.**

Examples Multiply.

5. $-3 \cdot (-5) = 15$

6. $-5.2(-10) = 52$

7. $-8.8 \times (-3.5) = 30.8$

8. $\left(-\frac{3}{4}\right) \cdot \left(-\frac{5}{2}\right) = \frac{15}{8}$

The commutative and associative laws hold for all real numbers.

> **For any real numbers a, b, and c,**
>
> $\qquad a \cdot b = b \cdot a$ $\qquad$ **(The Commutative Law of Multiplication)**
>
> $\qquad a \cdot (b \cdot c) = (a \cdot b) \cdot c$ $\quad$ **(The Associative Law of Multiplication)**

The number 1 is special. It is called the *multiplicative identity*.

> **For any real number a, $a \cdot 1 = a$. (The number 1 is the *multiplicative identity*.)**

The number 1 can be named in many ways with fractional notation. Here are some examples.

$$\frac{3}{3}, \quad \frac{-5}{-5}, \quad \frac{1}{1}, \quad \frac{x}{x}, \quad \frac{3/4}{3/4}$$

(Any number, except zero, divided by itself is 1.)

Examples   Multiply.

9. $\dfrac{2}{3} \cdot \dfrac{5}{5} = \dfrac{10}{15}$ $\qquad$ Since we multiplied by 1, we know that $\dfrac{2}{3} = \dfrac{10}{15}$.

10. $\dfrac{-4}{5} \cdot \dfrac{-1}{-1} = \dfrac{4}{-5}$ $\qquad$ Since we multiplied by 1, we know that

$$\frac{-4}{5} = \frac{4}{-5}.$$

DIVISION

The definition of division parallels the one for subtraction.

> **The quotient p/q is defined to be the number (if it exists) that when multiplied by q gives p.**

Using this definition, we can see how to handle signs when dividing.

Examples

11. $\dfrac{10}{-2} = -5$, because $-5 \cdot (\,-2\,) = 10$

12. $\dfrac{-32}{4} = -8$, because $-8 \cdot \boxed{4} = -32$

13. $\dfrac{-25}{-5} = 5$, because $5 \cdot (\boxed{-5}) = -25$

> **In division, if both numbers are negative, the answer is positive. If one is positive and one is negative, the answer is negative.**

Examples Divide.

14. $\dfrac{40}{-4} = -10$

15. $\dfrac{-10}{5} = -2$

16. $\dfrac{-10}{-40} = \dfrac{1}{4}$, or 0.25

17. $\dfrac{-10}{-3} = \dfrac{10}{3}$, or $3.33\overline{3}$

DIVISION BY ZERO

We do not divide by zero. Let's see why. By the definition of division, $n/0$ would be some number that when multiplied by 0 gives n. But when any number is multiplied by 0 the result is 0. The only possibility for n would be 0.

Consider $0/0$. We might say that it is 5 because $5 \cdot 0 = 0$. We might also say that it is -8 because $-8 \cdot 0 = 0$. In fact, $0/0$ could be any number at all. So we agree not to divide by 0.

Examples Which of the following divisions are possible?

18. $\dfrac{7}{0}$ Not possible; division by 0.

19. $\dfrac{0}{7}$ Possible; the quotient is 0 because $0 \cdot 7 = 0$.

20. $\dfrac{4}{x - x}$ Not possible; $x - x = 0$ for any x.

DIVISION AND RECIPROCALS

Two numbers whose product is 1 are called *reciprocals* (or *multiplicative inverses*) of each other.

Every nonzero real number a has a reciprocal $1/a$. The reciprocal of a positive number is positive. The reciprocal of a negative number is negative.

Examples   Find the reciprocal of each number.

21. $\dfrac{4}{5}$ The reciprocal is $\dfrac{5}{4}$, because $\dfrac{4}{5} \cdot \dfrac{5}{4} = 1$.

22. 8   The reciprocal is $\dfrac{1}{8}$, because $8 \cdot \dfrac{1}{8} = 1$.

23. $-\dfrac{2}{3}$ The reciprocal is $-\dfrac{3}{2}$, because $-\dfrac{2}{3} \cdot \left(-\dfrac{3}{2}\right) = 1$.

24. 0.25   The reciprocal is $\dfrac{1}{0.25}$ or 4, because $0.25 \times 4 = 1$.

(To find $\dfrac{1}{0.25}$, divide 1 by 0.25.)

> Remember that the reciprocal of a negative number is negative. Do *not* change sign when taking the reciprocal of a number.

The number 1 and reciprocals can be used to explain division with fractional notation. We write fractional notation to express the division, and then multiply by 1.

Example 25   Divide: $\dfrac{2}{5} \div \dfrac{7}{3}$.

We first write fractional notation for the division: $\dfrac{\frac{2}{5}}{\frac{7}{3}}$.

We then find the reciprocal of the denominator, $\frac{3}{7}$. Next, we multiply by 1, as follows:

$$\frac{\dfrac{2}{5}}{\dfrac{7}{3}} = \frac{\dfrac{2}{5}}{\dfrac{7}{3}} \times \frac{\dfrac{3}{7}}{\dfrac{3}{7}} \qquad \text{Multiplying by 1}$$

$$= \frac{\dfrac{2}{5} \times \dfrac{3}{7}}{\dfrac{7}{3} \times \dfrac{3}{7}} = \frac{\dfrac{6}{35}}{1} = \frac{6}{35}. \qquad \begin{array}{l}\text{Multiplying numerators and} \\ \text{denominators and simplifying}\end{array}$$

In the example above we wound up multiplying the dividend by the reciprocal of the divisor. This always happens.

> **To do the division $a/b \div c/d$ we can do the multiplication $a/b \times d/c$ (i.e., we can multiply by the reciprocal of the divisor).**

Examples Divide by multiplying by the reciprocal of the divisor.

26. $\dfrac{1}{4} \div \dfrac{3}{5} = \dfrac{1}{4} \times \dfrac{5}{3} = \dfrac{5}{12}$

27. $\dfrac{2}{3} \div \left(-\dfrac{4}{9} \right) = \dfrac{2}{3} \times \left(-\dfrac{9}{4} \right) = -\dfrac{18}{12}, \quad \text{or} \quad -\dfrac{3}{2}$

To explain the simplification in Example 27, we can use the idea of multiplying by 1.

Example 28 Simplify: $-\dfrac{18}{12}$.

We first factor numerator and denominator:

$$-\frac{18}{12} = -\frac{3 \cdot 6}{2 \cdot 6}.$$

Next, we factor the fraction, with one factor equal to 1:

$$-\frac{3 \cdot 6}{2 \cdot 6} = -\frac{3}{2} \cdot \frac{6}{6}.$$

Then we leave off the factor of 1, to get $-\frac{3}{2}$.

EXERCISE SET 1.3

Multiply.

1. $3(-7)$ **2.** $5(-8)$ **3.** $-2 \cdot 4$ **4.** $-5 \cdot 9$

5. $-8(-2)$ **6.** $-7(-3)$ **7.** $-9 \cdot 14$ **8.** $-8 \cdot 17$

9. $-6(-5.7)$ **10.** $-7(-6.1)$ **11.** $-4.2(-6.3)$ **12.** $-7.4(-9.6)$

13. $-3\left(-\dfrac{2}{3}\right)$ **14.** $-5\left(-\dfrac{3}{5}\right)$ **15.** $-3(-4)(5)$ **16.** $-6(-8)(9)$

17. $-\dfrac{3}{5} \cdot \dfrac{4}{7}$ **18.** $-\dfrac{5}{4} \cdot \dfrac{11}{3}$ **19.** $-\dfrac{9}{11} \cdot \left(-\dfrac{11}{9}\right)$ **20.** $-\dfrac{13}{7} \cdot \left(-\dfrac{5}{2}\right)$

21. $-\dfrac{2}{3} \cdot \left(-\dfrac{2}{3}\right) \cdot \left(-\dfrac{2}{3}\right)$ **22.** $-\dfrac{4}{5} \cdot \left(-\dfrac{4}{5}\right) \cdot \left(-\dfrac{4}{5}\right)$

Divide.

23. $\dfrac{-8}{4}$ **24.** $\dfrac{-16}{2}$ **25.** $\dfrac{56}{-8}$ **26.** $\dfrac{63}{-7}$

27. $\dfrac{-77}{-11}$ **28.** $\dfrac{-48}{-6}$ **29.** $\dfrac{-5.4}{-18}$ **30.** $\dfrac{-8.4}{-12}$

Which of these divisions are possible?

31. $\dfrac{9}{0}$ **32.** $\dfrac{8}{0}$ **33.** $\dfrac{0}{16}$

34. $\dfrac{0}{28}$ **35.** $\dfrac{9}{y-y}$ **36.** $\dfrac{2x-2x}{2x-2x}$

Find the reciprocal of each.

37. $\dfrac{3}{4}$ **38.** $\dfrac{9}{10}$ **39.** $-\dfrac{7}{8}$ **40.** $-\dfrac{5}{6}$

41. 15 **42.** -75 **43.** 4.4 **44.** -5.5

Divide.

45. $\dfrac{2}{7} \div \left(-\dfrac{11}{3}\right)$ **46.** $\dfrac{3}{5} \div \left(-\dfrac{6}{7}\right)$ **47.** $-\dfrac{10}{3} \div \left(-\dfrac{2}{15}\right)$ **48.** $-\dfrac{12}{5} \div \left(-\dfrac{3}{10}\right)$

49. $18.6 \div (-3.1)$ **50.** $39.9 \div (-13.3)$ **51.** $(-75.5) \div (-15.1)$

52. $(-12.1) \div (-0.11)$

☆
──

Compute.

53. ▦ $-80{,}397 \times (-583)$ **54.** ▦ -0.56004×0.003

55. The reciprocal of an electric resistance is called *conductance*. When two resistors are connected in parallel, the conductance is the sum of the conductances, $1/r_1 + 1/r_2$. Find the conductance of two resistors of 10 ohms and 5 ohms when connected in parallel.

Compute.

56. $\dfrac{(-7 - 3) \div (-5)}{(4 - 8)(-2 + 5)}$ **57.** $|-0.09 \div 0.3| \cdot |0.3 - 0.9|$ **58.** $\left[\dfrac{1}{2} \div \left(-\dfrac{2}{3}\right)\right] \cdot \left(\dfrac{1}{4} - 4\right)$

Place parentheses in each expression so that it has the given value.

59. $3 \cdot 3 + 10 - 8; \ 15$ **60.** $7 \cdot 6 - 1 + 4 \cdot 3 - 12; \ -1$ **61.** $20 \div 2 \cdot 5 - 2 + 14; \ 44$

──

1.4 THE DISTRIBUTIVE LAWS AND THEIR USE

SOME AGREEMENTS

When a letter can represent various numbers, we call it a *variable*. When we write two or more letters together we agree that this means multiplication. We make the same agreement when variables and numerals occur together.

Examples

1. Prt means $P \cdot r \cdot t$.

2. $5xy$ means $5 \cdot x \cdot y$.

We also agree that in an expression like $(4 \cdot 10) + (9 \cdot 2)$, we can omit the parentheses. Thus $4 \cdot 10 + 9 \cdot 2$ means $(4 \cdot 10) + (9 \cdot 2)$. In other words, we do the multiplications first. We make a similar agreement for subtraction. That is, $4 \cdot 10 - 9 \cdot 2$ means $(4 \cdot 10) - (9 \cdot 2)$. The multiplications are to be done first.

Example 3   Evaluate the following expression when $x = 4$ and $y = 5$. Then simplify.

$$\begin{aligned} 8x + 3y &= 8 \cdot 4 + 3 \cdot 5 \quad &&\text{Substituting} \\ &= 32 + 15 \quad &&\text{Multiplying} \\ &= 47 \quad &&\text{Adding} \end{aligned}$$

THE DISTRIBUTIVE LAWS

Compare.

a) $5\,(4 + 8) = 5 \cdot 12 = 60$

$5 \cdot 4 + 5 \cdot 8 = 20 + 40 = 60$

b) $-4\,(9 + 5) = -4 \cdot 14 = -56$

$-4 \cdot 9 + (-4) \cdot 5 = -36 + (-20) = -56$

We can either add and then multiply or multiply and then add.

> *The Distributive Law of Multiplication over Addition:*
> **For any numbers a, b, and c, $a(b + c) = ab + ac$.**

> Note that we cannot omit the parentheses in the expression $a(b + c)$.
> If we did, we would have $ab + c$, which by our agreement means $(ab) + c$.

There is another distributive law.

> *The Distributive Law of Multiplication over Subtraction:*
> **For any numbers a, b, and c, $a(b - c) = ab - ac$.**

We can subtract and then multiply, or we can multiply and then subtract.

MULTIPLYING

The distributive laws are the basis of multiplying in algebra, as well as in arithmetic. In the following examples, note that we multiply each number or letter inside the parentheses by the factor outside.

Examples Multiply.

4. $4\,(x - 2) = 4 \cdot x - 4 \cdot 2 = 4x - 8$

5. $b\,(s - t + f) = \boxed{b}\;s - \boxed{b}\;t + \boxed{b}\,f$

6. $-3(y + 4) = \boxed{-3}\,\cdot y + (\,\boxed{-3}\,)\cdot 4 = -3y - 12$

7. $-2x(y - 1) = \boxed{-2x}\,\cdot y - (\,\boxed{-2x}\,)\cdot 1 = -2xy + 2x$

FACTORING

The reverse of multiplying is called *factoring*. To *factor* an expression means to rewrite it as a product.

Examples Factor.

8. $\boxed{8}\;x + \boxed{8}\;y = \boxed{8}\,(x + y)$

9. $\boxed{c}\;x - \boxed{c}\;y = \boxed{c}\,(x - y)$

The parts of an expression such as $3x + 4y - 2z$ are called the *terms* of the expression. In this case the terms are $3x$, $4y$, and $-2z$.

Whenever the terms of an expression have a factor in common, we can "remove" that factor using the distributive laws. We proceed as in Examples 8 and 9, but we may have to factor some of the terms first in order to display the common factor.

Example 10 Factor: $4x + 8$.

$$4x + 8 = 4x + \boxed{4 \cdot 2} \qquad \text{Factoring 8 to display the common factor}$$

$$= \boxed{4}\,(x + 2) \qquad \text{Using the distributive law}$$

Generally, we try to factor out the largest factor common to all the terms. In the following example, we might factor out 3, but there is a larger factor common to the terms, 9. So we factor out the 9.

Example 11 Factor: $9x + 27y$.

$$9x + 27y = 9x + 9 \cdot (3y) = 9(x + 3y)$$

We often have to supply a factor of 1 when factoring out a common factor, as in the next example, which is from a formula about simple interest.

Example 12   Factor: $P + Prt$.

$$P + Prt = P \cdot \boxed{1} + Prt \qquad \text{Writing } P \text{ as a product of } P \text{ and } 1$$

$$= P(\boxed{1} + rt) \qquad \text{Using the distributive law}$$

It is a common error to omit this 1. If you do that, when you factor out P, you might leave out an entire term.

COLLECTING LIKE TERMS

If two terms have the same letter, or letters, we say that they are *like terms*, or *similar terms*. If two terms have no letter at all but are just numbers, they are similar terms. We can simplify by *collecting** like terms, using the distributive laws.

Examples   Collect like terms.

13. $3\,\boxed{x} + 5\,\boxed{x} = (3 + 5)\,\boxed{x} = 8\,\boxed{x}$

14. $\boxed{x} - 3\,\boxed{x} = 1 \cdot \boxed{x} - 3 \cdot \boxed{x} = (1 - 3)\,\boxed{x} = -2\,\boxed{x}$

15. $2x + 3y - 5x - 2y$

$$= 2x + 3y + (-5x) + (-2y) \qquad \text{Subtraction is the same as adding an inverse.}$$

$$= 2x + (-5x) + 3y + (-2y) \qquad \text{Using the commutative and associative laws}$$

$$= (2 - 5)x + (3 - 2)y \qquad \text{Using the distributive law}$$

$$= -3x + y \qquad \text{Adding}$$

16. $3x + 2x + 5 + 7 = (3 + 2)x + (5 + 7) = 5x + 12$

You need not write the intermediate steps.

The following example shows one way in which factoring is useful for solving practical problems.

------

* Also called "combining" like terms.

Example 17   A certain amount of money is invested. This is called *principal*, and we will represent this amount by P. The investment pays 7% interest per year. *Interest* is the money you are paid for your investment. Find an expression for the value of the investment one year later.

$$\text{Principal} + \text{Interest} = P + 7\% \cdot P$$
$$= 1 \cdot P + 0.07P$$
$$= (1 + 0.07)P$$
$$= 1.07P$$

EXERCISE SET 1.4

Evaluate each expression when $x = -2$, $y = 3$, and $z = -4$.

1. $xy + x$

2. $xy - y$

3. $xz + yz$

4. $xy - xz$

5. $yz + xy$

6. $xz + xy$

The expression $P + Prt$ gives the value of an account of P dollars principal, invested at a rate r (in percent) for a time t (in years). Find the value of an account under the following conditions.

7. $P = \$120$
   $r = 6\%$
   $t = 1$ yr

8. $P = \$500$
   $r = 8\%$
   $t = \frac{1}{2}$ yr

Multiply.

9. $3(a + 1)$

10. $8(x + 1)$

11. $4(x - y)$

12. $9(a - b)$

13. $-5(2a + 3b)$

14. $-2(3c + 5d)$

15. $2a(b - c + d)$

16. $5x(y - z + w)$

17. $2\pi r(h + 1)$

18. $P(1 + rt)$

19. $\frac{1}{2}h(a + b)$

20. $\pi r(1 + s)$

Factor.

21. $8x + 8y$

22. $7a + 7b$

23. $9p - 9$

24. $12x - 12$

25. $7x - 21$

26. $6y - 36$

27. $xy + x$

28. $ab + a$

29. $2x - 2y + 2z$

30. $3x + 3y - 3z$

31. $3x + 6y - 3$

32. $4a + 8b - 4$

33. $ab + ac - ad$ **34.** $xy - xz + xw$ **35.** $\pi rr + \pi rs$ **36.** $\frac{1}{2}ah + \frac{1}{2}bh$

Collect like terms.

37. $4a + 5a$ **38.** $9x + 3x$ **39.** $8b - 11b$ **40.** $9c - 12c$

41. $14y + y$ **42.** $13x + x$ **43.** $12a - a$ **44.** $15x - x$

45. $t - 9t$ **46.** $x - 6x$ **47.** $5x - 3x + 8x$

48. $3x - 11x + 2x$ **49.** $5x - 8y + 3x$ **50.** $9a - 10b + 4a$

51. $7c + 8d - 5c + 2d$ **52.** $12a + 3b - 5a + 6b$ **53.** $4x - 7 + 18x + 25$

54. $13p + 5 - 4p + 7$ **55.** $13x + 14y - 11x - 47y$ **56.** $17a + 17b - 12a - 38b$

☆ ───────────────────────────────

Evaluate each expression when $a = -1$, $b = 7$, $c = -6$, and $d = \frac{1}{2}$.

57. $\dfrac{d(ab - c)}{2a}$ **58.** $\dfrac{ad + bd}{b - a} - \dfrac{ac}{d}$

1.5 MULTIPLYING BY -1 AND SIMPLIFYING

MULTIPLYING BY -1

What happens when we multiply a number by -1?

Examples

    1. $-1 \cdot 9 = -9$ **2.** $-1 \cdot (-6) = 6$ **3.** $-1 \cdot 0 = 0$

> **For any number a, $-1 \cdot a = -a$. (Negative 1 times a is the additive inverse of a.)***

From this fact, we know that we can replace $-$ by -1 or the reverse, in any expression. In that way, we can rename additive inverses.

Examples

    4. $-(3x) = -1(3x)$ Replacing $-$ by -1

           $= (-1 \cdot 3)x$ Using associativity

           $= -3x$ Multiplying

───────────

* Or, *changing the sign* is also the same as multiplying by -1.

5. $-(-9y) = -1(-9y)$ Replacing $-$ by -1

$= [-1(-9)]y$ Using associativity

$= 9y$ Multiplying

Examples Find the additive inverse and name it without parentheses.

6. $4 + x$. The inverse is $-(4 + x)$.

$-(4 + x) = -1(4 + x)$ Replacing $-$ by -1

$= -1 \cdot 4 + (-1) \cdot x$ Multiplying

$= -4 + (-x)$ Replacing $-1 \cdot x$ by $-x$

$= -4 - x$ Adding an inverse is the same as subtracting.

7. $3x - 2y + 4$. The inverse is $-(3x - 2y + 4)$.

$-(3x - 2y + 4) = -1(3x - 2y + 4)$

$= -1 \cdot 3x - (-1)2y + (-1)4$

$= -3x + [-(-2y)] + (-4)$

$= -3x + 2y - 4$

8. $a - b$. The inverse is $-(a - b)$.

$-(a - b) = -1(a - b) = -1 \cdot a - (-1) \cdot b$

$= -a + [-(-1)b] = -a + b = b - a$

> Example 8 illustrates something that you should remember because it is a shortcut. The inverse of an expression $a - b$ is $b - a$; that is, $-(a - b) = b - a$.

The above examples show we can rename an additive inverse by multiplying every term by -1. We could also say that we change the sign of every term inside the parentheses. Thus we can skip some steps.

Example 9 Rename without parentheses: $-(-9t + 7z - \frac{1}{4}w)$.

$-\left(-9t + 7z - \frac{1}{4}w\right) = 9t - 7z + \frac{1}{4}w$ Changing the sign of every term

TERMS

What do we mean by the *terms* of an expression? When there are only addition signs, it is easy to tell, but if there are some subtraction signs, we might have reason to wonder. If there are subtraction signs, we can always rename using addition signs.

Example 10   What are the terms of $3x - 4y + 2z$?

$$3x - 4y + 2z = 3x + (-4y) + 2z \qquad \text{Renaming using additions}$$

Thus the terms are $3x$, $-4y$, and $2z$.

REMOVING CERTAIN PARENTHESES

In some expressions commonly encountered in algebra there are parentheses preceded by subtraction signs. These parentheses can be removed by changing the sign of *every* term inside.

Examples   Remove parentheses and simplify.*

11. $6x - (4x + 2) = 6x + [-(4x + 2)]$ Subtracting is adding the additive inverse.

$$= 6x + (-4x) + (-2) \qquad \text{Renaming the inverse}$$
$$= 6x - 4x - 2 \qquad \text{Subtracting is adding the inverse.}$$
$$= 2x - 2 \qquad \text{Collecting like terms}$$

12. $3y - 4 - (9y - 7) = 3y - 4 - 9y + 7$
$$= 6y + 3, \quad \text{or} \quad 3 - 6y$$

> If parentheses are preceded by an addition sign, *no* signs are changed when they are removed. For example, $2x + (5x - 7) = 2x + 5x - 7$.

We now consider subtracting an expression consisting of several terms preceded by a number.

---

* *Simplify* means to rewrite using a minimal amount of symbolism.

Examples   Remove parentheses and simplify.

13. $x - 3(x + y) = x + [-3(x + y)]$ Subtracting is adding the
                                                            inverse.

$$= x + [-3x - 3y]$$ Multiplying -3 by $x + y$

$$= x - 3x - 3y$$ Removing parentheses

$$= -2x - 3y$$ Collecting like terms

14. $3y - 2(4y - 5) = 3y + [-2(4y - 5)]$ Subtraction is adding
                                                              the inverse.

$$= 3y + [-8y + 10]$$ Multiplying -2 by $4y - 5$

$$= 3y - 8y + 10$$ Removing parentheses

$$= -5y + 10$$ Collecting like terms

A common error is to forget to change this sign. *Remember*: When
multiplying like this, change the sign of *every* term inside the parentheses.

PARENTHESES WITHIN PARENTHESES

When parentheses occur within parentheses, we may make them of
different shapes, such as [ ] (also called "brackets") and { } (usually
called "braces"). All of these have the same meaning. When parentheses occur within parentheses, computations in the *innermost* ones
are to be done first.

Example 15   Simplify: $5 - \{6 - [3 - (7 + 3)]\}$.

$$5 - \{6 - [3 - (7 + 3)]\} = 5 - \{6 - [3 - 10]\}$$ Adding $7 + 3$

$$= 5 - \{6 - (-7)\}$$ Subtracting $3 - 10$

$$= 5 - 13$$ Subtracting $6 - (-7)$

$$= -8$$

Example 16   Simplify: $7 - [3(2 - 5) - 4(2 + 3)]$.

$$7 - [3(2 - 5) - 4(2 + 3)] = 7 - [3(-3) - 4(5)]$$ Doing the
                                                                        calculations
                                                                        in the
                                                                        innermost
                                                                        parentheses

$$= 7 - [-9 - 20] = 7 - [-29]$$

$$= 36$$

When expressions with parentheses contain variables, we still work from the inside out when simplifying.

Example 17   Simplify: $6y - \{4[3(y - 2) - 4(y + 2)] - 3\}$.

$6y - \{4[3y - 6 - 4y - 8] - 3\}$ Multiplying to remove the innermost parentheses

$6y - \{4[-y - 14] - 3\}$ Collecting like terms in the inner parentheses

$6y - \{-4y - 56 - 3\}$ Multiplying to remove the inner parentheses

$6y - \{-4y - 59\}$ Collecting like terms

$6y + 4y + 59$ Removing parentheses

$10y + 59$ Collecting like terms

EXERCISE SET 1.5

Rename each additive inverse without parentheses.

1. $-(-4b)$ **2.** $-(-5x)$ **3.** $-(a + 2)$ **4.** $-(b + 9)$

5. $-(b - 3)$ **6.** $-(x - 8)$ **7.** $-(t - y)$ **8.** $-(r - s)$

Find the additive inverse.

9. $a + b + c$

10. $x + y + z$

11. $8x - 6y + 13$

12. $9a - 7b + 24$

13. $-2c + 5d - 3e + 4f$

14. $-4x + 8y - 5w + 9z$

What are the terms of the following?

15. $4a - 5b + 6$

16. $5x - 9y + 12$

17. $2x - 3y - 2z$

18. $5a - 7b - 9c$

Simplify.

19. $a + (2a + 5)$

20. $x + (5x + 9)$

21. $4m - (3m - 1)$

22. $5a - (4a - 3)$

23. $3d - 7 - (5 - 2d)$

24. $8x - 9 - (7 - 5x)$

25. $-2(x + 3) - 5(x - 4)$

26. $-9(y + 7) - 6(y - 3)$

27. $5x - 7(2x - 3)$

28. $8y - 4(5y - 6)$

29. $9a - [7 - 5(7a - 3)]$

30. $12b - [9 - 7(5b - 6)]$

31. $5\{-2 + 3[4 - 2(3 + 5)]\}$

32. $7\{-7 + 8[5 - 3(4 + 6)]\}$

33. $2y + \{7[3(2y - 5) - (8y + 7)] + 9\}$

34. $7b - \{6[4(3b - 7) - (9b + 10)] + 11\}$

☆ ————————————————————————————————

35. 🖩 Remove parentheses and simplify.

$(0.0079x - 0.000843y) - (-0.00793x - 0.000853y)$

Simplify.

36. $-\{-[-(-9)]\}$

37. $\dfrac{2}{3}[2(x + y) + 4(x + 4y)]$

38. $-4[3(x - y - z) - 3(2x + y - 5z)]$

39. $[-(7a - b) - (a + 5b)] - \left[2\left(a + \dfrac{1}{2}b\right) + 3\left(7a - \dfrac{5}{3}b\right)\right]$

40. $0.01\{0.1(x - 2y) - [0.001(3x + y) - (0.2x - 0.1y)]\} - (x - y)$

———————————————————————————————————————

1.6 EXPONENTIAL NOTATION

WHOLE-NUMBER EXPONENTS

Exponential notation is a shorthand device. For $3 \cdot 3 \cdot 3 \cdot 3$ we write 3^4. The latter is called *exponential notation*. In 3^4 the number 3 is called the *base* and the 4 is called the *exponent*.

> **Exponential notation a^n, where n is an integer greater than 1, means**
>
> $$\underbrace{a \cdot a \cdot a \cdots a \cdot a}_{n \text{ factors.}}$$

Caution! a^n does *not* mean to multiply n times. For example, 3^2 means $3 \cdot 3$, or 9. We multiply only once.

Examples   Write exponential notation.

1. $7 \cdot 7 \cdot 7$ *Answer:* 7^3

2. xxx *Answer:* x^3

3. $2x \cdot 2x \cdot 2x \cdot 2x$ *Answer:* $(2x)^4$

Examples   Rewrite without exponents.

4. 5^2 *Answer:* $5 \cdot 5$, or 25

5. $(4y)^2$ *Answer:* $4y \cdot 4y$, or $16yy$

Examples   Simplify.

6. $(4x)^2 = 4x \cdot 4x = 16x^2$

7. $(-3y)^3 = (-3y)(-3y)(-3y) = -27y^3$

In general, an exponent tells how many times the base occurs as a factor. What happens when the exponent is 1 or 0? We cannot have the base occurring as a factor 1 time or 0 times. Look for a pattern.

$$10^3 = 10 \cdot 10 \cdot 10 = 1000$$
$$10^2 = 10 \cdot 10 = 100$$
$$10^1 = ?$$
$$10^0 = ?$$

In order for the pattern to continue, 10^1 would have to be 10 and 10^0 would have to be 1. We shall *agree* that exponents of 1 and 0 have that meaning.

> **For any number a, we agree that a^1 means a.**
> **For any nonzero number a, we agree that a^0 means 1.**

Caution! a^0 does *not* mean a times 0! It means 1. We will see later why we do not allow the base a to be zero.

Examples   Rewrite each of the following without an exponent.

8. 4^1 *Answer:* 4   9. $(-9y)^1$ *Answer:* $-9y$

10. 6^0 *Answer:* 1   11. $(-37.4)^0$ *Answer:* 1

NEGATIVE INTEGERS AS EXPONENTS

How shall we define negative integers as exponents? Look for a pattern.

$$10^2 = 100 \qquad 10^1 = 10 \qquad 10^0 = 1$$
$$10^{-1} = \qquad 10^{-2} = \qquad 10^{-3} =$$

In order for the pattern to continue, 10^{-1} would have to be $\frac{1}{10}$ and 10^{-2} would have to be $\frac{1}{100}$. This leads to the following agreement.

> **If n is any integer, a^{-n} is given the meaning $1/a^n$. In other words, a^n and a^{-n} are reciprocals.**

Examples   Rewrite without using a negative exponent.

12. 7^{-3} *Answer:* $\frac{1}{7^3}$

13. $(-2)^{-3}$ *Answer:* $\frac{1}{(-2)^3}$, or $\frac{1}{-8}$

> *Caution!* A negative exponent does *not* necessarily indicate that an answer is negative! For example, 3^{-2} means $\frac{1}{3^2}$, which is $\frac{1}{9}$.

Examples   Rewrite using a negative exponent.

14. $\frac{1}{5^2}$ *Answer:* 5^{-2} **15.** $\frac{1}{(-7)^4}$ *Answer:* $(-7)^{-4}$

EXERCISE SET 1.6

Write exponential notation.

1. $4 \cdot 4 \cdot 4 \cdot 4 \cdot 4$ **2.** $6 \cdot 6 \cdot 6$ **3.** $5 \cdot 5 \cdot 5 \cdot 5 \cdot 5 \cdot 5$

4. $x \cdot x \cdot x \cdot x$ **5.** $mmmm$ **6.** $ttttt$

7. $3a \cdot 3a \cdot 3a \cdot 3a$ **8.** $5x \cdot 5x \cdot 5x \cdot 5x \cdot 5x$

9. $5 \cdot 5 \cdot c \cdot c \cdot c \cdot d \cdot d \cdot d \cdot d$ **10.** $2 \cdot 2 \cdot 2 \cdot r \cdot r \cdot r \cdot r \cdot t \cdot t$

Rewrite without exponents.

11. 2^5 **12.** 7^3 **13.** $(-3)^4$ **14.** $(-8)^2$ **15.** x^4

16. y^6 **17.** $(4b)^3$ **18.** $(3x)^4$ **19.** $(ab)^4$ **20.** $(xyz)^3$

21. 5^1 **22.** $(\sqrt{6})^1$ **23.** $\left(\frac{7}{8}\right)^1$ **24.** $\left(\frac{9}{7}\right)^1$ **25.** $(\sqrt{8})^0$

26. $(-4)^0$ **27.** $(3z)^0$ **28.** $(5xy)^0$

Rewrite without using negative exponents.

29. 6^{-3} **30.** 8^{-4} **31.** 9^{-5}

32. 16^{-2} **33.** $(-11)^{-1}$ **34.** $(-4)^{-3}$

Rewrite using negative exponents.

35. $\dfrac{1}{3^4}$ **36.** $\dfrac{1}{9^2}$ **37.** $\dfrac{1}{10^3}$

38. $\dfrac{1}{12^5}$ **39.** $\dfrac{1}{(-16)^2}$ **40.** $\dfrac{1}{(-8)^6}$

☆

---

41. How can you place $+$ and $\times$ in the blanks to make a true sentence?

    $1 \underline{\quad} 2 \underline{\quad} 3 \underline{\quad} 4 \underline{\quad} 5 \underline{\quad} 6 \underline{\quad} 7 \underline{\quad} 8 \underline{\quad} 9 = 100$

Simplify.

42. $(-2)^0 - (-2)^3 - (-2)^{-1} + (-2)^4 - (-2)^{-2}$

43. $2(6^1 \cdot 6^{-1} - 6^{-1} \cdot 6^0)$ **44.** $\dfrac{(-8)^{-2} \cdot (8 - 8^0)}{2^{-6}}$

45. $\left[\dfrac{1}{(-3)^{-2}} - (-3)^1 \right] \cdot \left[(-3)^2 + (-3)^{-2} \right]$

46. Evaluate when $x = 1$ and $y = -2$.   **47.** True or false.

    $(x - y)(x^{y-x} - y^{x-y})$ $-5^2 = (-5)^2$

1.7 PROPERTIES OF EXPONENTS

MULTIPLICATION AND DIVISION

To see how to multiply, or simplify, in an expression such as $a^3 \cdot a^2$, recall the definition of exponential notation.

    $a^3 \cdot a^2$

means

    $(a \cdot a \cdot a) \cdot (a \cdot a)$

or

    $a^5.$

The exponent in a^5 is the *sum* of those in $a^3 \cdot a^2$. In general, the exponents are added when we multiply, but note that the base must be the same in all parts.

Let us consider a case in which one exponent is positive and one is negative.

$$b^5 \cdot b^{-2} = b \cdot b \cdot b \cdot b \cdot b \cdot \frac{1}{b \cdot b} \qquad \text{By the definition of exponents}$$

$$= \frac{b \cdot b}{b \cdot b} \cdot b \cdot b \cdot b$$

$$= b^3$$

Adding exponents again gives the correct result. Let us now consider a case in which both exponents are negative.

$$c^{-3} \cdot c^{-2} = \frac{1}{c \cdot c \cdot c} \cdot \frac{1}{c \cdot c} \qquad \text{By the definition of exponents}$$

$$= \frac{1}{c \cdot c \cdot c \cdot c \cdot c} \qquad \text{Multiplying}$$

$$= c^{-5} \qquad \text{Simplifying}$$

Again, adding exponents gives the correct result. This is always the case, even if exponents are negative or zero.

> **In multiplication with exponential notation, we can add exponents if the bases are the same.**
>
> $$a^m a^n = a^{m+n}$$

Examples Multiply and simplify.

1. $x^4 \cdot x^3 = x^{4+3} = x^7$

2. $4^5 \cdot 4^{-3} = 4^{5+(-3)} = 4^2$

3. $(-2)^{-3}(-2)^7 = (-2)^{-3+7} = (-2)^4 = 16$

4. $(8x^4 y^{-2})(-3x^{-3} y) = 8 \cdot (-3) \cdot x^4 \cdot x^{-3} \cdot y^{-2} \cdot y^1$

$$= -24x^{4-3} y^{-2+1}$$

$$= -24xy^{-1} \quad \text{or} \quad -\frac{24x}{y}$$

> The answer $\dfrac{-24x}{y}$ would also be correct here.

There is no standard way of simplifying in cases like Example 4. In some cases it is better to have all exponents positive, but in some it is not. We give answers in two ways. You should also be able to give answers in two ways.

We consider a division.

$$\frac{8^5}{8^3} \quad \text{means} \quad \frac{8 \cdot 8 \cdot 8 \cdot 8 \cdot 8}{8 \cdot 8 \cdot 8}. \quad \text{This simplifies to } 8 \cdot 8, \text{ or } 8^2.$$

We can obtain the result by subtracting exponents. This is always the case, even if exponents are negative or zero.

> **In division with exponential notation, we can subtract exponents if the bases are the same.**
>
> $$\frac{a^m}{a^n} = a^{m-n}$$

Examples   Divide and simplify.

5. $\dfrac{5^7}{5^3} = 5^{7-3} = 5^4$ Subtracting exponents

6. $\dfrac{5^7}{5^{-3}} = 5^{7-(-3)} = 5^{7+3}$ Subtracting exponents (adding an inverse)

$$= 5^{10}$$

7. $\dfrac{9^{-2}}{9^5} = 9^{-2-5} = 9^{-7}, \quad \text{or} \quad \dfrac{1}{9^7}$

8. $\dfrac{7^{-4}}{7^{-5}} = 7^{-4-(-5)}$

$$= 7^{-4+5} = 7^1 = 7$$

9. $\dfrac{16x^4y^7}{-8x^3y^9} = \dfrac{16}{-8} \cdot \dfrac{x^4}{x^3} \cdot \dfrac{y^7}{y^9}$

$$= -2xy^{-2}, \quad \text{or} \quad -\dfrac{2x}{y^2}$$

The answer $\dfrac{-2x}{y^2}$ would also be correct here.

10. $\dfrac{14x^7y^{-3}}{4x^5y^{-5}} = \dfrac{14}{4} \cdot \dfrac{x^7}{x^5} \cdot \dfrac{y^{-3}}{y^{-5}}$

$$= \dfrac{7}{2}x^2y^2$$

In exercises such as Examples 1–10 above, it may help to think as follows: After writing the base, write the top exponent. Then write a subtraction sign. Then write the bottom exponent. Then do the subtraction. For example,

$$\frac{x^{-3}}{x^{-5}} = x^{-3-(-5)}$$

Writing the base and the top exponent   Writing a subtraction sign   Writing the bottom exponent

We do not define 0^0. Now we can see why. Note the following. We know that 0^0 is equal to 0^{1-1}. But 0^{1-1} is also equal to $0/0$. We have already seen that we must leave $0/0$ undefined, so we also leave 0^0 undefined.

RAISING POWERS TO POWERS

Consider the expression $(5^2)^4$. This means

$$5^2 \cdot 5^2 \cdot 5^2 \cdot 5^2, \quad \text{or} \quad 5^8.$$

We can obtain the result by multiplying the exponents.

Consider $(8^{-2})^3$. This means

$$\frac{1}{8^2} \cdot \frac{1}{8^2} \cdot \frac{1}{8^2}, \quad \text{or} \quad \frac{1}{8^6}, \quad \text{which is} \quad 8^{-6}.$$

Again, we could obtain the result by multiplying the exponents. This works, in general, whether the exponents are positive, negative, or zero.

To raise a power to a power we can multiply exponents.

$$(a^m)^n = a^{m \cdot n}$$

Examples Simplify.

11. $(3^5)^7 = 3^{5 \cdot 7} = 3^{35}$

12. $(x^{-5})^4 = x^{-5 \cdot 4} = x^{-20}$

We consider cases in which there are several factors inside the parentheses. Consider $(5^2 \cdot 7^3)^4$. This means

$$(5^2 \cdot 7^3) \cdot (5^2 \cdot 7^3) \cdot (5^2 \cdot 7^3) \cdot (5^2 \cdot 7^3), \quad \text{or} \quad 5^8 \cdot 7^{12}.$$

Again, we can obtain the result by multiplying the exponents. This is true, in general, for positive, negative, or zero exponents.

> **To raise an expression with several factors to a power, raise each factor to the power, multiplying exponents.**
>
> $$(a^m b^n)^p = a^{m \cdot p} \cdot b^{n \cdot p}$$

Examples   Simplify.

13. $(3x^2 y^{-2})^3 = 3^3 (x^2)^3 (y^{-2})^3 = 3^3 x^6 y^{-6}, \quad \text{or} \quad 27x^6 y^{-6}, \quad \text{or} \quad \dfrac{27x^6}{y^6}$

14. $(5x^3 y^{-5} z^2)^4 = 5^4 (x^3)^4 (y^{-5})^4 (z^2)^4,$

$\quad \text{or} \quad 625 x^{12} y^{-20} z^8, \quad \text{or} \quad \dfrac{625 x^{12} z^8}{y^{20}}$

We now consider raising a quotient to a power. Let's take as an example $\left(\dfrac{3^2}{5^3}\right)^4$. This means

$$\left(\frac{3^2}{5^3}\right) \cdot \left(\frac{3^2}{5^3}\right) \cdot \left(\frac{3^2}{5^3}\right) \cdot \left(\frac{3^2}{5^3}\right), \quad \text{or} \quad \frac{3^8}{5^{12}}.$$

Once more, we see that we can obtain the result by multiplying the exponents. This is true, in general, for positive, negative, or zero exponents.

> **To raise a quotient to a power, raise both numerator and denominator to the power, multiplying exponents.**
>
> $$\left(\frac{a^m}{b^n}\right)^p = \frac{a^{m \cdot p}}{b^{n \cdot p}}$$

Examples   Simplify.

15. $\left(\dfrac{x^2}{y^{-3}}\right)^{-5} = \dfrac{x^{2 \cdot (-5)}}{y^{-3 \cdot (-5)}} = \dfrac{x^{-10}}{y^{15}}, \quad \text{or} \quad \dfrac{1}{x^{10} y^{15}}$

16. $\left(\dfrac{2x^3 y^{-2}}{3y^4}\right)^5 = \dfrac{(2x^3 y^{-2})^5}{(3y^4)^5} = \dfrac{2^5 x^{3 \cdot 5} y^{-2 \cdot 5}}{3^5 y^{4 \cdot 5}}$

$$= \dfrac{32 x^{15} y^{-10}}{243 y^{20}} = \dfrac{32 x^{15}}{243 y^{30}}, \quad \text{or} \quad \dfrac{32}{243} x^{15} y^{-30}$$

EXERCISE SET 1.7

Multiply and simplify.

1. $5^6 \cdot 5^3$ **2.** $6^2 \cdot 6^6$ **3.** $8^{-6} \cdot 8^2$ **4.** $9^{-5} \cdot 9^3$

5. $8^{-2} \cdot 8^{-4}$ **6.** $9^{-1} \cdot 9^{-6}$ **7.** $b^2 \cdot b^{-5}$ **8.** $a^4 \cdot a^{-3}$

9. $a^{-3} \cdot a^4 \cdot a^2$ **10.** $x^{-8} \cdot x^5 \cdot x^3$ **11.** $(2x)^3 (3x)^2$ **12.** $(9y)^2 (2y)^3$

13. $(14m^2 n^3)(-2m^3 n^2)$ **14.** $(6x^5 y^{-2})(-3x^2 y^3)$

15. $(-2x^{-3})(7x^{-8})$ **16.** $(6x^{-4} y^3)(-4x^{-8} y^{-2})$

Divide and simplify.

17. $\dfrac{6^8}{6^3}$ **18.** $\dfrac{7^9}{7^4}$ **19.** $\dfrac{4^3}{4^{-2}}$ **20.** $\dfrac{5^8}{5^{-3}}$

21. $\dfrac{10^{-3}}{10^6}$ **22.** $\dfrac{12^{-4}}{12^8}$ **23.** $\dfrac{9^{-4}}{9^{-6}}$ **24.** $\dfrac{2^{-7}}{2^{-5}}$

25. $\dfrac{a^3}{a^{-2}}$ **26.** $\dfrac{y^4}{y^{-5}}$ **27.** $\dfrac{9a^2}{(-3a)^2}$ **28.** $\dfrac{24a^5 b^3}{-8a^4 b}$

29. $\dfrac{-24x^6 y^7}{18x^{-3} y^9}$ **30.** $\dfrac{14a^4 b^{-3}}{-8a^8 b^{-5}}$ **31.** $\dfrac{-18x^{-2} y^3}{-12x^{-5} y^5}$ **32.** $\dfrac{-14a^{14} b^{-5}}{-18a^{-2} b^{-10}}$

Simplify.

33. $(4^3)^2$ **34.** $(5^4)^5$ **35.** $(8^4)^{-3}$ **36.** $(9^3)^{-4}$

37. $(6^{-4})^{-3}$ **38.** $(7^{-8})^{-5}$ **39.** $(3x^2 y^2)^3$ **40.** $(2a^3 b^4)^5$

41. $(-2x^3 y^{-4})^{-2}$ **42.** $(-3a^2 b^{-5})^{-3}$ **43.** $(-6a^{-2} b^3 c)^{-2}$ **44.** $(-8x^{-4} y^5 z^2)^{-4}$

45. $\left(\dfrac{4^{-3}}{3^4}\right)^3$ **46.** $\left(\dfrac{5^2}{4^{-3}}\right)^{-3}$ **47.** $\left(\dfrac{2x^3 y^{-2}}{3y^{-3}}\right)^3$ **48.** $\left(\dfrac{-4x^4 y^{-2}}{5x^{-1} y^4}\right)^{-4}$

☆

Simplify.

49. $\dfrac{(2^{-2})^{-4} \cdot (2^3)^{-2}}{(2^{-2})^2 \cdot (2^5)^{-3}}$ **50.** $\{[(8^{-2})^3]^{-4}\}^5 \cdot [(8^0)^{-2}]^6$

51. $\left[\dfrac{(-3x^{-2} y^5)^{-3}}{(2x^4 y^{-8})^{-2}}\right]^2$ **52.** $\left[\left(\dfrac{a^{-2}}{b^7}\right)^{-3} \cdot \left(\dfrac{a^4}{b^{-3}}\right)^2\right]^{-1}$

TEST OR REVIEW—CHAPTER 1

1. Find decimal notation for $-\dfrac{13}{5}$.

2. Find fractional notation for 3.26.

Find percent notation.

3. 0.036

4. $\dfrac{3}{16}$

Find decimal notation.

5. 5.71%

6. $\dfrac{3}{5}\%$

7. Insert < or > to make a true sentence.

-4.5 _____ -8.7

Simplify.

8. $\left|\dfrac{7}{8}\right|$

9. $|-13.4|$

10. $|0|$

Add.

11. $7 + (-9)$

12. $-5.3 + (-7.8)$

13. $-\dfrac{5}{2} + \left(-\dfrac{7}{4}\right)$

Find $-a$ when a is as given. (In other words, find the additive inverse of each number.)

14. 8

15. -13

16. 0

Subtract.

17. $-6 - (-5)$

18. $-18.2 - 11.5$

19. $\dfrac{19}{4} - \left(-\dfrac{3}{2}\right)$

Multiply.

20. $(-4.1)(8.2)$

21. $-\dfrac{4}{5}\left(-\dfrac{15}{16}\right)$

22. $-6(-4)(-11)$

Divide.

23. $\dfrac{-10}{2}$
24. $\dfrac{-75}{-5}$
25. $\dfrac{7}{-5}$

Divide.

26. $-\dfrac{4}{3} \div \dfrac{5}{7}$
27. $-\dfrac{11}{3} \div \left(-\dfrac{7}{6}\right)$
28. $39.9 \div (-26.6)$

Multiply.

29. $-2(3a - 4b)$
30. $3\pi r(s + 1)$

Factor.

31. $ab - ac + 2ad$
32. $2ah + h$

Collect like terms.

33. $6y - 8x + 4y + 3x$
34. $4a - 7 + 17a + 21$

35. Find the additive inverse of $-9x + 7y - 22$.

Simplify.

36. $-3(x + 2) - 4(x - 5)$
37. $9y - 5(4y - 6)$

Simplify.

38. $4x - [6 - 3(2x - 5)]$
39. $3a - 2[5(2a - 5) - (8a + 4)] + 10$

Multiply or divide and simplify.

40. $(3a^4b^{-2})(-2a^5b^{-3})$
41. $\dfrac{-12x^3y^{-4}}{8x^7y^{-6}}$

Simplify.

42. $(-3a^{-3}b^2c)^{-4}$
43. $\left(\dfrac{-3x^3y^{-3}}{5x^{-2}y^5}\right)^{-3}$

2

LINEAR EQUATIONS

2.1 SOLVING EQUATIONS

EQUATIONS

By an *equation* we mean a number sentence with = for its verb.

Examples

1. $1 + 10 = 11$ is an equation. It is true.

2. $7 - 8 = 9 - 13$ is an equation. It is false.

3. $x - 9 = 3$ is an equation. It is neither true nor false until we determine what x represents.

What is the meaning of an equation?

An equation *says* that the symbols on either side of the equals sign represent the same thing.

Example 4 $7 - 8 = 9 - 13$ says that $7 - 8$ and $9 - 13$ represent the same number. In this case, the equation is false.

SOLUTIONS OF EQUATIONS

If an equation contains a variable it may be neither true nor false. Some replacements may make it true and some may make it false.

The replacements making an equation true are called its *solutions.* When we find them, no matter how, we say that we have *solved* the equation. There are various procedures for solving equations.

Example 5   The number 5 makes the equation $2x + 3 = 13$ true, because if we substitute 5 into the left side we get $2 \cdot 5 + 3$, or 13.

THE ADDITION PRINCIPLE

One of the principles we use in solving equations concerns adding. Consider the equation $a = b$. It says that a and b represent the same number. Suppose this is true and then add a number c to a. We will get the same answer if we add c to b, because a and b are the same number.

> **The Addition Principle.** **If $a = b$ is true, then $a + c = b + c$ is true for any number c.**

When we use the addition principle, we sometimes say that we "add the same number on both sides of an equation."

Example 6 Solve: $x + 6 = -15$.

$$x + 6 + (\,-6\,) = -15 + (\,-6\,) \qquad \text{Using the addition}$$

principle, we added -6.

$$\left. \begin{array}{r} x + 0 = -15 + (-6) \\ x = -21 \end{array} \right\} \quad \text{Simplifying}$$

Check:

$$\begin{array}{c|c} x + 6 = -15 \\ \hline -21 + 6 & -15 \\ -15 & \end{array}$$

The solution is -21.

In Example 6, why did we add -6? Because we wanted the variable x alone on one side of the equation. Then the other side shows what calculations to do to find the answer. When we added the inverse of 6, that got rid of the 6 on the left.

Example 7 Solve: $y - 4.7 = 13.9$.

$$y - 4.7 + \boxed{4.7} = 13.9 + \boxed{4.7} \qquad \text{Using the addition}$$
$$y + 0 = 13.9 + 4.7 \qquad \text{principle}$$
$$y = 13.9 + 4.7$$
$$y = 18.6 \qquad \text{Doing the calculation}$$
$$13.9 + 4.7$$

The number 18.6 checks, so it is the solution.

THE MULTIPLICATION PRINCIPLE

Suppose an equation $a = b$ is true and we multiply a by a number c. We will get the same answer if we multiply b by c. This is because a and b are the same number.

> *The Multiplication Principle.* **If an equation $a = b$ is true, then an equation $a \cdot c = b \cdot c$ is true for any number c.**

Example 8   Solve: $4x = 9$.

$$\frac{1}{4} \cdot 4x = \frac{1}{4} \cdot 9$$

Using the multiplication principle, we multiply by $\frac{1}{4}$, the reciprocal of 4.

$$1 \cdot x = \frac{9}{4}$$

$$x = \frac{9}{4}$$

Simplifying

Check:

$$\frac{4x = 9}{4 \cdot \frac{9}{4} \mid 9}$$

$$9 \mid$$

The solution is $\frac{9}{4}$.

In Example 8, why did we choose to multiply by $\frac{1}{4}$? Because we wanted x alone on one side of the equation. We multiplied by the reciprocal of 4. Then we got $1 \cdot x$, which simplified to x. This got rid of the 4 on the left.

Example 9   Solve: $22 = -\dfrac{4}{5}x$.

$$-\frac{5}{4} \cdot 22 = -\frac{5}{4} \cdot \left(-\frac{4}{5}\right)x$$

Multiplying by $-\frac{5}{4}$, the reciprocal of $-\frac{4}{5}$

$$-\frac{55}{2} = 1 \cdot x$$

$$-\frac{55}{2} = x$$

Check:

$$\frac{22 = -\frac{4}{5}x}{22 \mid -\frac{4}{5} \cdot \left(-\frac{55}{2}\right)}$$

$$\mid 22$$

The solution is $-\frac{55}{2}$.

USING THE PRINCIPLES TOGETHER

Let's see how we can use the addition and multiplication principles together.

Example 10 Solve: $3x - 4 = 13$.

$$3x - 4 + \boxed{4} = 13 + \boxed{4}$$ Using the addition principle, we add 4.

$$3x + (-4) + 4 = 13 + 4$$ Subtraction is done by adding an inverse.

$$3x = 17$$ Simplifying

$$\tfrac{1}{3} \cdot 3x = \boxed{\tfrac{1}{3}} \cdot 17$$ Using the multiplication principle, we multiply by $\tfrac{1}{3}$.

$$x = \tfrac{17}{3}$$

Check:

$$
\begin{array}{c|c}
3x - 4 = 13 & \\
\hline
3 \cdot \frac{17}{3} - 4 & 13 \\
17 - 4 & \\
13 &
\end{array}
$$

The solution is $\tfrac{17}{3}$, or $5\tfrac{2}{3}$.

In Example 10 we used the addition principle first. That usually makes the work simpler.

> There are various ways to solve equations. With the method we are developing, the idea is to find an equation so simple that we can *see* what the solution is. Then we check to see if that solution is also a solution of the original equation.

When there are fractions or decimals in an equation, it helps to eliminate them at the outset. We do that by using the multiplication principle. The process is called *clearing* of decimals or fractions.

Example 11 Clear the equation $\tfrac{3}{4}x + \tfrac{1}{2} = \tfrac{3}{2}$ of fractions.

We multiply on both sides by the least common multiple of the denominators—in this case, 4.

$$4\left(\tfrac{3}{4}x + \tfrac{1}{2}\right) = \boxed{4} \cdot \tfrac{3}{2}$$ Multiplying by 4

$$4 \cdot \tfrac{3}{4}x + 4 \cdot \tfrac{1}{2} = 4 \cdot \tfrac{3}{2}$$ Using the distributive law

$$3x + 2 = 6$$ Simplifying

Example 12   Clear the equation $12.4 - 5.12x = 3.14$ of decimals.

We multiply on both sides by a power of ten to get rid of the decimal points.

$$100\ (12.4 - 5.12x) = \boxed{100}\ (3.14) \qquad \text{Multiplying by 100}$$

$$100 \times 12.4 - 100 \times 5.12x = 100 \times 3.14 \qquad \text{Using the distributive law}$$

$$1240 - 512x = 314 \qquad \text{Simplifying}$$

When there are like terms in an equation we combine them on each side. Then if there are still like terms on opposite sides, we get them on the same side using the addition principle.

Example 13   Solve: $8x + 6 - 2x = -4x - 14$.

$$6x + 6 = -4x - 14 \qquad \text{Combining like terms on the left}$$

$$6x + 4x + 6 = -14 \qquad \text{Adding } 4x \text{ on both sides}$$

$$10x + 6 = -14 \qquad \text{Combining like terms on the left}$$

$$10x = -20 \qquad \text{Adding } -6 \text{ on both sides}$$

$$\frac{1}{10} \cdot 10x = \frac{1}{10} \cdot (-20) \qquad \text{Multiplying on both sides by } \tfrac{1}{10}$$

$$x = -2 \qquad \text{Simplifying}$$

Check:

$$
\begin{array}{c|c}
\multicolumn{2}{c}{8x + 6 - 2x = -4x - 14} \\
\hline
8(-2) + 6 - 2(-2) & -4(-2) - 14 \\
-16 + 6 + 4 & 8 - 14 \\
-6 & -6
\end{array}
$$

The solution is -2.

Here is a summary of the method of solving equations, as used in the preceding examples:

1. **Clear of fractions or decimals if that is needed.**
2. **Collect like terms on both sides of the equation, if necessary.**
3. **Get all like terms with letters on one side and all other terms on the other side.**
4. **Collect like terms again, if necessary.**
5. **Use the multiplication principle to get the letter alone on one side.**

EXERCISE SET 2.1

Solve, using the addition principle. Don't forget to check.

1. $x + 5 = 14$ **2.** $y + 7 = 19$ **3.** $x - 18 = -22$ **4.** $y - 19 = -26$

5. $-8 + y = 15$ **6.** $-9 + t = 17$ **7.** $-12 + z = -51$ **8.** $-37 + x = -89$

9. $p - 296 = 839$ **10.** $z - 149 = 573$

Solve, using the multiplication principle. Don't forget to check.

11. $5x = 20$ **12.** $3x = 21$ **13.** $-4x = 88$ **14.** $-11y = 121$

15. $4 = 24t$ **16.** $8 = 40y$ **17.** $-3x = -96$ **18.** $-8y = -120$

19. $48y = -288$ **20.** $39t = -273$

Solve, using the principles together. Don't forget to check.

21. $4x - 12 = 60$ **22.** $4x - 6 = 70$ **23.** $5x - 10 = 45$

24. $6z - 7 = 11$ **25.** $9t + 4 = -104$ **26.** $5x + 7 = -108$

27. $-\dfrac{7}{3}x + \dfrac{2}{3} = -18$ **28.** $-\dfrac{9}{2}y + 4 = -\dfrac{91}{2}$ **29.** $\dfrac{6}{5}x + \dfrac{4}{10}x = \dfrac{32}{10}$

30. $\dfrac{9}{5}y + \dfrac{4}{10}y = \dfrac{66}{10}$ **31.** $0.9y - 0.7y = 4.2$ **32.** $0.8t - 0.3t = 6.5$

33. $8x + 48 = 3x - 12$ **34.** $15x + 20 = 8x - 22$ **35.** $7y - 1 = 23 - 5y$

36. $3x - 15 = 15 - 3x$ **37.** $4x - 3 = 5 + 12x$ **38.** $9t - 4 = 14 + 15t$

39. $5 - 4a = a - 13$ **40.** $8 - 5x = x - 16$

41. $3m - 7 = -7 - 4m - m$ **42.** $5x - 8 = -8 + 3x - x$

43. $5x + 3 = 11 - 4x + x$ **44.** $6y + 20 = 10 + 3y + y$

☆

Solve. Remember to check. (With a calculator, you need not clear of decimals.)

45. ▦ $0.0008x = 0.0000056$ **46.** ▦ $43.008z = 1.201135$

Solve.

47. $-\dfrac{3}{4}x + \dfrac{1}{8} = -2$ **48.** $y - \dfrac{1}{3}y - 15 = 0$

49. $\dfrac{a}{5} - \dfrac{a}{25} = 3.1$ **50.** $\dfrac{3x}{2} + \dfrac{5x}{3} - \dfrac{13x}{6} - \dfrac{2}{3} = \dfrac{5}{6}$

51. $8b - 11 - 3b + 1 - b = 15 - b$ **52.** $10 - x + 2 = 2x - 6 - 9x + 7$

53. $\dfrac{11}{2}x + \dfrac{1}{2} - \dfrac{3}{4}x - x - \dfrac{5}{8} = 2x + \dfrac{7}{8} - 4x - x - \dfrac{1}{4}$

2.2 MORE ON SOLVING EQUATIONS

EQUATIONS CONTAINING PARENTHESES

Equations containing parentheses can often be solved by first multiplying to remove parentheses and then proceeding as before.

Example 1   Solve: $3(7 - 2x) = 14 - 8(x - 1)$.

$$21 - 6x = 14 - 8x + 8 \qquad \text{Multiplying to remove parentheses}$$
$$21 - 6x = 22 - 8x \qquad \text{Combining like terms}$$
$$8x - 6x = 22 + (-21) \qquad \text{Adding } -21 \text{ and also } 8x$$
$$2x = 1 \qquad \text{Combining like terms}$$
$$x = \tfrac{1}{2} \qquad \text{Multiplying by } \tfrac{1}{2}$$

Check:

$$
\begin{array}{c|c}
\multicolumn{2}{c}{3(7 - 2x) = 14 - 8(x - 1)} \\
\hline
3(7 - 2 \cdot \tfrac{1}{2}) & 14 - 8(\tfrac{1}{2} - 1) \\
3(7 - 1) & 14 - 8(-\tfrac{1}{2}) \\
3 \cdot 6 & 14 + 4 \\
18 & 18
\end{array}
$$

The solution is $\tfrac{1}{2}$.

On the right side of the equation, note that $14 - 8(x - 1) = 14 - 8x + 8$.

THE PRINCIPLE OF ZERO PRODUCTS

Equations can have more than one solution. We now look at an
equation-solving principle that will help to solve some such equations.

When we multiply two numbers, the product will be zero if one of the
factors is zero. Furthermore, if a product is zero, then one of the factors
must be zero. This gives us another principle for solving.

The Principle of Zero Products.   **For any numbers a and b,
if $a \cdot b = 0$, then $a = 0$ or $b = 0$; and
if $a = 0$ or $b = 0$, then $a \cdot b = 0$.**

Example 2   Solve: $(x + 4)(x - 2) = 0$.

Here we have a product that is zero. The equation will become true when either factor is zero. Hence it is true when

$$x + 4 = 0 \quad \text{or} \quad x - 2 = 0.$$

Now, solving the equations separately, we get

$$x = -4 \quad \text{or} \quad x = 2.$$

There are two solutions, -4 and 2. This is so because when x is replaced by either number, the sentence becomes true.

Check:

We show that -4 is a solution.   We show that 2 is also a solution.

$(x + 4)(x - 2) = 0$	
$(-4 + 4)(-4 - 2)$	0
$0 \cdot (-6)$	
	0

$(x + 4)(x - 2) = 0$	
$(2 + 4)(2 - 2)$	0
$6 \cdot 0$	
	0

> **To solve using the principle of zero products:**
>
> 1. **You *must* have a product on one side of the equation and 0 on the other.**
> 2. **Set each factor of the product equal to zero. This gives several simple equations.**
> 3. **Solve the simple equations.**

Example 3   Solve: $7x(4x + 2) = 0$.

$$7x(4x + 2) = 0$$

$7x = 0 \quad \text{or} \quad 4x + 2 = 0$ Using the principle of zero products.

$x = 0 \quad \text{or} \qquad 4x = -2$ We solve the equations separately.
$x = 0 \quad \text{or} \qquad x = -\frac{1}{2}$ For $7x = 0$ we multiply by $\frac{1}{7}$. For $4x = -2$ we multiply by $\frac{1}{4}$.

The solutions are 0 and $-\frac{1}{2}$.

SPECIAL CASES

There are equations with no solution.

Example 4   Solve: $-8x + 5 = 14 - 8x$.

$$-8x + 5 = 14 - 8x$$
$$5 = 14 \qquad \text{Adding } 8x, \text{ we get a false equation.}$$

No matter what number we try for x we get a false sentence. Thus the equation has no solution.

There are equations for which any real number is a solution.

Example 5 Solve: $-8x + 5 = 5 - 8x$.

$$-8x + 5 = 5 - 8x$$
$$5 = 5 \qquad \text{Adding } 8x, \text{ we get a true equation.}$$

Replacing x by any real number gives a true sentence. Thus any real number is a solution.

EXERCISE SET 2.2

Solve.

1. $2(x + 6) = 8x$ **2.** $3(y + 5) = 8y$

3. $80 = 10(3t + 2)$ **4.** $27 = 9(5y - 2)$

5. $180(n - 2) = 900$ **6.** $210(x - 3) = 840$

7. $5y - (2y - 10) = 25$ **8.** $8x - (3x - 5) = 40$

9. $7(3x + 6) = 11 - (x + 2)$ **10.** $9(2x + 8) = 20 - (x + 5)$

11. $\frac{1}{8}(16y + 8) - 17 = -\frac{1}{4}(8y - 16)$ **12.** $\frac{1}{6}(12t + 48) - 20 = -\frac{1}{8}(24t - 144)$

13. $(x + 2)(x - 5) = 0$ **14.** $(x + 4)(x - 8) = 0$

15. $(y - 8)(y - 9) = 0$ **16.** $(t - 3)(t - 7) = 0$

17. $(2x - 3)(3x - 2) = 0$ **18.** $(3y - 4)(4y - 1) = 0$

19. $m(m - 8) = 0$ **20.** $p(p - 5) = 0$

21. $0 = (2x + 8)(3x - 9)$ **22.** $0 = (4x + 16)(5x - 10)$

☆

Solve.

23. $x(x - 1)(x + 2) = 0$ **24.** $y(y - 4)(y + 2) = 0$

25. ▦ $4.62(x + 6.01) = 7.92x$ **26.** ▦ $4.83y - (2.61y - 9.88) = 23.65$

27. $2x - 4 - (x + 1) - 3(x - 2) = 6(2x - 3) - 3(6x - 1) - 8$

28. $5(x - 3) + 4(x + 6) = 9\left(\frac{1}{3}x + 1\right)$ **29.** $2x - 10 - \left(6x + \frac{1}{2}\right) = 13 - 4x$

30. $\frac{1}{7}(a - 3)(7a + 4) = 0$ **31.** $24\left(\frac{x}{6} - \frac{1}{3}\right) = x - 24$

32. $0.5(x - 2) - 2(x - 5) = 0.4(x - 5) - 5(x - 2)$

2.3 SOLVING PROBLEMS

TRANSLATING AND SOLVING

The first step in solving a problem is to translate to mathematical language. Very often this means translating to an equation. Drawing a picture usually helps a great deal. We solve the equation and then check to see if we have a solution to the problem.

Example 1   A 28-foot rope is cut into two pieces. One piece is 3 feet longer than the other. How long are the pieces?

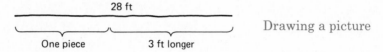

Drawing a picture

The picture can help in translating. Here is one way to do it.

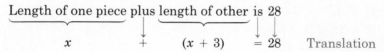

Translation

We used x for the length of one piece and $x + 3$ for the length of the other.

Now we solve:
$$x + (x + 3) = 28$$
$$2x + 3 = 28$$
$$2x = 25$$
$$x = \tfrac{25}{2}, \quad \text{or} \quad 12\tfrac{1}{2}.$$

Do we have an answer to the *problem*? If one piece is $12\tfrac{1}{2}$ ft long, the one that is 3 ft longer must be $15\tfrac{1}{2}$ ft long. The lengths of the pieces add up to 28 ft, so this checks.

> *Steps to Use in Solving a Problem*
> 1. **Translate to an equation. (Always draw a picture if it makes sense to do so.) Carefully keep track of what each variable you use stands for.**
> 2. **Solve the equation.**
> 3. **Check the answer in the original problem.**

Example 2   Five plus twice a number is seven times the number. What is the number?

This time it does not make sense to draw a picture.

①   5 plus $\underbrace{\text{twice a number}}$ is seven times $\underbrace{\text{the number.}}$

$$5 \ + \qquad 2x \qquad = \quad 7 \quad \cdot \qquad x \qquad \text{Translation}$$

We have used x to represent the unknown number. Notice that "is" translates to $=$.

②   Now solve: $5 + 2x = 7x$

$$5 = 7x - 2x$$
$$5 = 5x$$
$$1 = x.$$

③   Now check: Twice 1 is 2. If we add 5, we get 7. This is $7 \cdot 1$. This checks so the answer to the *problem* is 1.

Example 3   What percent of 76 is 13.68?

$$y \qquad \% \qquad \times \ 76 \ = \ 13.68 \qquad \text{Translating* (a picture is not needed here)}$$

$$y \times (0.01) \times 76 = 13.68$$
$$0.76y = 13.68$$
$$y = \frac{13.68}{0.76}, \quad \text{or} \quad 18$$

Solving

Check: $18\% \cdot 76 = 0.18 \times 76$
$$= 13.68$$

The answer checks. The solution of the problem is 18%.

> When solving a problem, remember these two things.
>
> **1.** Read the problem carefully, more than once.
> **2.** Make a list of the known information.

Example 4   One angle of a triangle is five times as large as the first angle. The measure of the third angle is $2°$ less than that of the first angle. How large are the angles?

---

* Notice that *what* translates to a variable, *of* translates to a multiplication sign, and *is* translates to an equals sign.

We draw a picture. We use x for the
measure of the first angle. The
second is five times as large so its
measure will be $5x$. The third angle
has a measure $2°$ less than that of
the first angle so its measure will
be $x - 2$.

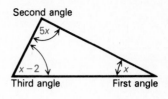

Now, to translate, we need to recall a geometric fact. The measures of
the angles of any triangle add up to $180°$.

$$
\underbrace{\text{Measure of}}_{\text{first angle}} \;\text{plus}\; \underbrace{\text{Measure of}}_{\text{second angle}} \;\text{plus}\; \underbrace{\text{Measure of}}_{\text{third angle}} \;\text{is}\; 180°
$$

$$
x \quad + \quad 5x \quad + \quad (x - 2) \quad = \quad 180
$$

Now we solve:
$$
\begin{aligned}
x + 5x + (x - 2) &= 180 \\
7x - 2 &= 180 \\
7x &= 182 \\
x &= 26
\end{aligned}
$$

The solution of the equation is 26.

The angles will then have measures as follows:

First angle	$x = 26°$,
Second angle	$5x = 130°$,
Third angle	$x - 2 = 24°$.

These add up to $180°$ so they give an answer to the *problem*.

> *Important!* *Write down* what your letters represent. In Example 4, you
> should write that $x =$ the measure of the first angle. Then write that
> $5x =$ the measure of the second angle and $x - 2 =$ the measure of the
> third angle.

Example 5 It has been found that the world record for the 10,000-
meter run has been decreasing steadily since 1940. The record is 30.18
minutes minus 0.12 times the number of years since 1940. If the record
continues to decrease in this way, what will it be in 1985?

$$
\underbrace{\text{Record}}_{R} \;\text{is}\; \underbrace{30.18}_{= 30.18} \;\text{minutes minus}\; \underbrace{0.12}_{-\;0.12} \;\text{times}\; \underbrace{\text{number of years since 1940}}_{t}
$$

$$
R = 30.18 - 0.12t
$$

In 1985, t will be 45, so we have

$R = 30.18 - 0.12 \cdot 45.$ This is the translation.

Solving, we get 24.78. This checks in the problem, so 24.78 minutes is the answer. (In this case it is only a *prediction*.)

Example 6   On December 17, 1974, the pilots of Pan American Airlines shocked the business world by taking a pay cut of 11% to a new salary of $48,950 per year. What was their former salary?

$$\underbrace{\text{Former salary}}\ \underbrace{\text{minus}}\ \underbrace{\text{11%}}\ \underbrace{\text{of}}\ \underbrace{\text{former salary}}\ \underbrace{\text{is}}\ \underbrace{\text{new salary}}$$

$$x \qquad - \qquad 11\% \cdot \qquad x \qquad = \qquad 48{,}950 \quad \text{Translation}$$

We have used x to represent the former salary. We solve.

$$x - 11\% \cdot x = 48{,}950$$
$$1x - 0.11x = 48{,}950 \qquad \text{Replacing 11\% by 0.11}$$
$$\left.\begin{array}{l} (1 - 0.11)x = 48{,}950 \\ 0.89x = 48{,}950 \end{array}\right\} \quad \text{Collecting like terms}$$
$$x = 55{,}000 \qquad \text{Multiplying by } \frac{1}{0.89}$$

Check: 11% of 55,000 is 6050. Subtracting this from 55,000 we get 48,950. This checks, so the former salary was $55,000. This is the answer to the *problem*.

Example 7   The sum of two consecutive integers is 35. What are the integers? (Consecutive integers are next to each other, such as 4 and 5.)

$$\underbrace{\text{First integer}}\ +\ \underbrace{\text{second integer}}\ = 35$$

$$x \qquad + \qquad (x + 1) \qquad = 35 \qquad \text{Translation}$$

Since the integers are consecutive, we know that one of them is 1 greater than the other; thus we call one of them x and the other $x + 1$.

Solve: $x + (x + 1) = 35$
$$2x + 1 = 35$$
$$2x = 34$$
$$x = 17$$

Check: The answers are 17 and 18. These are both integers and consecutive. Their sum is 35, so the answers check in the problem.

Example 8   The perimeter of a rectangle (distance around it) is 322 meters (m). The length is 25 m greater than the width. Find the dimensions.

We first draw a picture.

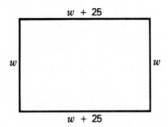

We have used w to represent the width. Then $w + 25$ represents the length.

$$\underbrace{\text{Twice the width}} + \underbrace{\text{twice the length}} = 322$$

$$2w \qquad + \qquad 2(w + 25) \quad = 322 \qquad \text{Translation}$$

Next we solve and get $w = 68$ (width) and $w + 25 = 93$ (length). Finally we check: The length is 25 more than the width. The perimeter is $2 \cdot 68 + 2 \cdot 93$, which is 322. The answer checks. The dimensions of the rectangle are 68 m and 93 m.

EXERCISE SET 2.3

For these exercises you may find the list of formulas on the inside front cover helpful.

Translate to equations and solve.

1. A 12-in. piece of tubing is cut into two pieces. One piece is 4 in. longer than the other. How long are the pieces?

2. A 30-foot piece of wire is cut into two pieces. One piece is 2 feet longer than the other. How long are the pieces?

3. A piece of wire four meters long is cut into two pieces so that one piece is two-thirds as long as the other. Find the length of each piece.

4. A piece of rope five meters long is cut into two pieces so that one piece is three-fifths as long as the other. Find the length of each piece.

5. A babysitting service charges $2.50 per day plus $1.75 per hour. What is the cost of a seven-hour babysitting job?

6. The cost of renting a rug shampooer is $3.25 per hour plus $2.75 for the shampoo. Find the cost of shampooing if the time involved is 3.5 hours.

7. Klunker Car Rental charges $16 per day plus 15¢ per mile. Find the cost of renting a car for a one-day trip of 290 miles.

8. A phone company charges 30¢ per long-distance call plus 20¢ per minute. Find the cost of an 18-minute long-distance call.

9. Eight plus five times a number is seven times the number. What is the number?

10. Six plus three times a number is four times the number. What is the number?

11. Five more than three times a number is the same as ten less than six times the number. What is the number?

12. Six more than nine times a number is the same as two less than ten times the number. What is the number?

interest

13. A pro shop in a bowling alley drops the price of bowling balls 24% to a sale price of $34.20. What was the former price?

14. An appliance store drops the price of a certain type of television 18% to a sale price of $410. What was the former price?

15. Money is borrowed at 9% simple interest. After one year $708.50 pays off the loan. How much was originally borrowed?

16. Money is borrowed at 7% simple interest. After one year $856 pays off the loan. How much was originally borrowed?

17. The second angle of a triangle is three times the first, and the third is 12° less than twice the first. Find the measures of the angles.

18. The second angle of a triangle is four times the first, and the third is 5° more than twice the first. Find the measures of the angles.

19. The perimeter of a college basketball court is 96 m and the length is 14 m more than the width. What are the dimensions?

20. The perimeter of a certain soccer field is 310 m. The length is 65 m more than the width. What are the dimensions?

21. Find three consecutive integers such that the sum of the first, twice the second, and three times the third is 80.

22. Find two consecutive even integers such that two times the first plus three times the second is 76.

23. After a person gets a 20% raise in salary the new salary is $9600. What was the old salary? (*Hint:* What number plus 20% of that number is 9600?)

24. A person gets a 17% raise, bringing the salary to $21,645. What was the salary before the raise?

25. The total cost for tuition plus room and board at State University is $2584. Tuition costs $704 more than room and board. What is the tuition fee?

26. The cost of a private pilot course is $1275. The flight portion costs $625 more than the ground school portion. What is the cost of each?

27. A student's scores on five tests are 93%, 89%, 72%, 80%, and 96%. What must the student score on the sixth test so that the average will be 88%?

28. The yearly changes in the population census of a city for three consecutive years are, respectively, 20% increase, 30% increase, and 20% decrease. What is the total percent change from the beginning to the end of the third year, to the nearest percent?

29. Three numbers are such that the second is 6 less than 3 times the first, and the third is 2 more than $\frac{2}{3}$ the second. The sum of the three numbers is 172. Find the largest number.

30. An appliance store is having a sale on 13 TV models. They are displayed left to right in order of increasing prices. The price of each TV differs by $20 from that of each adjacent TV. For the price of the TV at the extreme right a customer can buy both the second and the seventh models. What is the price of the least expensive TV?

31. A tank at a marine exhibit contains 2000 gallons of seawater. The seawater is 7.5% salt. How many gallons, to the nearest gallon, of freshwater should be added to the tank so that the mixture contains only 7% salt?

2.4 FORMULAS

SOLVING FORMULAS

A formula is a kind of recipe, or rule, for doing a certain kind of calculation. Formulas are often given by equations. For example, the formula

$$C = \pi D$$

tells us that to find the circumference of a circle, we should multiply the diameter by π. Suppose that we know the circumference and want to find the diameter. To do this, we can get D alone on one side of the equation. We "solve" for D.

Example 1 Solve the circle formula for D.

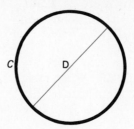

$$C = \pi D \qquad \text{We want this letter alone.}$$

$$\frac{1}{\pi} \cdot C = \frac{1}{\pi} \cdot \pi D \qquad \text{Multiplying by } \frac{1}{\pi}$$

$$\frac{C}{\pi} = D$$

When we solve a formula for a given letter, we do the same things we would do to solve any equation. To see this, compare Example 1 with the following.

Example 1A Solve the equation $3 = 2x$ for x.

$$3 = 2x$$

$$\frac{1}{2} \cdot 3 = \frac{1}{2} \cdot 2x \qquad \text{Multiplying by } \frac{1}{2}$$

$$\tfrac{3}{2} = x$$

> **The idea when solving a formula for a certain letter is to get that letter alone on one side of the equation. To do that, proceed as in solving any equation.**

Example 2 Solve for b: $\quad A = \dfrac{5}{2}(b - 20)$.

$$A = \frac{5}{2}(b - 20)$$

$$\frac{2}{5}A = b - 20 \qquad \text{Multiplying by } \frac{2}{5}$$

$$\frac{2}{5}A + 20 = b \qquad \text{Adding 20}$$

Example 3 Solve for r: $H = 2r + 3m$.

$$H = 2r + 3m \qquad \text{We want this letter alone.}$$

$$H - 3m = 2r \qquad \text{Adding } -3m$$

$$\frac{H - 3m}{2} = r \qquad \text{Multiplying by } \frac{1}{2}$$

Compare Example 3 with an ordinary equation with numbers.

Example 3A Solve for x: $11 = 2x + 3 \cdot 5$.

$$11 = 2x + 3 \cdot 5$$

$$11 - 3 \cdot 5 = 2x \qquad \text{Adding } -3 \cdot 5$$

$$\frac{11 - 3 \cdot 5}{2} = x \qquad \text{Multiplying by } \frac{1}{2}$$

In this case we can simplify further, whereas in Example 3 we could not.

Example 4 The formula $A = P + Prt$ tells how much a principal P, in dollars, will earn at simple interest at a rate r in t years. Solve the formula for P.

$$A = P + Prt \qquad \text{We want this letter alone.}$$

$$A = P(1 + rt) \qquad \text{Factoring (or collecting like terms)}$$

$$A \cdot \frac{1}{1 + rt} = P(1 + rt) \cdot \frac{1}{1 + rt} \qquad \text{Multiplying by } \frac{1}{1 + rt}$$

$$\frac{A}{1 + rt} = P \qquad \text{Simplifying}$$

Compare Example 4 with an equation in which there are numbers.

Example 4A Solve $7 = x + 4 \cdot 5 \cdot x$ for x.

$$7 = x(1 + 4 \cdot 5) \qquad \text{Factoring, or collecting like terms}$$

$$7 \cdot \frac{1}{1 + 4 \cdot 5} = x(1 + 4 \cdot 5) \cdot \frac{1}{1 + 4 \cdot 5} \qquad \text{Multiplying by } \frac{1}{1 + 4 \cdot 5}$$

$$\frac{7}{1 + 4 \cdot 5} = x \qquad \text{Simplifying}$$

In this case we can simplify further, to $\frac{1}{21} = x$, whereas in Example 4 we could not simplify further.

EXERCISE SET 2.4

Solve.

1. $A = lw$, for l
(an area formula)

2. $A = lw$, for w

3. $W = EI$, for I
(an electricity formula)

4. $W = EI$, for E

5. $F = ma$, for m
(a physics formula)

6. $F = ma$, for a

7. $I = Prt$, for t
(an interest formula)

8. $I = Prt$, for P

9. $E = mc^2$, for m
(a relativity formula)

10. $E = mc^2$, for c^2

11. $P = 2l + 2w$, for l
(a perimeter formula)

12. $P = 2l + 2w$, for w

13. $c^2 = a^2 + b^2$, for a^2
(a geometry formula)

14. $c^2 = a^2 + b^2$, for b^2

15. $A = \pi r^2$, for r^2
(a circle formula)

16. $A = \pi r^2$, for π

17. $W = \frac{11}{2}(h - 40)$, for h

18. $C = \frac{5}{4}(F - 32)$, for F
(a temperature formula)

19. $V = \frac{4}{3}\pi r^3$, for r^3 (a volume formula)

20. $V = \frac{4}{3}\pi r^3$, for π

21. $A = \frac{1}{2}h(a + b)$, for h (an area formula)

22. $A = \frac{1}{2}h(a + b)$, for b

23. $F = \dfrac{mv^2}{r}$, for m (a physics formula)

24. $F = \dfrac{mv^2}{r}$, for v^2

☆

25. In Exercise 7, you solved the formula $I = Prt$ for t. Now use it to find how long it will take a deposit of \$75 to earn \$3 interest when invested at 5% simple interest.

26. In Exercise 8, you solved the formula $I = Prt$ for P. Now use it to find how much principal would be needed to earn \$6 in two-thirds of a year at 6% simple interest.

Solve.

27. $s = v_i t + \frac{1}{2}at^2$, for a

28. $A = \pi rs + \pi r^2$, for s

29. $\dfrac{P_1 V_1}{T_1} = \dfrac{P_2 V_2}{T_2}$, for V_1

30. $\dfrac{P_1 V_1}{T_1} = \dfrac{P_2 V_2}{T_2}$, for T_2

2.5 GRAPHS

POINTS AND ORDERED PAIRS

On a number line each point is the graph of a number. On a plane each point is the graph of a number pair. We use two perpendicular number lines, called *axes*. The point where they cross is called the *origin* and is labeled 0. The arrows show the positive directions.

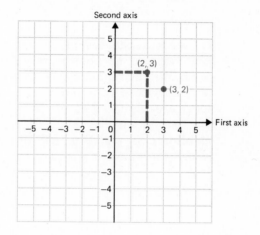

PLOTTING POINTS

Note that (2, 3) and (3, 2) give different points. These are called *ordered pairs* of numbers since it makes a difference which number comes first.

Example 1 Plot the point $(-4, 3)$.

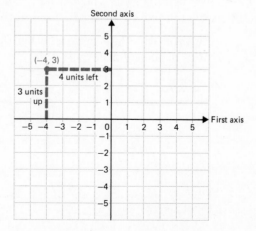

The first number, -4, tells us the distance in the first, or horizontal, direction. We go 4 units *left*. The second number tells us the distance in the second, or vertical, direction. We go 3 units *up*.

The numbers in an ordered pair are called *coordinates*. In $(-4, 3)$, the *first coordinate* is -4 and the *second coordinate** is 3.

QUADRANTS

The axes divide the plane into four regions, as shown, called *quadrants*. In region I (the *first* quadrant) both coordinates of a point are positive. In region II (the *second* quadrant) the first coordinate is negative and the second coordinate is positive.

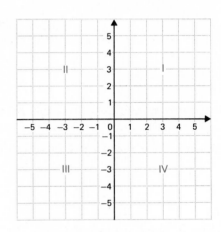

SOLUTIONS OF EQUATIONS

If an equation has two variables, its solutions are pairs of numbers. We usually take the variables in alphabetical order, hence we get ordered pairs of numbers for solutions.

Example 2   Determine whether the following ordered pairs are solutions of the equation $y = 3x - 1$: $(-1, -4), (7, 5)$.

$$y = 3x - 1$$

-4	$3(-1) - 1$
-4	$-3 - 1$
	-4

We substitute -1 for x and -4 for y (alphabetical order of variables).

---

* The first coordinate is sometimes called the *abscissa* and the second coordinate is called the *ordinate*.

The equation becomes true: $(-1, -4)$ is a solution.

$$\frac{y = 3x - 1}{5 \mid 3 \cdot 7 - 1}$$
$$5 \mid 21 - 1$$
$$\mid 20$$

We substitute.

The equation becomes false: $(7, 5)$ is not a solution.

GRAPHING EQUATIONS OF THE TYPE $y = mx$

To *graph* an equation means to make a drawing of its solutions. If an equation has a graph that is a line, we can graph it by plotting a few points and then drawing a line through them.

Example 3 Graph the equation $y = x$.

We will use alphabetical order. Thus the first axis is the x-axis and the second axis is the y-axis.

Next, we find some ordered pairs that are solutions of the equation. In this case, it is easy. Here are a few:

$$(0, 0), \quad (1, 1), \quad (5, 5), \quad (-1, -1), \quad (-6, -6).$$

Now we plot these points. We can see that if we were to plot a million solutions, the dots that we draw would merge into a solid line. We see the pattern, so we can draw the line with a ruler. The line is the graph of the equation $y = x$. We label the line $y = x$ on the graph paper.

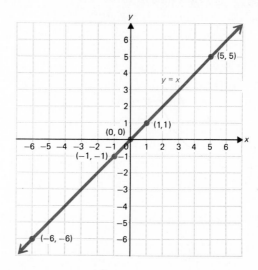

Example 4   Graph the equation $y = 2x$.

We find some ordered pairs that are solutions. This time we shall keep them in a table. To find an ordered pair, we can choose *any* number for x and then determine y. For example, if we choose 3 for x, then $y = 2(3)$ (substituting into the equation of the line), or 6. We make some negative choices for x, as well as positive ones. If a number takes us off of the graph paper, we usually do not use it.

x	y
0	0
1	2
3	6
-2	-4
-3	-6

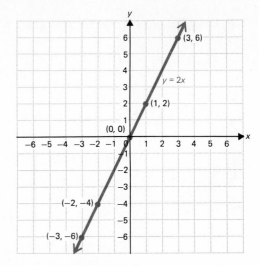

Next, we plot these points. If we had enough of them they would make a solid line. We can draw the line with a ruler, and we label it $y = 2x$.

Example 5   Graph the equation $y = -5x$.

To find an ordered pair, we choose any number for x and then determine y. For example, if we choose 3 for x, we get $y = -5(3)$, or -15.

When we choose -2 for x, we get $y = -5(-2)$, or 10. We find several ordered pairs, plot them, and draw the line.

x	y
3	-15
-2	10
0	0
1	-5

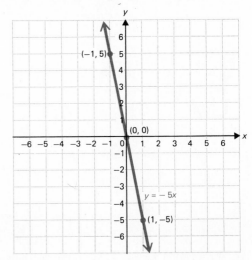

EXERCISE SET 2.5

Plot the following points.

1. $A(5, 3)$, $B(2, 4)$, $C(0, 2)$, $D(0, -6)$, $E(3, 0)$, $F(-2, 0)$, $G(1, -3)$, $H(-5, 3)$, $J(-4, 4)$

2. $A(3, 5)$, $B(1, 5)$, $C(0, 4)$, $D(0, -4)$, $E(5, 0)$, $F(-5, 0)$, $G(1, -5)$, $H(-7, 4)$, $J(-5, 5)$

3. Plot the points $M(2, 3)$, $N(5, -3)$, and $P(-2, -3)$. Draw $\overline{MN}$, $\overline{NP}$, and $\overline{MP}$. ($\overline{MN}$ means the line segment from M to N). What kind of geometric figure is formed? What is its area?

4. Plot the points $Q(-4, 3)$, $R(5, 3)$, $S(2, -1)$, and $T(-7, -1)$. Draw $\overline{QR}$, $\overline{RS}$, $\overline{ST}$, and $\overline{TQ}$. What kind of figure is formed? What is its area?

5. Determine whether $(10, 3)$ is a solution of $3x - 7y = 9$.

6. Determine whether $(-1, 1)$ is a solution of $5y + 2x = -3$.

7. Determine whether $(4, -2)$ is a solution of $4y - 5x = 7 + 3x$.

8. Determine whether $(8, -5)$ is a solution of $3x + 7y = 5 - 2x$.

Use graph paper. For each equation, make a table of values for x and y. Then graph the equation.

9. $y = 5x$

10. $y = \dfrac{1}{3}x$

11. $y = -4x$

12. $y = -3x$

13. $y = 0.25x$

14. $y = -1.5x$

☆

15. If $(-10, -2)$, $(-3, 4)$, and $(6, 4)$ are the coordinates of three of the four vertices of a parallelogram, what are the coordinates of the fourth vertex?

16. Graph $y = 6x$, $y = 3x$, $y = \frac{1}{2}x$, $y = -6x$, $y = -3x$, $y = -\frac{1}{2}x$ on the same set of axes and compare the slants of the lines. What does the number in front of the x tell you about the slant of the line?

17. Which of the following equations have $(-\frac{1}{3}, \frac{1}{4})$ as a solution?

a) $-\dfrac{3}{2}x - 3y = -\dfrac{1}{4}$

b) $8y - 15x = \dfrac{7}{2}$

c) $0.16y = -0.09x + 0.1$

d) $2(-y + 2) - \dfrac{1}{4}(3x - 1) = 4$

2.6 GRAPHS OF LINEAR EQUATIONS

GRAPHS OF EQUATIONS OF THE TYPE $y = mx + b$

We know that the graph of any equation $y = mx$ is a straight line through the origin. What will happen if we add a number b on the right-hand side to get an equation $y = mx + b$?

Example 1   Graph $y = x + 2$ and compare it with the graph of $y = x$.

We first make a table of values.

x	y (or $x + 2$)
0	2
1	3
5	7
-1	1
-2	0
-7	-5

Then graph and compare. The graph of $y = x + 2$ is a line. It looks just like the graph of $y = x$, but it is moved up 2 units.

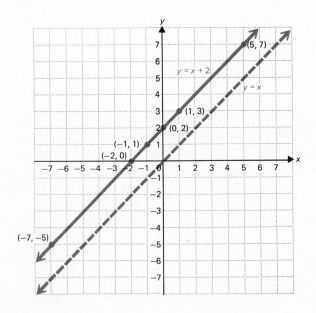

Example 2 Graph $y = 2x - 3$ and compare it with the graph of $y = 2x$.

We first make a table of values. Then graph and compare.

x	y (or $2x - 3$)
0	-3
1	-1
3	3
-2	-7

The graph of $y = 2x - 3$ is a line. It looks just like the graph of $y = 2x$, but it is moved down 3 units.

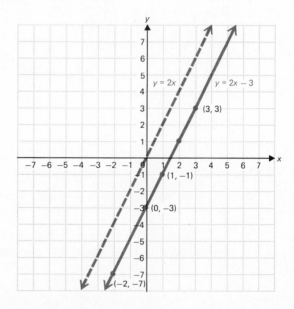

The graph of an equation $y = mx$ is a line through the origin. Any equation $y = mx + b$ also has a graph that is a line. It is parallel to $y = mx$ but moved up or down. The graph of $y = mx + b$ goes through the point $(0, b)$. This point is called the *y-intercept*. We also refer to the number b as the y-intercept.

Example 3 Find the y-intercept: $y = -5x + 4$.

The y-intercept is $(0, 4)$, or simply 4.

Example 4   Find the y-intercept: $y = 5.3x - 12$.

The y-intercept is $(0, -12)$, or simply -12.

We now look at the number m in the equation $y = mx + b$. The number m is called the *slope*, and it shows how the line slants. The greater the slope, the steeper the line slants, from left to right. If the slope is negative, the line slants downward from left to right.

Example 5   Determine the slope and y-intercept for $y = \frac{2}{3}x + 4$ and draw its graph.

The *y-intercept* is the point $(0, 4)$, so we know one point on the graph. The *slope* is the number $\frac{2}{3}$. It tells us that from *any* point of the graph, the line slants *up* 2 units for every 3 *over* to the right. We start at the *y*-intercept. We go *up* 2 units and *over* 3. This gives us the point $(3, 6)$ and we can now draw the graph.

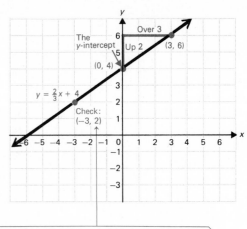

Important: As a check we choose some other value for x, say -3, and determine y (in this case, 2). We plot that point and see if it is on the line. If it is not, there has been some error.

In Example 5 the slope is $\frac{2}{3}$. We can find some other names for $\frac{2}{3}$ by multiplying by 1 and then find other points of the graph.

$$\frac{2}{3} = \frac{2}{3} \cdot \frac{2}{2} = \frac{4}{6}$$ Thus from any point of the graph we can find another point by going *up* 4 units and *over* 6.

$$\frac{2}{3} = \frac{2}{3} \cdot \frac{3}{3} = \frac{6}{9}$$ Thus we can find a point by going *up* 6 and *over* 9, from any point of the graph.

When numbers are negative, we reverse directions.

$$\frac{2}{3} = \frac{2}{3} \cdot \frac{-1}{-1} = \frac{-2}{-3}$$ Thus from any point of the graph we can find another point by going *down* 2 units and then *to the left* 3 units.

If the slope of a line is negative, it slants downward from left to right.

Example 6   Graph $y = -\dfrac{1}{2}x + 5$.

The y-intercept is $(0, 5)$. The slope is $-\frac{1}{2}$, or $\frac{-1}{2}$. From the y-intercept we go *down* 1 unit and to the right 2 units. That gives us the point $(2, 4)$. We can now draw the graph.

As a check, we rename the slope and find another point:

$$\frac{-1}{2} = \frac{-1}{2} \cdot \frac{-3}{-3} = \frac{3}{-6}.$$ Thus we can go *up* 3 units and then *to the left* 6 units. This gives the point $(-6, 8)$.

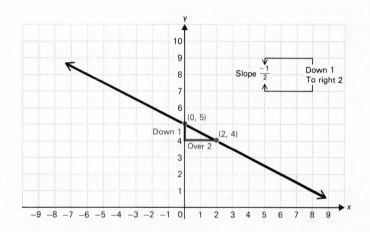

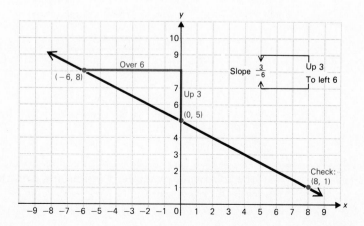

We can also check by choosing a number for x and finding y. If we choose 8 for x, we find that y is 1. Thus the point (8, 1) should be on the graph.

All the equations we have been graphing so far have graphs that are straight lines. Therefore, they are called *linear* equations. We know that any equation $y = mx + b$ is linear, but there are other equations that have straight lines for graphs.

> The *slope-intercept* **equation of a line is** $y = mx + b$.
> **Any equation** $y = mx + b$ **has a graph that is a straight line. It goes through the point (0,** b**) and has slope** m**.**

GRAPHS THAT ARE PARALLEL TO AN AXIS

Some equations have graphs that are parallel to one of the axes. These equations have a missing variable. In the following example, x is missing.

Example 7   Graph $y = 3$.

Since x is missing, any number for x will do. Thus all ordered pairs $(x, 3)$ are solutions. The graph is a line parallel to the x-axis.

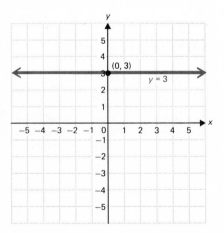

In the following example, y is missing and the graph turns out to be parallel to the y-axis.

Example 8   Graph $x = -2$.

Since y is missing, any number for y will do. Thus we know that all ordered pairs $(-2, y)$ are solutions. The graph is a line parallel to the y-axis.

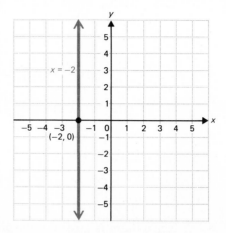

> **The graph of an equation $x = a$, where a is a constant, is a vertical line through the point $(a, 0)$.**
>
> **The graph of an equation $y = b$, where b is a constant, is a horizontal line through the point $(0, b)$.**

WHICH EQUATIONS ARE LINEAR?

A *linear* equation is one having a straight line for its graph. From the examples in this section, we know which equations are linear.

> **A *linear equation* is one that is equivalent to an equation $y = mx + b$ or an equation $x = c$, where m, b, and c are constants.**

Keep in mind that the constants m, b, and c can be 0. Therefore an equation $y = mx$ is linear. Also, the equation $y = 0$ is linear. (It has a slope of 0 and its y-intercept is also 0.)

Example 9   Show that the equation $2x - 3y = 6$ is linear. Determine the slope and the y-intercept.

To show that the equation is equivalent to an equation $y = mx + b$, we solve for y:

$$2x = 6 + 3y \quad \text{Adding } 3y$$

$$2x - 6 = 3y \quad \text{Adding } -6$$

$$\frac{1}{3}(2x - 6) = \frac{1}{3} \cdot 3y \quad \text{Multiplying by } \frac{1}{3}$$

$$\frac{2}{3}x - 2 = y$$

$$y = \frac{2}{3}x - 2 \quad \text{Reversing the equation}$$

Since we have an equation $y = mx + b$, we know that the graph is a straight line. The slope is $\frac{2}{3}$ and the y-intercept is -2.

Examples   Which of these equations are linear?

10. $5x + 10 = 0$

Since there is no y in the equation, we solve for x to see if we get an equation $x = c$:

$$5x = -10 \quad \text{Adding } -10$$

$$\frac{1}{5} \cdot 5x = \frac{1}{5}(-10) \quad \text{Multiplying by } \frac{1}{5}$$

$$x = -2.$$

Since we have an equation $x = c$, we know that the equation is linear. The graph is parallel to the y-axis, 2 units to the left of that axis.

11. $2x^2 = 3y - 4$

We solve for y:

$$2x^2 + 4 = 3y \quad \text{Adding } 4$$

$$\frac{1}{3}(2x^2 + 4) = \frac{1}{3} \cdot 3y \quad \text{Multiplying by } \frac{1}{3}$$

$$\frac{2}{3}x^2 + \frac{4}{3} = y, \quad \text{or} \quad y = \frac{2}{3}x^2 + \frac{4}{3}.$$

The equation is not linear because it has an x^2-term.

EXERCISE SET 2.6

Determine the slope and y-intercept.

1. $y = 4x + 5$

2. $y = 5x + 3$

3. $y = -2x - 6$

4. $y = -5x - 7$

5. $y = -\dfrac{6}{16}x - 0.2$

6. $y = \dfrac{15}{7}x + 2.2$

7. $y = 15.3x + 4.04$

8. $y = -0.314x + 18.9$

Determine the slope and y-intercept. Then draw a graph. Be sure to use a third point as a check.

9. $y = \dfrac{5}{2}x + 1$

10. $y = \dfrac{2}{5}x - 4$

11. $y = -\dfrac{5}{2}x - 4$

12. $y = -\dfrac{2}{5}x + 3$

13. $y = 2x - 5$

14. $y = -2x + 4$

$$\left(Hint: \text{Think of 2 as } \dfrac{2}{1}.\right)$$

15. $y = \dfrac{1}{3}x + 6$

16. $y = -3x + 6$

17. $y = -0.25x + 2$

18. $y = 1.5x - 3$

Graph.

19. $3y - 9 = 0$

20. $6y + 24 = 0$

21. $3x + 15 = 0$

22. $2x - 10 = 0$

Determine which equations are linear.

23. $3x + 5y + 15 = 0$

24. $5x - 3y = 15$

25. $3x - 12 = 0$

26. $16 = -4y$

27. $2x + 4y^2 = 19$

28. $3y - 5x^2 - 4 = 0$

29. $5x - 4xy = 12$

30. $3y = 7xy - 5$

☆

Determine the slope and y-intercept of each line.

31. $2x + 5y + 2 = 5x - 10y - 8$

32. $\dfrac{1}{8}y = -x - \dfrac{7}{16}$

33. $0.4y - 0.004x = -0.04$

34. $x = -\dfrac{7}{3}y - \dfrac{2}{11}$

2.7 ANOTHER LOOK AT LINEAR EQUATIONS AND GRAPHS

GRAPHING USING INTERCEPTS

In many cases we can graph a linear equation very quickly by finding the x-intercept and the y-intercept and then connecting those points. To do this, it is not necessary to solve the equation for either y or x.

Example 1   Graph $3x + 2y = 12$.

This equation is linear. (We could show that it is by solving for y.) To find the y-intercept, we let $x = 0$ and see what y must be. If we just cover up the $3x$ or ignore it, that amounts to letting x be 0. So we cover up $3x$ and see $2y = 12$. Then y must be 6. The y-intercept is $(0, 6)$.

To find the x-intercept, we can cover up the y-term and look at the rest of the equation. We have

$$3x = 12, \quad \text{or} \quad x = 4.$$

The x-intercept is $(4, 0)$.

We plot these points and draw the line, using a third point as a check.

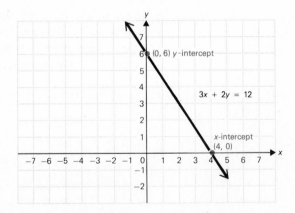

An x-intercept is a point $(a, 0)$. To find a, let $y = 0$ in the equation.

A y-intercept is a point $(0, b)$. To find b, let $x = 0$ in the equation.

We summarize the quickest procedure for graphing linear equations.

> *To Graph Linear Equations*
>   1. **Is there a variable missing? If so, solve for the other one. The graph will be a line parallel to an axis.**
>   2. **If no variable is missing, find the intercepts. Graph using the intercepts if that is feasible.**
>   3. **If the intercepts are too close together, choose another point farther from the origin.**
>   4. **In any case, use a third point as a check.**

FINDING SLOPE

If we know the coordinates of *any* two points of a line, we can find its slope. We divide the vertical distance (the rise) by the horizontal distance (the run). That is, slope = rise/run.

Example 2   Two points of a line are (1, 3) and (5, 6). Find the slope of the line.

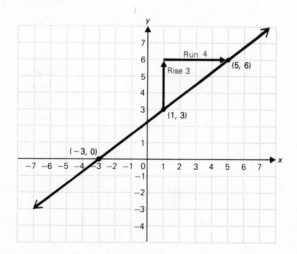

We graph the points and draw the line. Then we mark the rise (the distance *up*) and the run (the distance *over*) by drawing a right triangle. The rise is 3 (from 3 to 6) and the run is 4 (from 1 to 5).

$$\text{Slope} = \frac{\text{rise}}{\text{run}} = \frac{3}{4}$$

We can choose *any* two points of a line to find its slope.

Example 3   Find the slope of the line in Example 2 by using a different pair of points.

We can use *any* two points. Let's choose $(-3, 0)$ and $(5, 6)$.

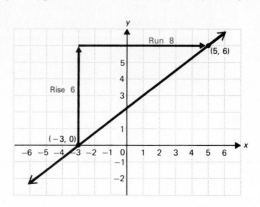

The rise is 6 (from 0 to 6) and the run is 8 (from -3 to 5).

$$\text{Slope} = \frac{\text{rise}}{\text{run}} = \frac{6}{8} = \frac{3}{4}$$

Example 4   Two points of a line are $(-5, -3)$ and $(6, 2)$. Find the slope of the line.

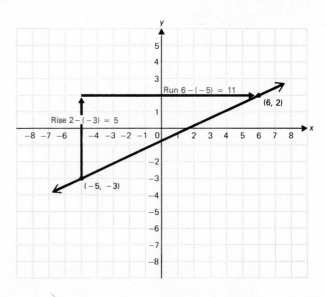

We graph the points and draw the line. We mark the rise and the run. We find them by subtracting.

$$\text{Rise} = 2 - (-3) = 5 \text{ (from } -3 \text{ to 2),}$$
$$\text{Run} = 6 - (-5) = 11 \text{ (from } -5 \text{ to 6),}$$
$$\text{Slope} = \frac{\text{rise}}{\text{run}} = \frac{5}{11}.$$

In Example 4, what will happen if we subtract coordinates in the opposite order? For the rise we get $-3 - 2$, or -5. For the run we get $-5 - 6 = -11$. For the slope we get

$$\text{Slope} = \frac{\text{rise}}{\text{run}} = \frac{-5}{-11} = \frac{5}{11}.$$

This is the same answer we got before. We can subtract either way, as long as we subtract in the same order both times.

> If (x_1, y_1) and (x_2, y_2) are two points of a line, the slope m of the line can be calculated as follows:
>
> $$m = \frac{y_2 - y_1}{x_2 - x_1} \quad \text{or} \quad \frac{y_1 - y_2}{x_1 - x_2}.$$

Examples Find the slope of the line through each pair of points.

5. $(5, -2)$ and $(-7, -5)$

$$\text{Slope} = \frac{y_2 - y_1}{x_2 - x_1} = \frac{-5 - (-2)}{-7 - 5} = \frac{-5 + 2}{-12}$$

$$= \frac{-3}{-12} = \frac{1}{4} \quad \text{or} \quad 0.25$$

We could also have subtracted in the opposite order, as follows:

$$\frac{y_1 - y_2}{x_1 - x_2} = \frac{-2 - (-5)}{5 - (-7)} = \frac{3}{12} = \frac{1}{4}$$

6. $(-4.2, 5.9)$ and $(8.4, -2.2)$

$$\text{Slope} = \frac{y_2 - y_1}{x_2 - x_1} = \frac{-2.2 - 5.9}{8.4 - (-4.2)} = \frac{-8.1}{12.6}$$

$$\approx -0.643 \qquad \approx \text{means "is approximately equal to."}$$

ZERO SLOPE AND LINES WITHOUT SLOPE

If two points have the same second coordinate, what about the slope of the line joining them? Since $y_2 = y_1$, we have

$$\frac{y_2 - y_1}{x_2 - x_1} = \frac{0}{x_2 - x_1} = 0.$$

Every horizontal line has a slope of 0.

Suppose two different points are on a vertical line. Then they have the same first coordinate. Since $x_2 = x_1$, we have

$$\frac{y_2 - y_1}{x_2 - x_1} = \frac{y_2 - y_1}{0}.$$

Since we cannot divide by 0 this is undefined.

Vertical lines do not have a slope.

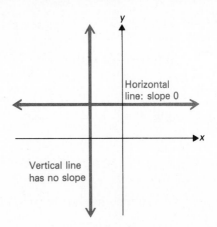

EXERCISE SET 2.7

Graph these equations, using intercepts and a third point as a check.

1. $5y = -15 + 3x$ **2.** $7x = 3y - 21$

3. $6x - 7 + 3y = 9x - 2y + 8$ **4.** $7x - 8 + 4y = 8y + 4x + 4$
(*Hint:* Collect like terms before graphing.)

5. $1.4y - 3.5x = -9.8$ **6.** $3.6x - 2.1y = 22.68$

For each pair of points, find the slope, if it exists, of the line containing them.

7. $(3, 8)$ and $(9, -4)$ **8.** $(17, -12)$ and $(-9, -15)$

9. $(-8, -7)$ and $(-9, -12)$ **10.** $(14, 3)$ and $(2, 12)$

11. $(-16.3, 12.4)$ and $(-5.2, 8.7)$ **12.** $(14.4, -7.8)$ and $(-12.5, -17.6)$

13. $(3.2, -12.8)$ and $(3.2, 2.4)$ **14.** $(-1.5, 7.6)$ and $(-1.5, 8.8)$

Which equations have zero slope and which have no slope?

15. $3x = 12 + y$ **16.** $5y - 12 = 3x$

17. $5x - 6 = 15$ **18.** $-12 = 4x - 7$

19. $5y = 6$ **20.** $19 = -6y$

21. $y - 6 = 14$ **22.** $3y - 5 = 8$

23. $12 - 4x = 9 + x$ **24.** $15 + 7x = 3x - 5$

25. $2y - 4 = 35 + x$ **26.** $2x - 17 + y = 0$

27. $3y + x = 3y + 2$ **28.** $x - 4y = 12 - 4y$

29. $3y - 2x = 5 + 9y - 2x$ **30.** $17y + 4x + 3 = 7 + 4x$

☆

31. Determine a so that the slope of the line through this pair of points has the given value.

$$(-2, 3a), (4, -a); \qquad m = -\frac{5}{12}$$

32. Find the slope of the line that contains the given pair of points.
   a) $(5b, -6c)(b, -c)$ b) $(b, d)(b, d + e)$
   c) $(c + f, a + d)(c - f, -a - d)$

33. A line contains the points $(-100, 4)$ and $(0, 0)$. List four more points of the line.

34. Find the slopes of the sides of the triangle formed by joining the midpoints of the triangle whose vertices are $A(-8, -2)$, $B(-2, 6)$, and $C(4, -10)$. *The midpoint formula:* If the endpoints of a segment are (x_1, y_1) and (x_2, y_2), then the coordinates of the midpoint are $\left(\dfrac{x_1 + x_2}{2}, \dfrac{y_1 + y_2}{2}\right).$

2.8 OTHER EQUATIONS OF LINES (Optional)

POINT-SLOPE EQUATIONS

Suppose we know one point on a line and also its slope. We can find an equation of the line. Suppose the known point is (x_1, y_1) and the slope is m. Then consider any other point (x, y) of the line.

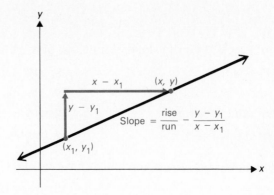

From this drawing we see that the slope of the line is $\dfrac{y - y_1}{x - x_1}$.

The slope is also m. This gives us the equation

$$\frac{y - y_1}{x - x_1} = m. \tag{1}$$

Multiplying by $x - x_1$ gives us the equation we are looking for.

> **The** *point-slope equation* **of a line is**
> $$y - y_1 = m(x - x_1).$$

Note that Equation (1) is not true when $x = x_1$ and $y = y_1$, but the point-slope equation is. Thus Equation (1) is not an equation of the entire line.

> Equation (1) is easy to remember. You can easily get the point-slope equation from it by multiplying by $x - x_1$.

Example 1   Find an equation of the line containing the point $(\frac{1}{2}, -1)$ with slope 5.

We substitute in the equation $y - y_1 = m(x - x_1)$:

$$y - (-1) = 5(x - \tfrac{1}{2}) \quad \text{Substituting}$$
$$y + 1 = 5x - \tfrac{5}{2} \quad \text{Simplifying}$$
$$y = 5x - \tfrac{7}{2}$$

> *Caution!* A common error here is forgetting to multiply *both* terms by 5.

Example 2 Find an equation of the line with x-intercept $(4, 0)$ and slope 2.

$$y - 0 = 2(x - 4) \qquad \text{Substituting}$$
$$y = 2x - 8 \qquad \text{Simplifying}$$

Vertical lines do not have point-slope equations, because they do not have slopes. All other lines do.

TWO-POINT EQUATIONS

Suppose we know two points of a nonvertical line. Call them (x_1, y_1) and (x_2, y_2). We can find the slope m:

$$m = \frac{y_2 - y_1}{x_2 - x_1}.$$

We substitute in the point-slope equation, and get the two-point equation. From this we can write an equation of a nonvertical line whenever we know two points on the line.

> **The** *two-point equation* **of a nonvertical line is**
>
> $$y - y_1 = \frac{y_2 - y_1}{x_2 - x_1}(x - x_1).$$

Example 3 Find an equation of the line containing the points $(2, 3)$ and $(1, -4)$.

Let us take $(2, 3)$ to be (x_1, y_1) and $(1, -4)$ to be (x_2, y_2). We substitute:

$$y - 3 = \frac{-4 - 3}{1 - 2}(x - 2)$$
$$y - 3 = 7(x - 2). \qquad \text{Simplifying}$$

If we wish, we can simplify further to get the slope-intercept equation:

$$y - 3 = 7x - 14$$
$$y = 7x - 11.$$

We can also take the points in opposite order. If we take $(1, -4)$ to be

(x_1, y_1) and $(2, 3)$ to be (x_2, y_2), we get

$$y - (-4) = \frac{3 - (-4)}{2 - 1}(x - 1)$$

$$y + 4 = 7(x - 1).$$

Simplifying further to get the slope-intercept equation, we obtain

$$y + 4 = 7x - 7$$
$$y = 7x - 11.$$

Note that in Example 3, what we did amounted to finding the slope and then substituting it into the *point-slope* equation. If you wish to think of it that way you need memorize only that formula, and not the *two-point* formula above.

PARALLEL AND PERPENDICULAR LINES

If two lines are vertical they are parallel. How can we tell whether nonvertical lines are parallel? The answer is simple: We look at their slopes.

Two nonvertical lines are parallel if they have the same slope.

If one line is vertical and another is horizontal, they are perpendicular. Otherwise, how can we tell whether two lines are perpendicular?

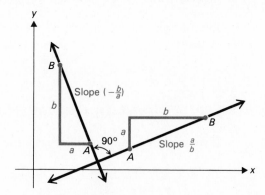

Consider a line $\overleftrightarrow{AB}$ as shown, with slope a/b. Then think of rotating the figure $90°$ to get a line perpendicular to $\overleftrightarrow{AB}$. For the new line the rise and run are interchanged, but the run is now negative. Thus the slope of the new line is $-b/a$. Let us multiply the slopes:

$$\frac{a}{b}\left(-\frac{b}{a}\right) = -1.$$

This is the condition under which lines will be perpendicular.

> **Two lines are perpendicular if the product of their slopes is −1. (If one line has slope m, the slope of a line perpendicular to it is $-1/m$. That is, we take the reciprocal and change the sign.)**

Example 4 Given the line $8y = 7x - 24$ and the point $(-1, 2)$.

a) Find an equation of a line containing the point and parallel to the line.

We find the slope:

$y = \frac{7}{8}x - 3.$ Solving for y

We use the point-slope equation:

$y - 2 = \frac{7}{8}[x - (-1)]$ or $y = \frac{7}{8}x + \frac{23}{8}$.   Substituting $\frac{7}{8}$ for the slope and $(-1, 2)$ for the point

b) Find an equation of the line containing the point and perpendicular to the line.

To find the slope we take the reciprocal of $\frac{7}{8}$ and then change the sign. The slope is $-\frac{8}{7}$.

$y - 2 = -\frac{8}{7}[x - (-1)]$ or $y = -\frac{8}{7}x + \frac{6}{7}$.   Substituting $-\frac{8}{7}$ for the slope and $(-1, 2)$ for the point

EXERCISE SET 2.8

Find an equation of the line having the given slope and containing the given point.

1. $m = 4$, $(3, 2)$

2. $m = 5$, $(5, 4)$

3. $m = -2$, $(4, 7)$

4. $m = -3$, $(7, 3)$

5. $m = 3$, $(-2, -4)$

6. $m = 1$, $(-5, -7)$

7. $m = -2$, $(8, 0)$

8. $m = -3$, $(-2, 0)$

9. $m = 0$, $(0, -7)$

10. $m = 0$, $(0, 4)$

Find an equation of the line containing each pair of points.

11. $(1, 4)$ and $(5, 6)$ **12.** $(2, 5)$ and $(4, 7)$

13. $(-1, -1)$ and $(2, 2)$ **14.** $(-5, -5)$ and $(7, 7)$

15. $(-2, 0)$ and $(0, 5)$ **16.** $(0, -3)$ and $(7, 0)$

17. $(-2, -3)$ and $(-4, -6)$ **18.** $(-4, -7)$ and $(-2, -1)$

19. $(0, 0)$ and $(5, 2)$ **20.** $(0, 0)$ and $(-3, 4)$

Write an equation of the line containing the given point and parallel to the given line.

21. $(3, 7)$, $x + 2y = 6$ **22.** $(0, 3)$, $2x - y = 7$

23. $(2, -1)$, $5x - 7y = 8$ **24.** $(-4, -5)$, $2x + y = -3$

Write an equation of the line containing the given point and perpendicular to the given line.

25. $(2, 5)$, $2x + y = 3$ **26.** $(4, 1)$, $x - 3y = 9$

27. $(3, -2)$, $3x + 4y = 5$ **28.** $(-3, -5)$, $5x - 2y = 4$

☆

29. Find an equation of the line containing $(4, -2)$ and parallel to the line containing $(-1, 4)$ and $(2, -3)$.

30. Find an equation of the line containing $(-1, 3)$ and perpendicular to the line containing $(3, -5)$ and $(-2, 7)$.

31. Use slopes to show that the triangle with vertices $(-2, 7)$, $(6, 9)$, and $(3, 4)$ is a right triangle.

32. Find an equation of the line containing each pair of points.

     a) $(-0.2, 0.7)(-0.7, -0.3)$

     b) $\left(\dfrac{1}{11}, \dfrac{1}{2}\right)\left(-\dfrac{10}{11}, -2\right)$

33. Which pairs of the following four equations represent perpendicular lines?
     (1) $7y - 3x = 21$
     (2) $-3x - 7y = 12$
     (3) $7y + 3x = 21$
     (4) $3y + 7x = 12$

34. Find a number a for which the graphs of $5y = ax + 5$ and $\frac{1}{4}y = \frac{1}{10}x - 1$ will be parallel.

35. Find a number k for which the graphs of $x + 7y = 70$ and $y + 3 = kx$ will be perpendicular.

36. Write an equation of the line that has y-intercept $(0, \frac{5}{7})$ and is parallel to the graph of $6x - 3y = 1$.

37. Write an equation of the line that has x-intercept $(-1.2, 0)$ and is perpendicular to the graph of $6x - 3y = 1$.

38. Write an equation of the line that has x-intercept $(-3, 0)$ and y-intercept $(0, \frac{2}{5})$.

2.9 APPLICATIONS OF LINEAR EQUATIONS (Optional)

FITTING EQUATIONS TO DATA

There are many situations in the world to which linear equations can be applied. How do we know when we have such a situation?

Example 1 Plot the following data and determine whether a linear equation gives an approximate fit.

Crickets are known to chirp faster when the temperature is higher. Here are some measurements made at different temperatures.

Temperature °C	6	8	10	15	20
Number of chirps per min	11	29	47	75	107

We make a graph with a T (temperature) axis and an N (number per minute) axis and plot these data. We see that they do lie approximately on a straight line, so we can use a linear equation in this situation.

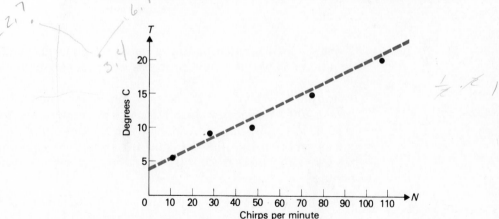

Example 2 Wind friction, or *resistance*, increases with speed. Here are some measurements made in a wind tunnel. Plot the data and determine whether a linear equation will give an approximate fit.

Velocity, km/h	10	21	34	40	45	52
Force of resistance, kg	3	4.2	6.2	7.1	15.1	29.0

We make a graph with a V (velocity) axis and an F (force) axis and plot the data. They do not lie on a straight line, even approximately. Therefore we cannot use a linear equation in this situation.*

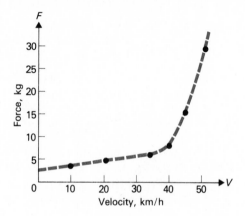

When we have a situation that a linear equation fits, we can find an equation and use it to solve problems and make predictions.

Example 3   When crickets chirp 40 times per minute, the temperature is 10°C. When they chirp 112 times per minute, the temperature is 20°C. We know that a linear equation fits this situation. **(a)** Find a linear equation that fits the data. **(b)** From your equation, find the temperature when crickets chirp 76 times per minute; 100 times per minute.

a) To find the equation, we use the two known ordered pairs (40, 10°) and (112, 20°). We call these *data points*. We find the two-point

---

* The situation in Example 2 might fit an equation that is not linear. We shall consider such problems later.

equation, using these data points.

$$T - T_1 = \frac{T_2 - T_1}{N_2 - N_1}(N - N_1)$$

$$T - 10 = \frac{20 - 10}{112 - 40}(N - 40) \qquad \text{Substituting}$$

$$T - 10 = \frac{10}{72}(N - 40) = \frac{5}{36}N - \frac{5}{36} \cdot 40$$

$$T = \frac{5}{36}N + \frac{40}{9}, \quad \text{or} \quad \frac{5N + 160}{36} \qquad \text{Solving for } T$$

b) Using this as a formula, we find T when $N = 76$.

$$T = \frac{5 \cdot 76 + 160}{36} = 15°$$

When $N = 100$, we get

$$T = \frac{5 \cdot 100 + 160}{36} \approx 18.3°. \quad \approx \text{means ``is approximately equal to.''}$$

Equations give decent results only within certain limits. A negative number of chirps per minute would be meaningless. Also, imagine what would happen to a cricket at $-40°C$ or at $100°C$!

EXERCISE SET 2.9

1. (*Life expectancy of females in the United States*). In 1950, the life expectancy of females was 72 years. In 1970, it was 75 years. Let E represent the life expectancy and t the number of years since 1950 ($t = 0$ gives 1950 and $t = 10$ gives 1960).

   a) Fit a linear equation to the data points. [They are (0, 72) and (20, 75).]

   b) Use the equation of (a) to predict the life expectancy of females in 1980; in 1985.

2. (*Life expectancy of males in the United States*). In 1950, the life expectancy of males was 65 years. In 1970, it was 68 years. Let E represent life expectancy and t the number of years since 1950.

   a) Fit a linear equation to the data points.

   b) Use the equation of (a) to predict the life expectancy of males in 1980; in 1985.

3. (*Weight vs. height*). It has been shown experimentally that a person's height H is related to that person's weight W by a linear equation. An average-size person who is 70 in. tall weighs 165 lb, and a person 67 in. tall weighs 145 lb.

    a) Fit a linear equation to the data points.

    b) Use the equation of (a) to estimate the height of an average-size person weighing 130 lb. (This equation is valid only for heights above 40 in.)

5. (*Records in the 1500-meter run*). In 1930, the record for the 1500-meter run was 3.85 minutes. In 1950, it was 3.70 minutes. Let R represent the record in the 1500-meter run and t the number of years since 1930.

    a) Fit a linear equation to the data points.

    b) Use the equation of (a) to predict the record in 1980; in 1984.

    c) When will the record be 3.3 minutes?

7. (*The cost of a taxi ride*). The cost of a taxi ride for 2 miles in Browntown is $1.75. For 3 miles the cost is $2.00.

    a) Fit a linear equation to the data.

    b) Use the equation to find the cost of a 7-mile ride.

4. (*Natural gas demand*). In 1950, natural gas demand in the United States was 20 quadrillion joules. In 1960, the demand was 22 quadrillion joules. Let D represent the demand for natural gas t years after 1950.

    a) Fit a linear equation to the data points.

    (b) Use the equation of (a) to predict the natural gas demand in 1980; in 2000.

6. (*Records in the 400-meter run*). In 1930, the record for the 400-meter run was 46.8 seconds. In 1970, it was 43.8 seconds. Let R represent the record in the 400-meter run and t the number of years since 1930.

    a) Fit a linear equation to the data points.

    b) Use the equation of (a) to predict the record in 1980; in 1990.

    c) When will the record be 40 seconds?

8. (*The cost of renting a car*). If you rent a car for one day and drive it 100 miles, the cost is $30. If you drive it 150 miles, the cost is $37.50.

    a) Fit a linear equation to the data points.

    b) Use the equation to find how much it will cost to rent the car for one day if you drive it 200 miles.

TEST OR REVIEW—CHAPTER 2

Solve.

1. $-9 + y = 13$

2. $-12x = -8$

3. $0.7x - 0.1 = 2.1 - 0.3x$

4. $5(3x + 6) = 6 - (x + 8)$

5. $(t + 1)(t - 3) = 0$

6. Solve $h = \dfrac{2}{11}W + 40$, for W.

7. Solve for y. Then determine the slope and y-intercept, and graph.

$$2x + 3y = 6$$

8. Graph using intercepts.

$$3y = 15 - 5x$$

9. Find the slope of the line containing the following points, if it exists.

$$(3, 5) \text{ and } (10, -4)$$

10. Find an equation of the line having the given slope and containing the given point.

$$m = 2, \quad (-2, -3)$$

11. Find an equation of the line containing the following pair of points.

$$(-2, 3) \text{ and } (-4, 6)$$

12. Find an equation of the line containing the given point and parallel to the given line.

$$(3, -1), 5x + 7y = 8$$

13. Find an equation of the line containing the given point and perpendicular to the given line.

$$(5, 2), \quad 3x + y = 5$$

Translate to equations and solve.

14. Five more than three times a number is five less than four times the number. Find the number.

15. The price of a saw went up 12% to $14.00. What was the price before the increase?

3

SYSTEMS OF EQUATIONS

3.1 SYSTEMS OF EQUATIONS IN TWO VARIABLES

IDENTIFYING SOLUTIONS

A pair of linear equations is called a *system* of linear equations. A solution of such a system is an ordered pair that makes *both* equations true.

Example 1 Determine whether (4, 7) is a solution of the system

$$x + y = 11$$
$$3x - y = 5$$

We use alphabetical order of the variables. Thus we replace x by 4 and y by 7.

$$
\begin{array}{c|c}
x + y = 11 & \\
\hline
4 + 7 & 11 \\
11 & \\
\end{array}
$$

$$
\begin{array}{c|c}
3x - y = 5 & \\
\hline
3 \cdot 4 - 7 & 5 \\
12 - 7 & \\
5 & \\
\end{array}
$$

The pair (4, 7) makes both equations true, so it is a solution of the *system*. We sometimes describe such a solution by saying that $x = 4$ and $y = 7$.

GRAPHING SYSTEMS OF EQUATIONS

Recall that the graph of an equation is a picture of its solution set. If the graph of an equation is a line, then every point on that line corresponds to an ordered pair that is a solution of the equation. If we graph a *system* of two linear equations, the point at which the lines intersect will be a solution of *both* equations.

Example 2 Solve this system graphically:

$$y - x = 1$$
$$y + x = 3.$$

All ordered pairs from line A are solutions of the first equation. All ordered pairs from line B are solutions of the second equation. The point of intersection has coordinates that make *both* equations true. The solution seems to be the point $(1, 2)$. Graphing is not perfectly accurate, so solving by graphing may give only approximate answers. The pair $(1, 2)$ does check, so it is a solution.

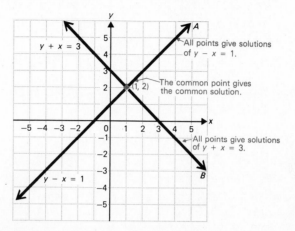

Some systems of equations have no solution. That happens when the graphs are parallel lines. Such equations are called *inconsistent*, because they are never *both* true at once. Some equations have only one solution, as in Example 2. Some have an infinite (unending) set of solutions. That happens when the equations have exactly the same graph. Such equations in two variables are called *dependent*. The three possibilities are shown in the following graphs.

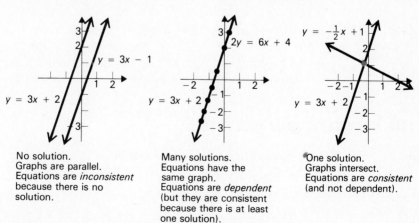

No solution.
Graphs are parallel.
Equations are *inconsistent* because there is no solution.

Many solutions.
Equations have the same graph.
Equations are *dependent* (but they are consistent because there is at least one solution).

One solution.
Graphs intersect.
Equations are *consistent* (and not dependent).

EXERCISE SET 3.1

Determine whether the given ordered pair is a solution of the system of equations. Remember to use alphabetical order of variables.

1. $(1, 2)$; $4x - y = 2$
 $10x - 3y = 4$

2. $(-1, -2)$; $2x + y = -4$
 $x - y = 1$

3. $(2, 5)$; $y = 3x - 1$
 $2x + y = 4$

4. $(-1, -2)$; $x + 3y = -7$
 $3x - 2y = 12$

5. $(1, 5)$; $x + y = 6$
 $y = 2x + 3$

6. $(5, 2)$; $a + b = 7$
 $2a - 8 = b$

7. $(2, -7)$; $3a + b = -1$
 $2a - 3b = -8$

8. $(2, 1)$; $3p + 2q = 5$
 $4p + 5q = 2$

Solve each system graphically. Be sure to check.

9. $x + y = 4$
 $x - y = 2$

10. $x - y = 3$
 $x + y = 5$

11. $2x - y = 4$
 $5x - y = 13$

12. $3x + y = 5$
 $x - 2y = 4$

13. $4x - y = 9$
 $x - 3y = 16$

14. $2y = 6 - x$
 $3x - 2y = 6$

15. $a = 1 + b$
 $b = -2a + 5$

16. $x = y - 1$
 $2x = 3y$

17. $2u + v = 3$
 $2u = v + 7$

18. $2b + a = 11$
 $a - b = 5$

19. $y = -\dfrac{1}{3}x - 1$
 $4x - 3y = 18$

20. $y = -\dfrac{1}{4}x + 1$
 $2y = x - 4$

21. $6x - 2y = 2$
 $9x - 3y = 1$

22. $y - x = 5$
 $2x - 2y = 10$

3.2 SOLVING SYSTEMS OF EQUATIONS

Graphing is not an efficient or accurate method for solving systems of equations. We now consider more efficient methods.

THE SUBSTITUTION METHOD

▸ **Example 1**   Solve this system:

$$x + y = 5$$
$$x = y + 1.$$

The second equation says that x and $y + 1$ name the same thing. In the first equation we can substitute $y + 1$ for x.

$$x + y = 5$$
$$(y + 1) + y = 5 \qquad \text{Substituting } y + 1 \text{ for } x$$

It can be very important to use parentheses here.

We solve this last equation, using methods learned earlier.

It is easy to be careless in removing these parentheses, especially if subtraction is involved.

$$(y + 1) + y = 5$$
$$2y + 1 = 5 \qquad \text{Removing parentheses and collecting like terms}$$
$$2y = 4 \qquad \text{Adding } -1 \text{ on both sides}$$
$$y = 2 \qquad \text{Multiplying on both sides by } \tfrac{1}{2}$$

We return to the original pair of equations, and substitute 2 for y in either of them. Calculation will be easier if we choose the second one.

$$x = y + 1$$
$$x = 2 + 1$$
$$x = 3$$

We obtain the ordered pair $(3, 2)$. We check to be sure it is a solution.

Check:

$$
\begin{array}{c|c}
x + y = 5 & x = y + 1 \\
\hline
3 + 2 \;\big|\; 5 & 3 \;\big|\; 2 + 1 \\
5 & 3
\end{array}
$$

Since $(3, 2)$ checks, it is the solution.

Suppose neither equation of a pair has a variable alone on one side. We then solve one equation for one of the variables.

Example 2 Solve this system:

$$2x + y = 6$$
$$3x + 4y = 4.$$

First solve one equation for one variable. We solve the first equation for y because that will make the work simpler:

$$y = 6 - 2x.$$

Then we substitute $6 - 2x$ for y in the second equation and solve for x:

$$3x + 4(6 - 2x) = 4 \qquad \text{Substituting } 6 - 2x \text{ for } y$$
$$3x + 24 - 8x = 4 \qquad \text{Multiplying to remove parentheses}$$
$$3x - 8x = 4 - 24$$
$$-5x = -20$$
$$x = 4.$$

Now we can substitute 4 for x in either equation and solve for y. We use the first equation.

$$2x + y = 6$$
$$2 \cdot 4 + y = 6$$
$$y = -2$$

We obtain $(4, -2)$. This checks so it is the solution.

THE ADDITION METHOD

The *addition method* for solving systems of equations makes use of the *addition principle*. Some systems are much easier to solve using the addition principle.

Consider this system:

$$2x - 3y = 0$$
$$-4x + 3y = -1.$$

The first equation says that $2x - 3y$ and 0 represent the same number. We use the addition principle with the second equation. We add the "same thing" on both sides. On the left we call it $2x - 3y$. On the right we call it 0:

$$-4x + 3y + (2x - 3y) = -1 + 0$$

or

$$-2x = -1.$$

Using this method we try to get an equation with only one variable. We often say that we have *eliminated* a variable.

Example 3 Solve this system:

$$2x - 3y = 0$$
$$-4x + 3y = -1.$$

We have

$$2x - 3y = 0$$
$$\underline{-4x + 3y = -1}$$
$$-2x + 0 = -1 \qquad \text{Adding}$$

We have made one variable "disappear." Now we solve this equation for x:

$$-2x = -1$$
$$x = \frac{1}{2}.$$

We substitute $\frac{1}{2}$ for x in the first equation and solve for y:

$$2 \cdot \frac{1}{2} - 3y = 0 \qquad \text{Substituting}$$

$$\left.\begin{array}{l} 1 - 3y = 0 \\[2mm] y = \dfrac{1}{3} \end{array}\right\} \qquad \text{Solving for } y$$

We obtain $(\frac{1}{2}, \frac{1}{3})$. This checks, so it is the solution.

Note in Example 3 that a term in one equation and a term in the other were additive inverses of each other. Thus their sum was 0. That enabled us to eliminate a variable.

In order to eliminate a variable we sometimes must multiply one or both of the equations by something before adding.

Example 4   Solve this system:

$$3x + 3y = 15$$
$$2x + 6y = 22.$$

If we add, we do not eliminate a variable. However, if the $3y$ in the first equation were $-6y$, we could. So we multiply by -2 on both sides of the first equation.

$$-6x - 6y = -30 \qquad \text{Multiplying by } -2 \text{ on both sides}$$
$$\underline{2x + 6y = 22}$$
$$-4x + 0 = -8 \qquad \text{Adding}$$
$$ x = 2 \qquad \text{Solving for } x$$

$$2 \cdot 2 + 6y = 22 \qquad \text{Substituting 2 for } x \text{ in second equation}$$
$$4 + 6y = 22$$
$$y = 3 \qquad \text{Solving for } y$$

We obtain (2, 3), or $x = 2$, $y = 3$. This checks, so it is the solution.

Sometimes we have to multiply twice to make two terms become additive inverses.

Example 5 Solve this system:

$$2x + 3y = 12$$
$$5x + 7y = 29.$$

From first equation:	$10x + 15y = 60$	Multiplying by 5
From second equation:	$-10x - 14y = -58$	Multiplying by -2
	$0 + y = 2$	Adding
	$y = 2$	Solving for y

We then have
$$2x + 3 \cdot 2 = 12 \qquad \text{Substituting 2 for } y \text{ in first equation}$$
$$2x = 6$$
$$x = 3 \qquad \text{Solving for } x$$

We obtain (3, 2), or $x = 3$, $y = 2$. This checks, so it is the solution.

When decimals or fractions occur, we *clear* before solving.

Example 6 Clear of decimals:
$$0.5x + 0.3y = 1.9$$
$$1.2x - 0.3y = 12.4.$$

$$5x + 3y = 19 \qquad \text{Multiplying on both sides by 10}$$
$$12x - 3y = 124 \qquad \text{Multiplying on both sides by 10}$$

We can now finish solving as in Example 4.

Example 7 Clear of fractions:
$$\tfrac{1}{2}x + \tfrac{2}{3}y = 1$$
$$\tfrac{3}{4}x - \tfrac{1}{3}y = 2.$$

We solve as follows:

$$6\left(\frac{1}{2}x + \frac{2}{3}y\right) = 6 \cdot 1 \qquad \text{Multiplying on both sides by 6}$$

$$12\left(\frac{3}{4}x - \frac{1}{3}y\right) = 12 \cdot 2 \qquad \text{Multiplying by 12}$$

We have multiplied each equation by the LCM of the denominators. Now we simplify.

First equation: $6 \cdot \frac{1}{2}x + 6 \cdot \frac{2}{3}y = 6$

$$3x + 4y = 6$$

Second equation: $12 \cdot \frac{3}{4}x - 12 \cdot \frac{1}{3}y = 24$

$$9x - 4y = 24$$

We can now finish solving as in Example 3.

WHICH METHOD IS BETTER?

The substitution method is usually better when a variable has a coefficient of 1, as in the following examples:

$$8x + 4y = 180 \qquad x + y = 561$$
$$-x + y = 3; \qquad x = 2y.$$

Both methods work. When in doubt, use the addition method.

EXERCISE SET 3.2

Solve using the substitution method.

1. $3x + 5y = 3$
  $x = 8 - 4y$

2. $2x - 3y = 13$
  $y = 5 - 4x$

3. $9x - 2y = 3$
  $3x - 6 = y$

4. $x = 3y - 3$
  $x + 2y = 9$

5. $5m + n = 8$
  $3m - 4n = 14$

6. $4x + y = 1$
  $x - 2y = 16$

7. $4x + 12y = 4$
  $-5x + y = 11$

8. $-3b + a = 7$
  $5a + 6b = 14$

Solve using the addition method.

9. $x + 3y = 7$
  $-x + 4y = 7$

10. $x + y = 9$
  $2x - y = -3$

11. $2x + y = 6$
  $x - y = 3$

12. $x - 2y = 6$
  $-x + 3y = -4$

13. $9x + 3y = -3$
$2x - 3y = -8$

14. $6x - 3y = 18$
$6x + 3y = -12$

15. $5x + 3y = 19$
$2x - 5y = 11$

16. $3x + 2y = 3$
$9x - 8y = -2$

17. $5r - 3s = 24$
$3r + 5s = 28$

18. $5x - 7y = -16$
$2x + 8y = 26$

19. $0.3x - 0.2y = 4$
$0.2x + 0.3y = 1$

20. $0.7x - 0.3y = 0.5$
$-0.4x + 0.7y = 1.3$

21. $\frac{1}{2}x + \frac{1}{3}y = 4$

$\frac{1}{4}x + \frac{1}{3}y = 3$

22. $\frac{2}{3}x + \frac{1}{7}y = -11$

$\frac{1}{7}x - \frac{1}{3}y = -10$

☆ _____

Solve.

23. 🖩 $3.5x - 2.1y = 106.2$
$4.1x + 16.7y = -106.28.$

24. $5x - 9y = 7$
$7y - 3x = -5$

25. $a - 2b = 16$
$b + 3 = 3a$

26. $3(a - b) = 15$
$4a = b + 1$

27. $1.3x - 0.2y = 12$
$0.4x + 17y = 89$

28. $x - \frac{1}{10}y = 100$

$y - \frac{1}{10}x = -100$

29. $\frac{1}{8}x + \frac{3}{5}y = \frac{19}{2}$

$-\frac{3}{10}x - \frac{7}{20}y = -1$

30. $\frac{x + y}{2} - \frac{x - y}{5} = 1$

$\frac{x - y}{2} + \frac{x + y}{6} = -2$

31. Solve for x and y in terms of a and b.

$5x + 2y = a$
$x - y = b$

32. Determine a and b for which $(-4, -3)$ will be a solution of the system.

$ax + by = -26$
$bx - ay = 7$

33. The points $(0, -3)$ and $(-\frac{3}{2}, 6)$ are two of the solutions of the equation $px - qy = -1$. Find p and q.

3.3 SOLVING PROBLEMS

To solve problems, we often translate to a system of equations in two variables. That can make the translation much easier than if we use just one equation.

Example 1 Eight times a certain number added to five times a second number is 184. The first number minus the second number is -3. Find the numbers.

There are two statements in the problem. We translate the first one.

$\underbrace{\text{8 times a certain number}}$ $\underbrace{\text{added to}}$ $\underbrace{\text{5 times a second number}}$ is 184.

$$8x \qquad\qquad + \qquad\qquad 5y \qquad\qquad = 184$$

Here x represents the first number and y represents the second. Now we translate the second statement, remembering to use x and y.

$\underbrace{\text{The first number}}$ minus $\underbrace{\text{the second number}}$ is -3.

$$x \qquad\qquad - \qquad\qquad y \qquad\qquad = -3$$

We now have a system of equations:

$$8x + 5y = 184$$
$$x - y = -3.$$

We solve this system by either substitution or addition. We'll use the substitution method this time.

$x = y - 3$ Adding y on both sides of the second equation

$8(y - 3) + 5y = 184$ Substituting into the first equation

$8y - 24 + 5y = 184$

$13y = 208$

$$y = \frac{208}{13}, \quad \text{or} \quad 16$$

> Remember to use parentheses when you substitute. Then be careful when you remove them.

We go back to the second of the original equations.

$x - 16 = -3$ Substituting the value of y we just obtained

$x = 13.$

Now we must check in the *original problem*. Since $8 \cdot 13 = 104$ and $5 \cdot 16 = 80$, the sum is 184. The difference between 13 and 16 is -3, so we have the answer.

Example 2   One day Glovers, Inc., sold 20 pairs of gloves. The cloth gloves brought $4.95 per pair and the pigskin gloves sold for $7.50 per pair. The company took in $137.25. How many of each kind were sold?

We let $x =$ the number of pairs of cloth gloves sold and $y =$ the number of pairs of pigskin gloves sold. The total was 20 pairs, so we have

$x + y = 20.$

The amount taken in for cloth gloves was $4.95x$. The amount taken in for pigskin gloves was $7.50y$. These amounts are in dollars. The total taken in was $137.25, so we have

$$4.95x + 7.50y = 137.25.$$

We can multiply on both sides by 100 to clear of decimals.

$$495x + 750y = 13{,}725.$$

We now have the following system of equations. We solve it, using the substitution method or the addition method. In this case, the substitution method is simpler.

$$x + y = 20$$
$$495x + 750y = 13{,}725$$

After we solve, we find that $x = 5$ and $y = 15$. This must be checked in the original problem. Remember, x is the number of cloth gloves sold and y is the number of pigskin gloves sold.

Money from cloth gloves: $\$4.95x = 4.95 \times 5 = \$24.75$

Money from pigskin gloves: $\$7.50y = 7.50 \times 15 = \underline{\$112.50}$

                                              Total $\$137.25$

Number of gloves $x + y = 5 + 15 = 20$.

The numbers check in the original problem, so the business sold 5 pairs of cloth gloves and 15 pairs of pigskin gloves.

Example 3   (*A mixture problem*)   Solution A is 2% alcohol. Solution B is 6% alcohol. A service station owner wants to mix the two to get 60 liters of a solution that is 3.2% alcohol. How many liters of each should the owner use?

We can keep information in a table. That will help in translating.

	Amount of solution	Percent of alcohol	Amount of alcohol in solution
A	x liters	2%	$2\%x$ .02 X
B	60 − X y liters	6%	6%y .06(60−x)
Mixture	60 liters	3.2%	0.032×60, or 1.92 liters

Note that we have used x for the number of liters of A and y for the number of liters of B. To get the amount of alcohol, we multiply by the percentages, of course. If we add x and y in the first column we get 60, and this gives us one equation:

$x + y = 60.$

If we add the amounts of alcohol in the third column we get 1.92, and this gives us another equation:

$2\%x + 6\%y = 1.92.$

After changing percents to decimals and clearing, we have this system:

$x + y = 60$
$2x + 6y = 192.$

Now we solve the system

$x + y = 60$
$2x + 6y = 192.$

$$\begin{array}{rl} -2x - 2y = -120 & \text{Multiplying in the first equation by } -2 \\ \underline{2x + 6y = 192} & \\ 0 + 4y = 72 & \text{Adding} \\ y = 18 & \text{Solving for } y \end{array}$$

$x + 18 = 60 \qquad$ Substituting into the first equation of the system

$x = 42$

Remember, $x =$ the number of liters of 2% solution; $y =$ the number of liters of 6% solution.

Total number of liters of mixture: $\quad x + y = 42 + 18 = 60$

Amount of alcohol: $\quad 2\% \times 42 + 6\% \times 18 = 0.02 \times 42 + 0.06 \times 18$
$$= 1.92 \text{ liters}$$

Percentage of alcohol in mixture: $\quad \dfrac{1.92}{60} = 0.032, \quad \text{or} \quad 3.2\%$

The numbers check in the original problem, so the answer is that the owner should use 42L of 2% alcohol and 18L of 6% alcohol.

Example 4 (*An interest problem*) Two investments are made totaling $4800. In the first year they yield $412 in simple interest. Part of the money is invested at 8% and the rest at 9%. Find the amount invested at each rate of interest.

Keeping information in a table will help in translating. The columns in the table come from the formula for simple interest: $I = Prt$.

	Principal	Rate of interest	Time	Interest
First investment	x	8%	1 yr	$8\%x$
Second investment	y	9%	1 yr	$9\%y$
Total	$4800			$412

Note that we have used x and y for the numbers of dollars invested. They total $4800, and this gives us one equation:

$$x + y = 4800.$$

Look at the last column. The interest, or *yield*, totals $412. This gives us a second equation:

$$8\%x + 9\%y = 412, \quad \text{or} \quad 0.08x + 0.09y = 412.$$

After we multiply on both sides to clear of decimals, we have

$$8x + 9y = 41{,}200.$$

We solve the system

$$x + y = 4800$$
$$8x + 9y = 41{,}200.$$

We find that $x = 2000$ and $y = 2800$. These numbers check in the problem, so $2000 is invested at 8% and $2800 at 9%.

Making a drawing is often helpful. Whenever it makes sense to do so, you should make a drawing before trying to translate.

Example 5   The perimeter of a rectangular field is 194 yards. The length is 2 more than four times the width. Find the length and the width.

First we make a drawing. We use l for length and w for width. We translate the first statement.

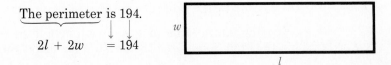

The perimeter is 194.

$$2l + 2w \quad = 194$$

We translate the second statement.

$$\underbrace{\text{The length}}_{l} \ \underbrace{\text{is}}_{=} \ \underbrace{2}_{2} \ \underbrace{\text{more than}}_{+} \ \underbrace{\text{four}}_{4} \ \underbrace{\text{times}}_{\cdot} \ \underbrace{\text{the width.}}_{w}$$

We now have the system of equations

$$2l + 2w = 194$$
$$l = 2 + 4w.$$

It is probably easier to use the substitution method, substituting $2 + 4w$ for l in the first equation.

$2(2 + 4w) + 2w = 194$	Substituting
$4 + 8w + 2w = 194$	Multiplying to remove parentheses on the left
$4 + 10w = 194$	Collecting like terms
$10w = 190$	Adding -4
$w = \dfrac{190}{10}, \quad \text{or} \quad 19$	Multiplying by $\dfrac{1}{10}$

Then we substitute 19 for w in the second equation:

$$l = 2 + 4 \cdot 19$$
$$= 2 + 76$$
$$= 78.$$

The numbers 78 and 19 check in the original problem, so we know that the length is 78 yards and the width is 19 yards.

EXERCISE SET 3.3

Solve.

1. The sum of a certain number and a second number is -42. The first number minus the second is 52. Find the numbers.

2. The sum of two numbers is -63. The first number minus the second is -41. Find the numbers.

3. The difference between two numbers is 16. Three times the larger number is nine times the smaller. What are the numbers?

4. The difference between two numbers is 11. Twice the smaller number plus three times the larger is 123. What are the numbers?

5. The perimeter of a rectangular field is 628 m. The length of the field exceeds its width by 6 m. Find the dimensions.

6. The perimeter of a lot is 750 ft. The width is one-fourth the length. Find the dimensions.

7. One day a store sold 30 sweatshirts. White ones cost $9.95 and yellow ones cost $10.50. In all, $310.60 worth of sweatshirts were sold. How many of each color were sold?

8. One week a business sold 40 scarves. White ones cost $4.95 and printed ones cost $7.95. In all, $282 worth of scarves were sold. How many of each kind were sold?

9. Soybean meal is 16% protein; corn meal is 9% protein. How many pounds of each should be mixed together to get a 350-pound mixture that is 12% protein?

10. A chemist has one solution of acid and water that is 25% acid and a second that is 50% acid. How many gallons of each should be mixed together to get 10 gallons of a solution that is 40% acid?

11. Two investments are made totaling $8800. For a certain year these investments yield $663 in simple interest. Part of the $8800 is invested at 7% and part at 8%. Find the amount invested at each rate.

12. Two investments are made totaling $15,000. For a certain year these investments yield $1432 in simple interest. Part of the $15,000 is invested at 9% and part at 10%. Find the amount invested at each rate.

13. For a certain year $3900 is received in interest from two investments. A certain amount is invested at 5% and $10,000 more than this is invested at 6%. Find the amount invested at each rate.

14. For a certain year $876 is received from two investments. A certain amount is invested at 7%, and $1200 more than this is invested at 8%. Find the amount invested at each rate.

15. Ann Teak is twice as old as her son. Ten years ago Ann was three times as old as her son. What are their present ages? (*Hint:* Ten years ago Ann was $x - 10$ years old and her son was $y - 10$ years old.)

16. Gerry Atric is twice as old as his youngest daughter. In eight years Gerry's age will be three times what his daughter's age was six years ago. What are their present ages?

17. John is 8 years older than his sister Sue. Four years ago Sue was $\frac{2}{3}$ as old as John. How old are they now?

18. Paula is 12 years older than her brother Bob. Four years from now Bob will be $\frac{2}{3}$ as old as Paula. How old are they now?

19. The perimeter of a rectangle is 86 inches. The length is 19 in. greater than the width. Find the length and the width.

20. The perimeter of a rectangle is 388 ft. The length is 82 ft greater than the width. Find the length and the width.

21. One day a store sold 45 pens, one kind at $8.50 and another kind at $9.75. In all, $398.75 was taken in. How many of each kind were sold?

22. At a club play, 117 tickets were sold. Adult tickets cost $1.25; children's tickets cost $0.75. In all, $129.75 was taken in. How many of each kind of ticket were sold?

23. $1150 is invested, part of it at 6% and part of it at 5%. The total yield was $65.75. How much was invested at each rate?

24. $27,000 is invested, part of it at 6% and part of it at 7%. The total yield was $1765. How much was invested at each rate?

☆

Solve.

25. An automobile radiator contains 16 liters of antifreeze and water. This mixture is 30% antifreeze. How much of this mixture should be drained and replaced with pure antifreeze so that there will be 50% antifreeze?

26. Mike and Barbara Jensen have been mathematics professors at a state university. In total they have 46 years service. Two years ago Mr. Jensen had taught 2.5 times as many years as Barbara. How long has each taught at the university?

27. The ten's digit of a two-digit positive integer is 2 more than three times the unit's digit. If the digits are interchanged, the new number is 13 less than half the given number. Find the integer.

28. The measure of one of two supplementary angles is 7° more than three times the other. Find the measure of the larger of the two angles.

29. A limited edition of a new hardback book published by a national historical society was offered for sale to only its membership. Orders were limited to not more than two books per member. The cost was one book for $12 or two books for $20. The society sold 880 books, and the total amount of money taken in was $9840. How many members ordered two books?

30. The numerator of a fraction is 12 more than the denominator. The sum of the numerator and the denominator is 5 more than three times the denominator. What is the reciprocal of the fraction?

31. Use a system of equations to solve the following problem. Two solutions of the equation $y = mx + b$ are $(1, 2)$ and $(-3, 4)$. Find m and b.

3.4 SYSTEMS OF EQUATIONS IN THREE VARIABLES

IDENTIFYING SOLUTIONS

A solution of a system of equations in three variables is an ordered triple that makes *all three* equations true.

Example 1   Determine whether $(\frac{3}{2}, -4, 3)$ is a solution of the system

$$
\begin{aligned}
4x - 2y - 3z &= 5 \\
-8x - y + z &= -5 \\
2x + y + 2z &= 5.
\end{aligned}
$$

We substitute $(\frac{3}{2}, -4, 3)$ into the equations, using alphabetical order.

$$
\begin{array}{c|c}
\multicolumn{2}{c}{4x - 2y - 3z = 5} \\
\hline
4 \cdot \frac{3}{2} - 2(-4) - 3 \cdot 3 & 5 \\
6 + 8 - 9 & \\
5 & \\
\end{array}
$$

$$
\begin{array}{c|c}
\multicolumn{2}{c}{-8x - y + z = -5} \\
\hline
-8 \cdot \frac{3}{2} - (-4) + 3 & -5 \\
-12 + 4 + 3 & \\
-5 & \\
\end{array}
$$

$$
\begin{array}{c|c}
\multicolumn{2}{c}{2x + y + 2z = 5} \\
\hline
2 \cdot \frac{3}{2} + (-4) + 2 \cdot 3 & 5 \\
3 - 4 + 6 & \\
5 & \\
\end{array}
$$

The triple makes *all three* equations true, so it is a solution.

SOLVING

The substitution method is not ordinarily convenient for solving systems of equations with three variables. We consider only the addition method.

Example 2   Solve:

$$
\begin{aligned}
x + y + z &= 4 &\quad (1) \\
x - 2y - z &= 1 &\quad (2) \\
2x - y - 2z &= -1. &\quad (3)
\end{aligned}
$$

a) We first use *any* two of the three equations to get an equation in two variables. Let us use equations (1) and (2) and add to eliminate z:

$$
\begin{array}{ll}
x + y + z = 4 & (1) \\
x - 2y - z = 1 & (2) \\
\hline
2x - y \quad\quad = 5 & (4)
\end{array}
$$

b) We use a different pair of equations and eliminate the *same variable*

17. $\dfrac{2}{x} - \dfrac{1}{y} + \dfrac{3}{z} = -1$

$\dfrac{1}{x} - \dfrac{1}{y} - \dfrac{2}{z} = -5$

$\dfrac{4}{x} + \dfrac{3}{y} - \dfrac{1}{z} = 1$

$\left(Hint: \text{Let } u = \dfrac{1}{x},\ v = \dfrac{1}{y},\ w = \dfrac{1}{z}. \right)$

18. $\begin{aligned} 2x - 4y + 10z &= 104 \\ -5x + 3y - 6z &= -63 \\ -3x - 10y + 9z &= 100 \end{aligned}$

19. $\begin{aligned} x - 2y &= 0 \\ 3x + z &= 1 \\ 5y - 2w &= 4 \\ 3z + w &= 1 \end{aligned}$

3.5 SOLVING PROBLEMS

Many problems can be solved by first translating to a system of three equations.

Example 1 The sum of three numbers is 4. The first number minus twice the second minus the third is 1. Twice the first number minus the second minus twice the third is -1. Find the numbers.

Let us call the three numbers x, y, and z. There are three statements in the problem. We translate:

The sum of the three numbers is 4.

$$x + y + z \qquad = 4$$

The first number minus twice the second minus the third is 1.

$$x \quad - \quad 2y \quad - \quad z \quad = 1$$

Twice the first number minus the second minus twice the third is -1.

$$2x \quad - \quad y \quad - \quad 2z \quad = -1$$

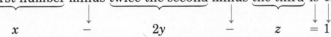

We now have a system of three equations:

$$\begin{aligned} x + y + z &= 4 \\ x - 2y - z &= 1 \\ 2x - y - 2z &= -1. \end{aligned}$$

We solved this system in Example 2 on p. 110 and found that $x = 2$, $y = -1$, and $z = 3$. These numbers check in the original problem, so we have a solution.

Example 2 In a triangle the largest angle is 70° greater than the smallest angle. The largest angle is twice as large as the remaining angle. Find the measure of each angle.

We first make a drawing, and label the angles. We let x represent the smallest angle, z the largest angle, and y the remaining angle.

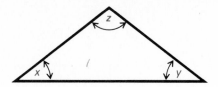

We know that the sum of the measures of the angles is 180° in any triangle. This gives us the equation

$x + y + z = 180.$

There are two statements in the problem. We translate them.

The largest angle is 70° greater than the smallest angle.

$z \qquad = 70 \qquad + \qquad x$

The largest angle is twice as large as the remaining angle.

$z \qquad = \qquad 2y$

We now have a system:

$$
\begin{array}{llll}
x + y + z = 180 & \qquad & x + y + z = 180 \\
x + 70 = z & \text{or} & x - z = -70 \\
2y = z & & 2y - z = 0.
\end{array}
$$

We solved this system in Example 4 on p. 113 and found that $x = 30$, $y = 50$, and $z = 100$. These numbers check, so we have a solution. The angles measure 30°, 50°, and 100°.

Example 3 In a factory there are three machines A, B, and C. When all three are running, they produce 222 suitcases per day. If A and B work but C does not, they produce 159 suitcases per day. If B and C work but A does not, they produce 147 suitcases per day. What is the daily production of each machine?

Let us use x, y, and z for the number of suitcases produced daily by the machines A, B, and C, respectively. There are three statements. We translate them.

When all three are running, they produce 222 suitcases per day.

$$x + y + z \qquad = \qquad 222$$

When A and B work, they produce 159 suitcases per day.

$$x + y \qquad = \qquad 159$$

When B and C work, they produce 147 suitcases per day.

$$y + z \qquad = \qquad 147$$

We now have a system of three equations:

$$x + y + z = 222$$
$$x + y \quad\ = 159$$
$$\quad\ y + z = 147.$$

We solve and get $x = 75$, $y = 84$, and $z = 63$. These numbers check, so the answer to the problem is that A produces 75 suitcases per day, B produces 84 suitcases per day, and C produces 63 suitcases per day.

Solving systems of three or more equations is important in many applications. Systems of equations arise very often in the use of statistics, for example, in such fields as the social sciences. They also come up in problems of business, science, and engineering.

EXERCISE SET 3.5

1. The sum of three numbers is 105. The third is 11 less than 10 times the second. Twice the first is 7 more than 3 times the second. Find the numbers.

2. The sum of three numbers is 57. The second is 3 more than twice the first. The third is 6 more than the first. Find the numbers.

3. The sum of three numbers is 5. The first number minus the second plus the third is 1. The first minus the third is 3 more than the second. Find the numbers.

4. The sum of three numbers is 26. Twice the first minus the second is 2 less than the third. The third is the second minus three times the first. Find the numbers.

5. In triangle ABC, the measure of angle B is 2° more than three times the measure of angle A. The measure of angle C is 8° more than the measure of angle A. Find the angle measures.

7. In triangle ABC, the measure of angle B is twice the measure of angle A. The measure of angle C is 80° more than that of angle A. Find the angle measures.

9. Paula sells magazines part time. On Thursday, Friday, and Saturday she sold $66 worth. On Thursday she sold $3 more than on Friday. On Saturday she sold $6 more than on Thursday. How much did she take in each day?

11. Linda has a total of 255 on three tests. The sum of the scores on the first and second tests exceeds her third score by 61. Her first score exceeds her second by 6. Find the three scores.

13. In a factory there are three polishing machines, A, B, and C. When all three of them are working, 5700 lenses can be polished in one week. When only A and B are working, 3400 lenses can be polished in one week. When only B and C are working, 4200 lenses can be polished in one week. How many lenses can be polished in a week by each machine?

15. When three pumps, A, B, and C, are running together, they can pump 3700 gallons per hour. When only A and B are running, 2200 gallons per hour can be pumped. When only A and C are running, 2400 gallons per hour can be pumped. What is the pumping capacity of each pump?

6. In triangle ABC, the measure of angle B is three times the measure of angle A. The measure of angle C is 30° greater than the measure of angle A. Find the angle measures.

8. In triangle ABC, the measure of angle B is three times that of angle A. The measure of angle C is 20° more than that of angle A. Find the angle measures.

10. John Gray picked strawberries on three days. He picked a total of 87 quarts. On Tuesday he picked 15 quarts more than on Monday. On Wednesday he picked 3 quarts fewer than on Tuesday. How many quarts did he pick each day?

12. Fred, Jane, and Mary made a total bowling score of 575. Fred's score was 15 more than Jane's. Mary's was 20 more than Jane's. Find the scores.

14. Sawmills A, B, and C can produce 7400 board feet of lumber per day. A and B together can produce 4700 board feet, while B and C together can produce 5200 board feet. How much can each mill produce by itself?

16. Three welders, A, B, and C, can weld 37 linear feet per hour when working together. A and B together can weld 22 linear feet per hour, while A and C together can weld 25 linear feet per hour. How much can each weld alone?

☆ _____

Solve.

17. Sarah's age is the sum of the ages of Janice and Jason. Janice's age is 2 years more than the sum of the ages of Jason and Steve. Jason is four times as old as Steve. The sum of the ages of all four is 42. How old is Sarah?

18. Hal gives Tom as many raffle tickets as Tom already has and gives Gary as many as Gary already has. In like manner, Tom then gives Hal and Gary as many tickets as each then has. Similarly, Gary gives Hal and Tom as many tickets as each then has. If each finally has 40 tickets, with how many tickets does Tom begin?

19. For a musical production at a university, the costs of tickets for students, adults, and children were $4.00, $5.50, and $1.50, respectively. The total number of children's and adult tickets sold was 30 more than half the number of student tickets sold. The number of adult tickets sold was 5 more than four times the number of children's tickets. How many of each type of ticket were sold if the total receipts from ticket sales was $13,162.50?

20. Find a three-digit positive integer such that the sum of all three digits is 14, the ten's digit is 2 more than the unit's digit, and, if the digits are reversed, the number is unchanged.

3.6 DETERMINANTS (Optional)

EVALUATING DETERMINANTS

The following is called a *determinant*:

$$\begin{vmatrix} a_1 & b_1 \\ a_2 & b_2 \end{vmatrix}.$$

To evaluate a determinant, we do two multiplications and subtract.

Example 1 Evaluate:

$$\begin{vmatrix} 2 & -5 \\ 6 & 7 \end{vmatrix}.$$

We multiply and subtract as follows:

$$\begin{vmatrix} 2 & -5 \\ 6 & 7 \end{vmatrix} = 2 \cdot 7 - 6 \cdot (-5) = 14 + 30 = 44.$$

Determinants are defined according to the pattern shown in Example 1.

> The determinant $\begin{vmatrix} a_1 & b_1 \\ a_2 & b_2 \end{vmatrix}$ is defined to mean $a_1b_2 - a_2b_1$.

> The value of a determinant is a *number*. In Example 1, the value is 44.

THIRD-ORDER DETERMINANTS

The determinants above are called *second-order*. A *third-order* determinant is defined as follows.

Note the minus sign here.

$$\begin{vmatrix} a_1 & b_1 & c_1 \\ a_2 & b_2 & c_2 \\ a_3 & b_3 & c_3 \end{vmatrix} = a_1 \begin{vmatrix} b_2 & c_2 \\ b_3 & c_3 \end{vmatrix} - a_2 \begin{vmatrix} b_1 & c_1 \\ b_3 & c_3 \end{vmatrix} + a_3 \begin{vmatrix} b_1 & c_1 \\ b_2 & c_2 \end{vmatrix}$$

↳Note that the a's come from the first column.

Note also that the second-order determinants above can be obtained by crossing out the row and column in which the a occurs.

For a_1: $\begin{vmatrix} a_1 & b_1 & c_1 \\ a_2 & b_2 & c_2 \\ a_3 & b_3 & c_3 \end{vmatrix}$ For a_2: $\begin{vmatrix} a_1 & b_1 & c_1 \\ a_2 & b_2 & c_2 \\ a_3 & b_3 & c_3 \end{vmatrix}$

For a_3: $\begin{vmatrix} a_1 & b_1 & c_1 \\ a_2 & b_2 & c_2 \\ a_3 & b_3 & c_3 \end{vmatrix}$

Example 2   Evaluate this third-order determinant.

$$\begin{vmatrix} -1 & 0 & 1 \\ -5 & 1 & -1 \\ 4 & 8 & 1 \end{vmatrix} = -1 \begin{vmatrix} 1 & -1 \\ 8 & 1 \end{vmatrix} - (-5) \begin{vmatrix} 0 & 1 \\ 8 & 1 \end{vmatrix} + 4 \begin{vmatrix} 0 & 1 \\ 1 & -1 \end{vmatrix}$$

We calculate as follows.

$$-1 \begin{vmatrix} 1 & 1 \\ 8 & 1 \end{vmatrix} - (-5) \begin{vmatrix} 0 & 1 \\ 8 & 1 \end{vmatrix} + 4 \begin{vmatrix} 0 & 1 \\ 1 & -1 \end{vmatrix}$$

$$= -1[1 \cdot 1 - 8(-1)] + 5(0 \cdot 1 - 8 \cdot 1) + 4[0 \cdot (-1) - 1 \cdot 1]$$

$$= -1(9) + 5(-8) + 4(-1)$$

$$= -9 - 40 - 4$$

$$= -53$$

SOLVING SYSTEMS BY DETERMINANTS

Here is a system of two equations in two variables:

$$a_1 x + b_1 y = c_1$$
$$a_2 x + b_2 y = c_2.$$

We form three determinants, which we call D, D_x, and D_y.

$$D = \begin{vmatrix} a_1 & b_1 \\ a_2 & b_2 \end{vmatrix} \qquad \text{In } D \text{ we have the coefficients of } x \text{ and } y.$$

$$D_x = \begin{vmatrix} c_1 & b_1 \\ c_2 & b_2 \end{vmatrix} \qquad \text{To form } D_x \text{ we use } D, \text{ but we replace the } x\text{-coefficients with the constants on the right side of the equations.}$$

$$D_y = \begin{vmatrix} a_1 & c_1 \\ a_2 & c_2 \end{vmatrix} \qquad \text{To form } D_y \text{ we use } D, \text{ but we replace the } y\text{-coefficients with the constants on the right.}$$

It is important that the replacement be done *without changing the order of the columns*. Then the solution of the system can be found as follows. This is known as *Cramer's rule*:

$$x = \frac{D_x}{D}, \qquad y = \frac{D_y}{D}.$$

Example 3 Solve, using Cramer's rule:

$$3x - 2y = 7$$
$$3x + 2y = 9.$$

We compute D, D_x, and D_y.

$$D = \begin{vmatrix} 3 & -2 \\ 3 & 2 \end{vmatrix} = 3 \cdot 2 - 3 \cdot (-2) = 6 + 6 = 12$$

$$D_x = \begin{vmatrix} 7 & -2 \\ 9 & 2 \end{vmatrix} = 7 \cdot 2 - 9(-2) = 14 + 18 = 32$$

$$D_y = \begin{vmatrix} 3 & 7 \\ 3 & 9 \end{vmatrix} = 3 \cdot 9 - 3 \cdot 7 = 27 - 21 = 6$$

Then

$$x = \frac{D_x}{D} = \frac{32}{12}, \quad \text{or} \quad \frac{8}{3}$$

and

$$y = \frac{D_y}{D} = \frac{6}{12} = \frac{1}{2}.$$

The solution is $(\frac{8}{3}, \frac{1}{2})$.

Cramer's rule for three equations is very similar to that for two.

$$a_1 x + b_1 y + c_1 z = d_1$$
$$a_2 x + b_2 y + c_2 z = d_2$$
$$a_3 x + b_3 y + c_3 z = d_3$$

$$D = \begin{vmatrix} a_1 & b_1 & c_1 \\ a_2 & b_2 & c_2 \\ a_3 & b_3 & c_3 \end{vmatrix}$$

$$D_x = \begin{vmatrix} d_1 & b_1 & c_1 \\ d_2 & b_2 & c_2 \\ d_3 & b_3 & c_3 \end{vmatrix}$$

$$D_y = \begin{vmatrix} a_1 & d_1 & c_1 \\ a_2 & d_2 & c_2 \\ a_3 & d_3 & c_3 \end{vmatrix}$$

D is again the determinant of the coefficients of x, y, and z. This time we have one more determinant, D_z. We get it by using D and replacing the z-coefficients by the constants on the right:

$$D_z = \begin{vmatrix} a_1 & b_1 & d_1 \\ a_2 & b_2 & d_2 \\ a_3 & b_3 & d_3 \end{vmatrix}.$$

The solution of the system is given by

$$x = \frac{D_x}{D}, \qquad y = \frac{D_y}{D}, \qquad z = \frac{D_z}{D}.$$

Example 4 Solve, using Cramer's rule:

$$x - 3y + 7z = 13$$
$$x + y + z = 1$$
$$x - 2y + 3z = 4.$$

We compute D, D_x, D_y, and D_z:

$$D = \begin{vmatrix} 1 & -3 & 7 \\ 1 & 1 & 1 \\ 1 & -2 & 3 \end{vmatrix} = -10; \qquad D_x = \begin{vmatrix} 13 & -3 & 7 \\ 1 & 1 & 1 \\ 4 & -2 & 3 \end{vmatrix} = 20;$$

$$D_y = \begin{vmatrix} 1 & 13 & 7 \\ 1 & 1 & 1 \\ 1 & 4 & 3 \end{vmatrix} = -6; \qquad D_z = \begin{vmatrix} 1 & -3 & 13 \\ 1 & 1 & 1 \\ 1 & -2 & 4 \end{vmatrix} = -24.$$

Then

$$x = \frac{D_x}{D} = \frac{20}{-10} = -2;$$

$$y = \frac{D_y}{D} = \frac{-6}{-10} = \frac{3}{5};$$

$$z = \frac{D_z}{D} = \frac{-24}{-10} = \frac{12}{5}.$$

The solution is $(-2, \frac{3}{5}, \frac{12}{5})$.

In Example 4 we would not have needed to evaluate D_z. Once we found x and y we could have substituted them into one of the equations to find z. In practice, we use determinants to find only two of the numbers; then we find the third by substituting into an equation.

In using Cramer's rule, we divide by D. If D should be 0, we could not do that. If $D = 0$ and at least one of the other determinants is not 0, then the system does not have a solution, and we say that it is *inconsistent*. If $D = 0$ and all of the other determinants are also 0, then there is an infinite set of solutions. In that case we say that the system is *dependent*.

EXERCISE SET 3.6

Evaluate.

1. $\begin{vmatrix} 2 & 7 \\ 1 & 5 \end{vmatrix}$ **2.** $\begin{vmatrix} 3 & 2 \\ 2 & -3 \end{vmatrix}$ **3.** $\begin{vmatrix} 6 & -9 \\ 2 & 3 \end{vmatrix}$ **4.** $\begin{vmatrix} 3 & 2 \\ -7 & 5 \end{vmatrix}$

5. $\begin{vmatrix} 0 & 2 & 0 \\ 3 & -1 & 1 \\ 1 & -2 & 2 \end{vmatrix}$ **6.** $\begin{vmatrix} 3 & 0 & -2 \\ 5 & 1 & 2 \\ 2 & 0 & -1 \end{vmatrix}$ **7.** $\begin{vmatrix} -1 & -2 & -3 \\ 3 & 4 & 2 \\ 0 & 1 & 2 \end{vmatrix}$

8. $\begin{vmatrix} 1 & 2 & 2 \\ 2 & 1 & 0 \\ 3 & 3 & 1 \end{vmatrix}$ **9.** $\begin{vmatrix} 3 & 2 & -2 \\ -2 & 1 & 4 \\ -4 & -3 & 3 \end{vmatrix}$ **10.** $\begin{vmatrix} 2 & -1 & 1 \\ 1 & 2 & -1 \\ 3 & 4 & -3 \end{vmatrix}$

Solve, using Cramer's rule.

11. $3x - 4y = 6$
$5x + 9y = 10$

12. $5x + 8y = 1$
$3x + 7y = 5$

13. $-2x + 4y = 3$
$3x - 7y = 1$

14. $5x - 4y = -3$
$7x + 2y = 6$

15. $3x + 2y - z = 4$
$3x - 2y + z = 5$
$4x - 5y - z = -1$

16. $3x - y + 2z = 1$
$x - y + 2z = 3$
$-2x + 3y + z = 1$

17. $2x - 3y + 5z = 27$
$x + 2y - z = -4$
$5x - y + 4z = 27$

18. $x - y + 2z = -3$
$x + 2y + 3z = 4$
$2x + y + z = -3$

19. $r - 2s + 3t = 6$
$2r - s - t = -3$
$r + s + t = 6$

20. $a - 3c = 6$
$b + 2c = 2$
$7a - 3b - 5c = 14$

☆
21. Show that an equation of the line through points (x_1, y_1) and (x_2, y_2) can be written

$$\begin{vmatrix} x & y & 1 \\ x_1 & y_1 & 1 \\ x_2 & y_2 & 1 \end{vmatrix} = 0.$$

Solve.

22. $\begin{vmatrix} y & -2 \\ 4 & 3 \end{vmatrix} = 44$ **23.** $\begin{vmatrix} 2 & x & -1 \\ -1 & 3 & 2 \\ -2 & 1 & 1 \end{vmatrix} = -12$ **24.** $\begin{vmatrix} m + 1 & -2 \\ m - 2 & 1 \end{vmatrix} = 27$

25. Write an expression equivalent to $3a + 5b$ using a determinant. Answers may vary.

3.7 BUSINESS AND ECONOMIC APPLICATIONS (Optional)

BREAK-EVEN ANALYSIS

When a company manufactures a product it invests money (this is its *cost*). When it sells the product it takes in money (called *revenue*). It hopes to make a profit. Profit is computed as follows:

Profit = Revenue − Cost

or

$$P = R - C.$$

If R is greater than C, the company makes a profit. If C is greater than R, the company has a loss. If $R = C$, the company breaks even.

There are two kinds of costs. First, there are costs like rent, insurance, buying machinery, and so on. These costs, which must be paid whether a product is produced or not, are called *fixed costs*. Then, when a product is being produced there are costs for labor, materials, marketing, and so on. These are called *variable costs*. They vary according to the amount of product being produced. Adding these together, we find the *total cost* of producing a product.

Example 1 Snowy Electronics, Inc., is planning to make a new kind of radio. Fixed costs the first year will be $90,000, and it will cost $15 to produce each radio (variable costs). Each radio will sell for $26. Determine the profit and break-even point.

a) We know that $P = R - C$, and that

C = (Fixed costs) plus (Variable costs)   or
$C = 90,000 + 15x,$

where x is the number of radios produced.

$R = 26x$ ($26 times the number of radios sold. We are assuming that all radios produced are sold.)

So we have

$P = R -$ C
$P = 26x - [90,000 + 15x],$ or
$P = 11x - 90,000.$

b) The break-even point occurs when revenue equals total cost. So we set $R = C$ and solve:

$$26x = 90{,}000 + 15x$$
$$11x = 90{,}000$$
$$x = 8181.8.$$

The firm will need to produce and sell 8182 radios in order to break even. (8181 radios produce a slight loss, and 8182 a slight profit.) It is interesting to graph the equations in a problem like this.

$C =$ (Fixed costs) plus (Variable costs)

$C = 90{,}000 + 15x$ (1)

$R = 26x$ (2) C and R are both in dollars.

Equation (1) has an intercept on the \$-axis of 90,000 and has a slope of 15.

Equation (2) has a graph that goes through the origin, with a slope of 26.

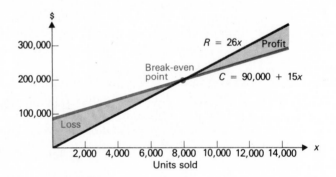

Profit occurs where the revenue is greater than the cost. Loss occurs where revenue is less than cost. The break-even point occurs where the graphs cross.

SUPPLY AND DEMAND

DEMAND VARIES WITH PRICE

As the price of coffee varied over a period of years, the amount sold varied. The table and graph both show how the amount *demanded* by consumers went down as the price went up.

Demand schedule	
Price, P, per kg	Quantity, Q, in millions of kg
$1	25
2	20
3	15
4	10
5	5

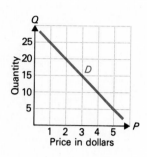

SUPPLY VARIES WITH PRICE

As the price of coffee varied, we see that the amount available varied. The amount of coffee on the market (the *supply*) went up as the price went up.

Supply schedule	
Price, P, per kg	Quantity, Q, in millions of kg
$2.00	5
2.50	10
3.00	15
3.50	20
4.00	25

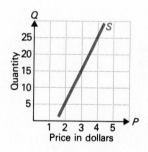

THE EQUILIBRIUM POINT

Let us look at the graphs together. We see that as supply increases, demand decreases. As supply decreases, demand increases. The point of intersection is called the *equilibrium point*. At that price, the amount that the seller will supply is the same amount that the consumer will buy.

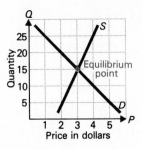

Any ordered pair of coordinates from this graph is (price, quantity), because the *first* axis is the *price* axis and the *second* axis is the *quantity* axis.

Example 2 Find the equilibrium point when demand D and supply S are given by

$$D = 1000 - 60P$$

and

$$S = 200 + 4P.$$

a) We first set $D = S$ and solve:

$$1000 - 60P = 200 + 4P$$
$$1000 - 64P = 200 \qquad \text{Adding } -4P$$
$$-64P = -800 \qquad \text{Adding } -1000$$

$$P = \frac{-800}{-64} = 12.5.$$

Thus the equilibrium price, P_E, is **$12.50** per unit.

b) To find the equilibrium quantity we substitute P_E into either D or S. We use S:

$$200 + 4P = 200 + 4(12.5)$$
$$= 200 + 50 = 250.$$

Thus the equilibrium quantity is 250 units, and the equilibrium point is (**$12.50**, 250).

EXERCISE SET 3.7

For each of the following, (a) find the profit, and (b) find the break-even point.

1. $C = 45x + 600,000$
  $R = 65x$

2. $C = 25x + 360,000$
  $R = 70x$

3. $C = 10x + 120,000$
  $R = 60x$

4. $C = 30x + 49,500$
  $R = 85x$

5. $C = 20x + 10,000$
  $R = 100x$

6. $C = 40x + 22,500$
  $R = 85x$

Find the equilibrium point.

7. $D = 2000 - 60P$
$S = 460 + 94P$

8. $D = 1000 - 10P$
$S = 250 + 5P$

9. $D = 760 - 13P$
$S = 430 + 2P$

10. $D = 800 - 43P$
$S = 210 + 16P$

11. $D = 7500 - 25P$
$S = 6000 + 5P$

12. $D = 8800 - 30P$
$S = 7000 + 15P$

☆

13. A manufacturer is planning a new type of lamp. For the first year, the fixed costs for setting up the production line are $22,500. Variable costs for producing each lamp are estimated to be $40. The sales department projects that 3000 lamps can be sold during the first year. The revenue from each lamp is to be $85.

a) Find the total cost C of producing x lamps.

b) Find the total revenue R from the sale of x lamps.

c) Find total profit P.

d) What profit or loss will the company realize if expected sales of 3000 lamps occur?

e) Find the break-even point.

14. A clothing firm is planning a new line of sport coats. For the first year, the fixed costs for setting up the production line are $10,000. Variable costs for producing each coat are $20. The sales department projects that 2000 coats can be sold during the first year. The revenue from each coat is to be $100.

a) Find the total cost C of producing x coats.

b) Find the total revenue R from the sale of x coats.

c) Find the total profit P.

d) What profit or loss will the company realize if expected sales of 2000 coats occur?

e) Find the break-even point.

3.8 INCONSISTENT AND DEPENDENT EQUATIONS (Optional)

INCONSISTENT EQUATIONS

> If a system of equations has a solution, we say that it is *consistent*. If a system does not have a solution, we say that it is *inconsistent*.

Example 1 Determine whether this system is consistent or inconsistent:

$$x - 3y = 1$$
$$-2x + 6y = 5.$$

We attempt to find a solution:

$$2x - 6y = 2 \qquad \text{Multiplying the first equation by 2}$$
$$\underline{-2x + 6y = 5}$$
$$0 = 7 \qquad \text{Adding}$$

The last equation says that $0 \cdot x + 0 \cdot y = 7$. There are no numbers x and y for which this is true, so there is no solution. The system is inconsistent.

Whenever we obtain a statement such as $0 = 7$, which is clearly false, we know that the system we are trying to solve is inconsistent.

Example 2 Determine whether this system is consistent or inconsistent:

$$y + 3z = 4 \qquad (1)$$
$$-x - y + 2z = 0 \qquad (2)$$
$$x + 2y + z = 1. \qquad (3)$$

We attempt to find a solution. If we add equations (2) and (3), we get

$$y + 3z = 1.$$

If we multiply this by -1 and then add the result to equation (1) we get

$$0 = 3$$

The system is inconsistent.

DEPENDENT EQUATIONS

Consider this system:

$$2x + 3y = 1$$
$$4x + 6y = 2.$$

If we multiply the first equation by 2, we get the second equation, so the two equations have the same solutions (are equivalent). The system of two equations is equivalent to only one of them. We call them *dependent*.

> **If a system of n linear equations is equivalent to a system of fewer than n of them, we say that the system is *dependent*. If such is not the case, we call the system *independent*.**

Example 3 Determine whether this system is dependent or independent:

$$2x + 3y = 1$$
$$4x + 6y = 2.$$

We attempt to solve:

$$-4x - 6y = -2 \qquad \text{Multiplying the first equation by } -2$$
$$\underline{4x + 6y = 2}$$
$$0 = 0. \qquad \text{Adding}$$

The last equation says that $0 \cdot x + 0 \cdot y = 0$. This is true for all numbers x and y. The system is dependent.

Whenever we get an obviously true sentence like this, we know that the system we are trying to solve is dependent.

Example 4 Determine whether this system is dependent or independent:

$$x + 2y + z = 1$$
$$x - y + z = 1$$
$$2x + y + 2z = 2.$$

We attempt to solve:

$$x + 2y + z = 1$$
$$-3y = 0 \qquad \text{Multiplying the first equation by } -1 \text{ and adding it to the second}$$
$$-3y = 0. \qquad \text{Multiplying the first equation by } -2 \text{ and adding it to the third}$$

It is now obvious that our system of three equations is equivalent to a system of two equations. The system is dependent.

When we attempt to solve a system and obtain two equations that are identical, or obviously equivalent, we know that the system is dependent. Or, if we obtain a statement obviously true, such as $3 = 3$, we know that the system is dependent.

EXERCISE SET 3.8

Determine whether these systems are consistent or inconsistent.

1. $x + 2y = 6$
$\ 2x = 8 - 4y$

2. $y - 2x = 1$
$\ 2x - 3 = y$

3. $y - x = 4$
$\ x + 2y = 2$

4. $y + x = 5$
$\ y = x - 3$

5. $x - 3 = y$
$\ 2x - 2y = 6$

6. $3y = x - 2$
$\ 3x = 6 + 9y$

7. $x + z = 0$
$\ x + y + 2z = 3$
$\ y + z = 2$

8. $x + y = 0$
$\ x + z = 1$
$\ 2x + y + z = 2$

9. $x + z = 0$
$\ x + y = 1$
$\ y + z = 1$

10. $2x + y = 1$
$\ x + 2y + z = 0$
$\ x + z = 1$

Determine whether these systems are dependent or independent.

11. $x - 3 = y$
$\ 2x - 2y = 6$

12. $3y = x - 2$
$\ 3x = 6 + 9y$

13. $y - x = 4$
$\ x + 2y = 2$

14. $y + x = 5$
$\ y = x - 3$

15. $2x + 3y = 1$
$\ x + 1.5y = 0.5$

16. $15x + 6y = 20$
$\ 7.5x - 10 = -3y$

17. $x + z = 0$
$\ x + y = 1$
$\ y + z = 1$

18. $2x + y = 1$
$\ x + 2y + z = 0$
$\ x + z = 1$

19. $x + y + z = 1$
$\ -x + 2y + z = 2$
$\ 2x - y = -1$

20. $y + z = 1$
$\ x + y + z = 1$
$\ x + 2y + 2z = 2$

TEST OR REVIEW—CHAPTER 3

1. Solve graphically.

$2x + y = 4$
$x - 3y = -5$

2. Solve.

$3x + y = -1$
$2x - 3y = -8$

3. The perimeter of a rectangular field is 710 meters. The length is 115 m greater than the width. Find the dimensions.

4. Solve.

$$2x + y + z = -3$$
$$2x - y + 2z = 0$$
$$3x - 2y + 2z = -3$$

5. A student scores a total of 248 points on three tests. The total of the first two scores is 66 more than the score on the third test. The score on the third test is 6 more than that on the second test. Find the scores.

6. Evaluate.

$$\begin{vmatrix} 1 & 0 & 2 \\ 3 & 1 & 5 \\ 1 & -2 & 2 \end{vmatrix}$$

7. Find the profit and the break-even point.

$$C = 15x + 90{,}000$$
$$R = 50x$$

8. Determine whether this system of equations is consistent or inconsistent.

$$2x + 2y = 0$$
$$4x + 4z = 4$$
$$2x + y + z = 2$$

9. Determine whether this system of equations is dependent or independent.

$$3x + 3z = 0$$
$$2x + 2y = 2$$
$$3y + 3z = 3$$

4

POLYNOMIALS AND FUNCTIONS

4.1 SOME PROPERTIES OF POLYNOMIALS

POLYNOMIALS

Expressions like these are called *polynomials in one variable:*

$$-7x + 5, \qquad 2y^2 + 5y - 3.$$

They show additions and subtractions. Each part to be added or sub-tracted is a number or a number times a variable to some whole-number power. A polynomial can also consist of just one of the parts. The following are polynomials:

$$5x^2, \qquad -8a, \qquad 5, \qquad 3t, \qquad 0.$$

Expressions like the following are called *polynomials in several variables:*

$$5x - xy^2 + 7y + 2, \qquad 15x^3y^2, \qquad \pi r^2 + 2\pi rh.$$

Evaluating polynomials is important in many applications.

Example 1   The cost, in cents per kilometer, of operating an auto-mobile at speed s (in km/h) is approximated by the polynomial $0.002s^2 - 0.21s + 15$. How much does it cost to operate at 80 km/h?

We substitute:

$$0.002(\;80\;)^2 - 0.21(\;80\;) + 15 = 11.$$

It costs 11¢ per kilometer.

Example 2   The *magic number* in baseball can be found by using the polynomial $G - P - L + 1$.

Here G is the number of games in the season, P is the number of games the leading team has played, and L is the number of games ahead in the loss column. Any combination of wins by the leading team and losses by the second-place team totaling the magic number results in the championship.

Find the magic number when there are 162 games in a season and team A has played 145 games and is in the lead by 5 games. The magic num-ber is

$$G - P - L + 1 = 162 - 145 - 5 + 1, \quad \text{or} \quad 13.$$

Example 3   If there are n teams in a league and they play each other twice in a season, the total number of games played is given by the polynomial

$$n^2 - n.$$

If the teams play each other only once, the total number of games is given by the polynomial

$$\frac{1}{2}n^2 - \frac{1}{2}n.$$

Find the number of games played when there are 12 teams in a league and they play each other **twice**.

The total number of games is

$$n^2 - n = 12^2 - 12 = 144 - 12, \quad \text{or} \quad 132.$$

Recall that we can subtract by adding an additive inverse. So any polynomial can be rewritten with only plus signs between the parts.

Examples

4. $4x - 3 = 4x + (-3)$

5. $7xy - 4x^2 = 7xy + (-4x^2)$

6. $-9x^3z - 7xyz^2 + y - 6x^3y^2 = -9x^3z + (-7xyz^2) + y + (-6x^3y^2)$

TERMS AND COEFFICIENTS

When a polynomial is written with plus signs between the parts, the parts are called its *terms*. The numbers in the terms, other than the exponents, are called *coefficients*.

Example 7 Identify the terms and the coefficients in

$$-3x^2 - 4x^3z + 14x^4y^3 - 9.$$

We rewrite with plus signs:

$$-3x^2 + (-4x^3z) + 14x^4y^3 + (-9).$$

Caution! Do not confuse terms (things to be added) with factors (things to be multiplied).

The terms are $-3x^2$, $-4x^3z$, $14x^4y^3$, and -9. The coefficients are -3, -4, 14, and -9.

If a coefficient is 0, we usually do not write the term. When there is a zero coefficient in a polynomial with one variable, we say that there is a *missing* term.

Example 8   In $9x^5 - 2x^3 + 5x - 9$ there is no term with x^4. We say that the "x^4-term is missing." The x^2-term is also missing.

DEGREES

The *degree of a term* is the sum of the exponents of the variables. The *degree of a polynomial* is the same as that of its term of highest degree.

Example 9   Determine the degree of each term and the degree of the polynomial:

$$6x^2 + 8x^2y^3 - 17xy + 2y + 3.$$

Term	$6x^2$	$8x^2y^3$	$-17xy$	$2y$	3
Degree	2	5	2	1	0

You might think of $2y$ as $2y^1$ and 3 as $3y^0$.

The degree of the polynomial is 5.

COLLECTING LIKE TERMS

Terms that have the same variables with the same exponents are called *like* terms (or *similar* terms). The following are examples of similar terms:

$$4x^2 \text{ and } -2x^2, \quad 5x^2y^3 \text{ and } 9x^2y^3, \quad 7y^6z^9 \text{ and } 23y^6z^9.$$

The following are examples of terms that are not similar:

$$x^2y \text{ and } xy^2, \quad 5xyz^2 \text{ and } 13xyz.$$

We can often simplify expressions by collecting like terms. The process is based on the distributive laws. Thus we know that the simplified expression and the original expression will name the same number for all replacements. That is, the expressions are *equivalent*.

Examples   Collect like terms.

10. $4x - 5x + 2xy + 3xy = (4 - 5)x + (2 + 3)xy$
$$= -1 \cdot x + 5xy, \quad \text{or} \quad -x + 5xy$$

11. $4x^3y + 5z - 4x^2 - 2x^3y + 5x^2 = 2x^3y + x^2 + 5z$

ASCENDING AND DESCENDING ORDER

We generally arrange polynomials in one variable so that the exponents decrease (descending order) or so that they increase (ascending order).

Examples

12. Arrange $-2x + x^2 + 5x^7 + 6x^5$ in descending order:

$5x^7 + 6x^5 + x^2 - 2x.$

13. Arrange $9x^{10} - 2 + 4x - 8x^2$ in ascending order:

$-2 + 4x - 8x^2 + 9x^{10}$

We usually write polynomials in one variable in descending order, although it is not incorrect to use ascending or any other order.

For polynomials in several variables we choose one of the variables and arrange the terms with respect to it.

Examples

14. Arrange $y^4 + 2 - 5x^2 + 3x^3y + 7xy$ in descending powers of x:

$3x^3y - 5x^2 + 7xy + y^4 + 2.$

15. Arrange $-17x^3y^2 + 4xy^3 + 13xy + 8$ in ascending powers of y:

$8 + 13xy - 17x^3y^2 + 4xy^3.$

MONOMIALS, BINOMIALS, AND TRINOMIALS

Polynomials with just one term are called *monomials*. Polynomials with just two terms are called *binomials*. Those with just three terms are called *trinomials*.

Example 16

Monomials	Binomials	Trinomials
$5x^2$	$3x - 5$	$5x^2 - 6x + 1$
-1	$-2y^2 + 8t$	$3pq - 4q^2 - 10$

EXERCISE SET 4.1

Rewrite each polynomial with plus signs and evaluate each when $x = -1$ and $y = 2$.

1. $8x^2 - 2x - 5$ **2.** $9y^3 - 3y^2 + 6$

3. $18xy - 8x^3 - y + 50$ **4.** $24y^3 - 9y^2 - xy - 21x$

Identify the terms and coefficients.

5. $5x^3 + 7x^2 - 3x - 9$ **6.** $8y^3 - 9y^2 + 12y + 11$

7. $-3xyz + 7x^2y^2 - 5xy^2z + 4xyz^2$ **8.** $-9abc + 19a^2bc - 8ab^2c + 12abc^2$

Determine the degree of each term and the degree of the polynomial.

9. $x^2 + 3y^5 - x^3y^4 - 7$ **10.** $y^3 + 2y^6 + x^2y - 9$

11. $x^5 + 3x^2y^4 - 5xy + 4x - 3$ **12.** $9y^6 + 2x^4y^4 - 8x^3y + 5x^2 - 9$

Collect like terms.

13. $6x^2 - 7x^2 + 3x^2$ **14.** $-2y^2 - 7y^2 + 5y^2$

15. $5x - 4y - 2x + 5y$ **16.** $4a - 9b - 6a + 3b$

17. $5a + 7 - 4 + 2a - 6a + 3$ **18.** $9x + 12 - 8 - 7x + 5x + 10$

19. $3a^2b + 4b^2 - 9a^2b - 6b^2$ **20.** $5x^2y^2 + 4x^3 - 8x^2y^2 - 12x^3$

21. $8x^2 - 3xy + 12y^2 + x^2 - y^2 + 5xy + 4y^2$

22. $a^2 - 2ab + b^2 + 9a^2 + 5ab - 4b^2 + a^2$

23. $4x^2y - 3y + 2xy^2 - 5x^2y + 7y + 7xy^2$

24. $3xy^2 + 4xy - 7xy^2 + 7xy + x^2y$

Arrange in descending order.

25. $x - 3x^2 + 1 + x^3$ **26.** $x^2 - 8x^4 + 9 - x^3$

27. $a - a^3 + 5a^5 - 9 + 6a^2$ **28.** $y^2 - 9y^4 + 18 - y + 7y^5$

Arrange in ascending order.

29. $2y^3 - y + y^4 - 7 + 3y^2$ **30.** $5y^4 - 4y^3 + 8y - 11 - 2y^2$

31. $x^2 - x + 5x^5 - 9x^3 + 18x^4$ **32.** $-6x^3 + 9x - 42 - 3x^2 + 5x^4$

Arrange in descending powers of y.

33. $x^2y^2 + x^3y - xy^3 + 1$ **34.** $x^3y - x^2y^2 + xy^3 + 6$

Arrange in ascending powers of x.

35. $-9x^3y + 3xy^3 + x^2y^2 + 2x^4$ **36.** $5x^2y^2 - 9xy + 8x^3y^2 - 5x^4$

☆

Collect like terms.

37. ▦ $0.00976x^2y^2 - 0.08054x^3y + 0.80149x^2y^2 + 0.00943x^3y$

38. ▦ $8{,}592{,}429xy^2z - 42{,}004x^2y^2 + 5{,}976{,}006x^2y^2 - 4{,}008{,}793xy^2z$

Evaluate each polynomial using the indicated numbers.

39. $3 - xy - x^2 - y^2$; $x = \dfrac{1}{2}, y = -\dfrac{1}{2}$ **40.** $\dfrac{1}{10{,}000}x^3 + \dfrac{1}{1000}x^2 - \dfrac{1}{100}x + \dfrac{1}{10}$; $x = 10$

41. $0.008z^4 - 0.002z^6$; $z = 0.1$ **42.** $a^2 + ab - 12b^2$; $a = 3t, b = t$

4.2 CALCULATIONS WITH POLYNOMIALS

ADDITION

The sum of two polynomials is a polynomial. To add polynomials we can write a plus sign between them and then collect like terms.

Example 1 Add: $-5x^3 + 3x - 5$ and $8x^3 + 4x^2 + 7$.

$(-5x^3 + 3x - 5) + (8x^3 + 4x^2 + 7)$

$= (-5 + 8)x^3 + 4x^2 + 3x + (-5 + 7) = 3x^3 + 4x^2 + 3x + 2$

We still arrange the terms in descending order.

The use of columns is often helpful. To do this we write the polynomials one under the other, writing like terms under one another and leaving spaces for missing terms. Let us do the addition in Example 1 using columns.

$$
\begin{array}{rrrr}
-5x^3 & & +\,3x & -\,5 \\
8x^3 & +\,4x^2 & & +\,7 \\
\hline
3x^3 & +\,4x^2 & +\,3x & +\,2 \\
\end{array}
$$

Example 2 Add: $4ax^2 + 4bx - 5$ and $3ax^2 + 5bx + 8$.

$$
\begin{array}{l}
4ax^2 + 4bx - 5 \\
3ax^2 + 5bx + 8 \\
\hline
7ax^2 + 9bx + 3 \\
\end{array}
$$

Although the use of columns is helpful for complicated examples, you should attempt to write only the answer when you can.

Example 3   Add.

$$(13x^3y + 3x^2y - 5y) + (x^3y + 4x^2y - 3xy + 3y)$$
$$= 14x^3y + 7x^2y - 3xy - 2y$$

ADDITIVE INVERSES

If the sum of two polynomials is 0, they are called *additive inverses* of each other. For example, $(3x - 2) + (-3x + 2) = 0$, so the additive inverse of $(3x - 2)$ is $(-3x + 2)$. In symbols,

The additive
inverse of $3x - 2$ is $-3x + 2$

$$-\quad (3x - 2) = -3x + 2.$$

Examples   Rewrite each additive inverse without parentheses.

4. $-(5x^3 + 2) = -1 \cdot (5x^3 + 2)$ Replacing $-$ by -1
$$= -1 \cdot 5x^3 + (-1) \cdot 2 \quad \text{Using the distributive law}$$
$$= -5x^3 + (-2)$$
$$= -5x^3 - 2$$

5. $-(9xy^2 - 4x^3y - 8x^4 + 7) = (-1)9xy^2 + (-1)(-4x^3y)$
$$+ (-1)(-8x^4) + (-1) \cdot 7$$
$$= -9xy^2 + 4x^3y + 8x^4 - 7$$

Renaming an additive inverse like this amounts to "changing the sign" of every term inside the parentheses.

Example 6   Rewrite without parentheses.

$$-(5xy^2 - 7x^3y - 8x + 4) = -5xy^2 + 7x^3y + 8x - 4$$

SUBTRACTION

To subtract one polynomial from another, we add the inverse of the subtrahend.

Example 7 Subtract.

$$(-9x^5 + 2x^2 + 4) - (2x^5 + 4x^3 - 3x^2)$$
$$= (-9x^5 + 2x^2 + 4) + (-2x^5 - 4x^3 + 3x^2) \qquad \text{Adding the inverse}$$
$$\text{of the subtrahend}$$

$$= -11x^5 - 4x^3 + 5x^2 + 4$$

After some practice, you will find that you can skip some steps, by mentally "changing the sign of each term" and then collecting like terms. Eventually, all you will write is the answer.

We can also use columns for subtraction. We change signs mentally in the subtrahend and then add.

Example 8 Subtract.

$$(4x^2y - 6x^3y^2 + x^2y^2) - (4x^2y + x^3y^2 + 3x^2y^3).$$

$$4x^2y - 6x^3y^2 \qquad\qquad + x^2y^2$$
$$4x^2y + \ \ x^3y^2 + 3x^2y^3 \longleftarrow \qquad \boxed{\text{Don't forget to change the sign of } \textit{every} \text{ term.}}$$
$$\overline{\qquad - 7x^3y^2 - 3x^2y^3 + x^2y^2}$$

As with addition, you should avoid the use of columns as much as possible. After sufficient experience, you should write only the answer.

EXERCISE SET 4.2

Add.

1. $3x^2 + 5y^2 + 6$ and
$2x^2 - 3y^2 - 1$

2. $9y^2 + 8y - 4$ and
$12y^2 - 5y + 8$

3. $2a + 3b - c$ and
$4a - 2b + 2c$

4. $5x - 4y + 2z$ and
$9x + 12y - 8z$

5. $a^2 - 3b^2 + 4c^2$ and
$-5a^2 + 2b^2 - c^2$

6. $x^2 - 5y^2 - 9z^2$ and
$-6x^2 + 9y^2 - 2z^2$

7. $x^2 + 2x - 3xy - 7$ and
$-3x^2 - x + 2xy + 6$

8. $3a^2 - 2b + ab + 6$ and
$-a^2 + 5b - 5ab - 2$

9. $7x^2y - 3xy^2 + 4xy$ and
$-2x^2y - xy^2 + xy$

10. $7ab - 3ac + 5bc$ and
$13ab - 15ac - 8bc$

11. $2r^2 + 12r - 11$ and
$6r^2 - 2r + 4$ and
$r^2 - r - 2$

12. $5x^2 + 19x - 23$ and
$-7x^2 - 11x + 12$ and
$-x^2 - 9x + 8$

Rename each additive inverse without parentheses.

13. $-(5x^3 - 7x^2 + 3x - 6)$

14. $-(8y^4 - 18y^3 + 4y - 9)$

Subtract.

15. $(8x - 4) - (-5x + 2)$

16. $(9y + 3) - (-4y - 2)$

17. $(-3x^2 + 2x + 9) - (x^2 + 5x - 4)$

18. $(-9y^2 + 4y + 8) - (4y^2 + 2y - 3)$

19. $(5a - 2b + c) - (3a + 2b - 2c)$

20. $(8x - 4y + z) - (4x + 6y - 3z)$

21. $(3x^2 - 2x - x^3) - (5x^2 - 8x - x^3)$

22. $(8y^2 - 3y - 4y^3) - (3y^2 - 9y - 7y^3)$

23. $(5a^2 + 4ab - 3b^2) - (9a^2 - 4ab + 2b^2)$

24. $(9y^2 - 14yz - 8z^2) - (12y^2 - 8yz + 4z^2)$

☆

25. ▦ Add: $15.36x^2 - 4.039y^2 + 1.002x$ and
$12.58x^2 + 13.144y^2 + 7.148x.$

26. ▦ Subtract: $(8035x - 1735) - (-4111x + 3892).$

Simplify.

27. $\dfrac{2}{3}(18a^2 - 33ab + 3b^2) - \dfrac{3}{4}(20a^2 + 44ab + 56b^2)$

28. $5(3 - x) + (5 - x - x^2) - 5(3 - 2x - x^3)$

29. $(0.06u^2 + 0.003v^2) - (v^2 - 0.6u^2) - (0.003u^2 - 6v^2)$

30. $\left(\dfrac{1}{2} - \dfrac{1}{3}y + \dfrac{1}{4}y^2\right) + \left(\dfrac{1}{2}y + \dfrac{1}{3}y^2 - \dfrac{1}{4}y^3\right) - \left(\dfrac{1}{2}y^2 - \dfrac{1}{3}y^3 - \dfrac{1}{4}y^4\right)$

4.3 MULTIPLYING POLYNOMIALS

MULTIPLYING MONOMIALS

To multiply monomials we multiply the coefficients. Then we multiply the variables. This process is based on the commutative and associative laws of multiplication.

Examples   Multiply and simplify.

1. $(-8x^4y^7)(5x^3y^2) = -8 \cdot 5 \cdot x^4 \cdot x^3 \cdot y^7 \cdot y^2$

$$= -40x^{4+3}y^{7+2} \qquad \text{Adding exponents}$$

$$= -40x^7y^9$$

2. $(3x^2yz^5)(-6x^5y^{10}z^2) = 3 \cdot (-6) \cdot x^2 \cdot x^5 \cdot y \cdot y^{10} \cdot z^5 \cdot z^2$

$$= -18x^7y^{11}z^7$$

You should try to work mentally, writing only the answer.

MULTIPLYING MONOMIALS AND BINOMIALS

The distributive laws are the basis for multiplying polynomials other than monomials.

Example 3   Multiply: $2x$ and $3x - 5$.

$$2x \cdot (3x - 5) = 2x \cdot (3x) - 2x \cdot (5) \qquad \text{Using a distributive law}$$

$$= 6x^2 - 10x \qquad \text{Multiplying the monomials}$$

Example 4   Multiply: $3y^2 + 4$ and $y - 2$.

$$(3y^2 + 4)\,(y - 2) \;=\; 3y^2\,(y - 2) \;+\; 4\,(y - 2) \qquad \begin{array}{l}\text{Using a}\\\text{distributive law}\end{array}$$

$$\qquad\qquad\qquad\text{(a)}\qquad\qquad\text{(b)}$$

We consider the two parts (a) and (b) separately.

a) $3y^2\,(y - 2) = 3y^2\,y - 3y^2\,(2) \qquad \text{Using a distributive law}$

$$= 3y^3 - 6y^2 \qquad \text{Multiplying the monomials}$$

b) $4\,(y - 2) = 4\,y - 4\,(2) \qquad \text{Using a distributive law}$

$$= 4y - 8 \qquad \text{Multiplying the monomials}$$

We replace the parts (a) and (b) in the original expression with their answers, and combine like terms if we can.

$$(3y^2 + 4)(y - 2) = (3y^3 - 6y^2) + (4y - 8)$$

$$= 3y^3 - 6y^2 + 4y - 8$$

MULTIPLYING ANY TWO POLYNOMIALS

Example 5   Multiply: $p + 2$ and $p^4 - 2p^3 + 3$.

By the distributive laws we have:

$$(\boxed{p + 2})(p^4 - 2p^3 + 3) = (\boxed{p + 2})(p^4) - (\boxed{p + 2})(2p^3)$$

$$+ (\boxed{p + 2})(3)$$

$$= p \cdot (\boxed{p^4}) + 2(\boxed{p^4}) - p \cdot (\boxed{2p^3})$$

$$- 2(\boxed{2p^3}) + p \cdot (\boxed{3}) + 2(\boxed{3})$$

$$= p^5 + 2p^4 - 2p^4 - 4p^3 + 3p + 6$$

$$= p^5 - 4p^3 + 3p + 6$$

In the last step we combined like terms.

Now we can see how to multiply any two polynomials.

> **To multiply two polynomials, multiply each term of one by every term of the other. Then add the results, combining like terms if possible.**

We can use columns for long multiplications. We multiply each term at the top by every term at the bottom, keeping like terms in columns. Then we add.

Example 6   Multiply: $5x^3 + x - 4$ and $-2x^2 + 3x + 6$.

$$
\begin{array}{l}
5x^3 + x - 4 \\
-2x^2 + 3x + 6 \\
\hline
-10x^5 - 2x^3 + 8x^2 \qquad \text{Multiplying by } -2x^2 \\
15x^4 + 3x^2 - 12x \qquad \text{Multiplying by } 3x \\
30x^3 + 6x - 24 \qquad \text{Multiplying by } 6 \\
\hline
-10x^5 + 15x^4 + 28x^3 + 11x^2 - 6x - 24 \qquad \text{Adding}
\end{array}
$$

EXERCISE SET 4.3

Multiply.

1. $2y^2$ and $5y$

2. $-3x^2$ and $2xy$

3. $5x$ and $-4x^2y$

4. $-3ab^2$ and $2a^2b^2$

5. $2x^3y^2$ and $-5x^2y^4$

6. $7a^2bc^4$ and $-8ab^3c^2$

7. $2x$ and $3 - x$

8. $4a$ and $a^2 - 5a$

9. $3ab$ and $a + b$

10. $2xy$ and $2x - 3y$

11. $5cd$ and $3c^2d - 5cd^2$

12. a^2 and $2a^2 - 5a^3$

13. $2x + 3$ and $3x - 4$

14. $2a - 3b$ and $4a - b$

15. $s + 3t$ and $s - 3t$

16. $y + 4$ and $y - 4$

17. $x - y$ and $x - y$

18. $a + 2b$ and $a + 2b$

19. $y + 8x$ and $2y - 7x$

20. $x + y$ and $x - 2y$

21. $a^2 - 2b^2$ and $a^2 - 3b^2$

22. $2m^2 - n^2$ and $3m^2 - 5n^2$

23. $x - 4$ and $x^2 + 4x + 16$

24. $y + 3$ and $y^2 - 3y + 9$

25. $x + y$ and $x^2 - xy + y^2$

26. $a - b$ and $a^2 + ab + b^2$

27. $a^2 + a - 1$ and $a^2 + 4a - 5$

28. $x^2 - 2x + 1$ and $x^2 + x + 2$

29. $4a^2b - 2ab + 3b^2$ and $ab - 2b + a$

30. $2x^2 + y^2 - 2xy$ and $x^2 - 2y^2 - xy$

☆

31. Express the area of this box as a polynomial. The box is rectangular with an open top, and dimensions as shown.

32. A box is to be made from a piece of cardboard 12 inches square. Corners are cut out and the sides are folded up. Express the volume of the box as a polynomial.

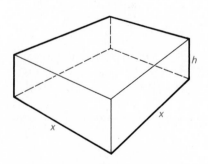

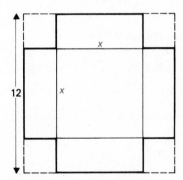

4.4 SPECIAL PRODUCTS OF POLYNOMIALS

We can multiply polynomials faster by learning some patterns.

PRODUCTS OF TWO BINOMIALS

To multiply any two polynomials we multiply each term of one by every term of the other. For two binomials we can think of it this way.

Multiply the first terms, then the outside terms, then the inside terms, then the last terms. We abbreviate this FOIL.

$$
\begin{array}{cccc}
& \text{F} & \text{O} & \text{I} & \text{L} \\
(A + B)(C + D) & = AC & + AD & + BC & + BD
\end{array}
$$

Examples   Multiply.

$$
\begin{array}{cccc}
\text{F} & \text{O} & \text{I} & \text{L}
\end{array}
$$

1. $(x + 5)(x - 8) = x^2 - 8x + 5x - 40$

$\qquad\qquad\qquad = x^2 - 3x - 40$ Combining like terms

2. $(p - 3q)(2p - 5q) = 2p^2 - 5pq - 6pq + 15q^2$

$\qquad\qquad\qquad\qquad = 2p^2 - 11pq + 15q^2$ Combining like terms

3. $(3xy + 2x)(x^2 + 2xy^2) = 3x^3y + 6x^2y^3 + 2x^3 + 4x^2y^2$

> Do this mentally if you can.

SQUARES OF BINOMIALS

Note the following:

$$
\begin{aligned}
(A + B)^2 &= (A + B)(A + B) \\
&= A^2 + AB + AB + B^2 \\
&= A^2 + 2AB + B^2
\end{aligned}
$$

$$
(A + B)^2 = A^2 + 2AB + B^2
$$
$$
(A - B)^2 = A^2 - 2AB + B^2
$$

The square of a binomial is the square of the first expression, plus or minus twice the product of the expressions, plus the square of the second expression.

> Memorize this rule. Then, when you are squaring a binomial, say the words to yourself. That will help keep you from making errors.

Examples Multiply.

$$(A - B)^2 = A^2 - 2\,A\,B + B^2$$

4. $(y - 5)^2 = y^2 - 2(5)(y) + 5^2$

$$= y^2 - 10y + 25$$

> Note that these signs are the same.

5. $(2x + 9)^2 = (2x)^2 + 2(2x)(9) + 9^2$

$$= 4x^2 + 36x + 81$$

Try to find such products mentally, and try to write only the answer.

Examples Multiply.

6. $(2x + 3y)^2 = (2x)^2 + 2(2x)(3y) + (3y)^2$

$$= 4x^2 + 12xy + 9y^2$$

> Remember to say the rule to yourself while multiplying.

7. $(3x^2 - 5xy^2)^2 = (3x^2)^2 - 2(3x^2)(5xy^2) + (5xy^2)^2$

$$= 9x^4 - 30x^3y^2 + 25x^2y^4$$

> A common error results from thinking that $(A + B)^2 = A^2 + B^2$, and thus leaving out the middle term. If you say the rule to yourself as you work, you will not make that mistake.

PRODUCTS OF SUMS AND DIFFERENCES

Note the following:

$$\begin{array}{cccc} \text{F} & \text{O} & \text{I} & \text{L} \end{array}$$
$$(A + B)(A - B) = A^2 - AB + AB - B^2$$
$$= A^2 - B^2.$$

> ### $(A + B)(A - B) = A^2 - B^2$
>
> **The product of the sum and difference of two expressions is the square of the first expression minus the square of the second.**

> Memorize this rule. Say it to yourself as you work.

Examples   Multiply.

$$(A + B)(A - B) = A^2 - B^2$$

8. $(y + 5)(y - 5) = y^2 - 5^2$
$$= y^2 - 25$$

9. $(3x - 2)(3x + 2) = (3x)^2 - 2^2$
$$= 9x^2 - 4$$

> If you say the rule as you work, you will not need to write a middle step. You can simply write the answer.

Examples   Multiply.

10. $(2xy^2 + 3x)(2xy^2 - 3x) = (2xy^2)^2 - (3x)^2$
$$= 4x^2y^4 - 9x^2$$

$$(\quad A \quad - B)(\quad A \quad + B) = \quad A^2 \quad - B^2$$

11. $(5y + 4 - 3x)(5y + 4 + 3x) = (5y + 4)^2 - (3x)^2$
$$= 25y^2 + 40y + 16 - 9x^2$$

> This multiplying could have been done by columns, but it is much faster using the present rule.

Example 12   Multiply: $(3xy^2 + 4y)(-3xy^2 + 4y)$.

$$(3xy^2 + 4y)(-3xy^2 + 4y) = (4y + 3xy^2)(4y - 3xy^2)$$
$$= (4y)^2 - (3xy^2)^2$$
$$= 16y^2 - 9x^2y^4$$

You should learn to multiply polynomials mentally, even when several kinds are mixed. First check to see what kinds of polynomials are to be multiplied. Then use the quickest method. In Exercise Set 4.4 the types are separated. In Exercise Set 4.4A they are mixed.

EXERCISE SET 4.4

Multiply.

1. $(a + 2)(a + 3)$

2. $(x + 5)(x + 8)$

3. $(y + 3)(y - 2)$

4. $(y - 4)(y + 7)$

5. $\left(b - \dfrac{1}{3}\right)\left(b - \dfrac{1}{2}\right)$

6. $\left(x - \dfrac{1}{2}\right)\left(x - \dfrac{1}{4}\right)$

7. $(2x + 9)(x + 2)$

8. $(3b + 2)(2b - 5)$

9. $(2x - 3y)(2x + y)$

10. $(2a - 3b)(2a - b)$

11. $(x + 3)^2$

12. $(y - 7)^2$

13. $\left(2a + \dfrac{1}{3}\right)^2$

14. $\left(3c - \dfrac{1}{2}\right)^2$

15. $(x - 2y)^2$

16. $(2s + 3t)^2$

17. $(2x^2 - 3y^2)^2$

18. $(3s^2 + 4t^2)^2$

19. $(a^2b^2 + 1)^2$

20. $(x^2y - xy^2)^2$

21. $(c + 2)(c - 2)$

22. $(x - 3)(x + 3)$

23. $(2a + 1)(2a - 1)$

24. $(3 - 2x)(3 + 2x)$

25. $(3m - 2n)(3m + 2n)$

26. $(3x + 5y)(3x - 5y)$

27. $(x^2 + yz)(x^2 - yz)$

28. $(2a^2 + 5ab)(2a^2 - 5ab)$

☆
Multiply. Assume that variables in exponents represent positive integers.

29. $(6y)^2 \left(-\dfrac{1}{3}x^2y^3\right)^3$

30. $[(a^{2n})^{2n}]^4$

31. $(-r^6s^2)^3 \left(-\dfrac{r^2}{6}\right)^2 (9s^4)^2$

32. $(z^{n^2})^{n^3}(z^{4n^3})^{n^2}$

33. $\left(-\dfrac{3}{8}xy^2\right)\left(-\dfrac{7}{9}x^3y\right)\left(-\dfrac{8}{7}x^2y\right)^2$

34. $(a^xb^{2y})\left(\dfrac{1}{2}a^{3x}b\right)^2$

35. $(-8s^3t)(2s^5 - 3s^3t^4 + st^7 - t^{10})$

36. $y^3z^n(y^{3n}z^3 - 4yz^{2n})$

37. $[(2x - 1)^2 - 1]^2$

38. $[x + y + 1][x^2 - x(y + 1) + (y + 1)^2]$

39. $[(a + b)(a - b)][5 - (a + b)][5 + (a + b)]$

40. $(y - 1)^6(y + 1)^6$

41. $(r^2 + s^2)^2(r^2 + 2rs + s^2)(r^2 - 2rs + s^2)$

42. $\left(3x^s - \dfrac{5}{11}\right)^2$

43. $(a - b + c - d)(a + b + c + d)$

44. $\left(\dfrac{2}{3}x + \dfrac{1}{3}y + 1\right)\left(\dfrac{2}{3}x - \dfrac{1}{3}y - 1\right)$

45. $[2(y - 3) - 6(x + 4)][5(y - 3) - 4(x + 4)]$

46. $\left(x - \dfrac{1}{7} \right)\left(x^2 + \dfrac{1}{7}x + \dfrac{1}{49} \right)$

47. $(4x^2 + 2xy + y^2)(4x^2 - 2xy + y^2)$

48. $(x^2 - 7x + 12)(x^2 + 7x + 12)$

49. $(x^a + y^b)(x^a - y^b)(x^{2a} + y^{2b})$

50. $\left[1 + \dfrac{1}{5}(x + 1)^2 \right]^2$

51. $[a - (b - 1)][(b - 1)^2 + a(b - 1) + a^2]$

52. $(x - 1)(x^2 + x + 1)(x^3 + 1)$

53. $\left[\left(\dfrac{1}{3}x^3 - \dfrac{2}{3}y^2 \right)\left(\dfrac{1}{3}x^3 + \dfrac{2}{3}y^2 \right) \right]^2$

54. $10(0.1x^4y - 0.01xy^4)^2$

55. $(x^{a-b})^{a+b}$

56. $(M^{x+y})^{x+y}$

EXERCISE SET 4.4A

Multiply. (First determine the quickest way to multiply; then proceed.)

1. $(4x - y)(2x - y)$

2. $\left(4r - \dfrac{1}{3} \right)^2$

3. $(2c + 7d)^2$

4. $(3x^2 + 2y)(3x^2 - 2y)$

5. $(-2d + 3c)(2d + 3c)$

6. $(5c - 2d)(2c - 3d)$

7. $(x^2 + 2y)(2x^2 - y)$

8. $(a - 2b)(a - 4b)$

9. $(2m^2 - n^2)(m^2 + 3n^2)$

10. $(5y^2 + 7w^3)^2$

11. $(3x - 4y)^2$

12. $(a^3 - 3b^2)(a^3 + 3b^2)$

13. $(a^2 + b)(a^2 - b)$

14. $(3y - 2)(2y + 5)$

15. $\left(\dfrac{1}{5} - x \right)\left(\dfrac{1}{5} + x \right)$

16. $(3a^2 + 2)^2$

17. $(2x^2 + 3y^2)(4x^2 - 5y^2)$

18. $(x^2 - 9)(x^2 + 9)$

19. $\left(\dfrac{1}{2}x - 6 \right)^2$

20. $(0.2x - 0.4y)(0.2x + 0.4y)$

21. $(x + 1)(x - 1)(x^2 + 1)$
Hint: First multiply $x + 1$ and $x - 1$.

22. $(y - 2)(y + 2)(y^2 + 4)$
Hint: First multiply $y - 2$ and $y + 2$.

23. $(a + b)(a - b)(a^2 + b^2)$

24. $(2x - y)(2x + y)(4x^2 + y^2)$

25. $(a + b + 1)(a + b - 1)$

26. $(m + n + 2)(m + n - 2)$

27. $(2x + 3y + 4)(2x + 3y - 4)$

28. $(3a - 2b + c)(3a - 2b - c)$

☆
_____

The amount to which $1000 will grow in 3 years, when interest is compounded annually, is given by the polynomial

$$1000r^3 + 3000r^2 + 3000r + 1000.$$

Here, r is the rate of interest.

29. Find the amount to which $1000 will grow in 3 years at 12%.

30. Find the amount to which $1000 will grow in 3 years at 15%.

_____

4.5 FACTORING

The reverse of multiplication is factoring. To *factor* an expression means to write an equivalent expression that is a product.

TERMS WITH COMMON FACTORS

Example 1   Factor out a common factor: $4y^2 - 8$.

$$4y^2 - 8 = \boxed{4} \cdot y^2 - \boxed{4} \cdot 2 = \boxed{4} \cdot (y^2 - 2)$$

If there is more than one common factor, we usually choose the one with the largest coefficient and the greatest exponent. Try to write the answer directly.

Example 2   Factor: $5x^4 - 20x^3$.

$$5x^4 - 20x^3 = \boxed{5x^3} \cdot x - \boxed{5x^3} \cdot 4 = \boxed{5x^3} (x - 4)$$

Example 3   Factor: $15y^5 + 12y^4 - 27y^3 - 3y^2$.

$$15y^5 + 12y^4 - 27y^3 - 3y^2 = 3y^2(5y^3 + 4y^2 - 9y - 1)$$

> Always factor out the largest factor common to the terms.

FACTORING BY GROUPING

Sometimes there is a common factor that is a binomial.

Example 4   Factor: $(a - b)(x + 5) + (a - b)(x - y^2)$.

$$(\boxed{a - b})(x + 5) + (\boxed{a - b})(x - y^2) = (\boxed{a - b})[(x + 5) + (x - y^2)]$$
$$= (a - b)(2x + 5 - y^2)$$

Example 5   Factor: $y^2 + 3y + 4y + 12$.

$$y^2 + 3y + 4y + 12 = y(\,y + 3\,) + 4(\,y + 3\,) = (y + 4)(\,y + 3\,)$$

In Example 5 we factored two parts of the expression. Then we factored as in Example 4. We use trial and error to see which terms are to be grouped.

Example 6   Factor: $4x^2 - 15 + 20x - 3x$.

$$4x^2 - 15 + 20x - 3x = 4x^2 + 20x - 3x - 15$$

$$= 4x(\,x + 5\,) - 3(\,x + 5\,)$$

$$= (4x - 3)(\,x + 5\,)$$

Example 7   Factor: $ax^2 + ay + bx^2 + by$.

$$ax^2 + ay + bx^2 + by = a(\,x^2 + y\,) + b(\,x^2 + y\,)$$

$$= (a + b)(\,x^2 + y\,)$$

DIFFERENCES OF SQUARES

To factor a difference of squares, we use the result established in the last section, in reverse.

$$A^2 - B^2 = (A + B)(A - B)$$

To factor a difference of two squares, write the square root of the first expression *plus* the square root of the second, times the square root of the first *minus* the square root of the second.

Example 8   Factor: $x^2 - 9$.

$$x^2 - 9 = x^2 - 3^2 = (x + 3)(x - 3)$$

You should memorize this rule, and then say it to yourself as you work.

Example 9   Factor: $25y^6 - 49x^2$.

$$A^2 \;-\; B^2 \;=\; (\,A \;+\; B\,)(\,A \;-\; B\,)$$
$$\downarrow \qquad \downarrow \qquad \quad \downarrow \quad \downarrow \quad \downarrow \quad \downarrow$$
$$25y^6 - 49x^2 = (5y^3 + 7x)(5y^3 - 7x)$$

Example 10   Factor: $5 - 5x^2y^6$.

Always look *first* for a factor common to the terms.

$$5 - 5x^2y^6 = 5(1 - x^2y^6)$$

$$= 5(1 + xy^3)(1 - xy^3)$$

Factoring the difference of squares

Example 11 Factor: $2x^4 - 8y^4$.

$$2x^4 - 8y^4 = 2(x^4 - 4y^4) \quad \text{Removing the common factor}$$
$$= 2(x^2 + 2y^2)(x^2 - 2y^2) \quad \text{Factoring the difference of squares}$$

Example 12 Factor: $16x^4y - 81y$.

$$16x^4y - 81y = y(16x^4 - 81) \quad \text{Removing the common factor}$$
$$= y(4x^2 + 9)(4x^2 - 9) \quad \text{Factoring the difference of squares}$$
$$= y(4x^2 + 9)(2x + 3)(2x - 3) \quad \text{Factoring } 4x^2 - 9, \text{ which is a difference of squares}$$

Always continue to factor as long as you can. That way you will be factoring *completely*.

Caution! A common error is trying to factor a *sum* of squares. $4x^2 + 9$ is a *sum* of squares and cannot be factored.

EXERCISE SET 4.5

Factor.

1. $4a^2 + 2a$

2. $6y^2 + 3y$

3. $3y^2 - 3y - 9$

4. $5x^2 - 5x + 15$

5. $6x^2 - 3x^4$

6. $8y^2 + 4y^4$

7. $4ab - 6ac + 12ad$

8. $8xy + 10xz - 14xw$

9. $10a^4 + 15a^2 - 25a - 30$

10. $12t^5 - 20t^4 + 8t^2 - 16$

11. $a(b - 2) + c(b - 2)$

12. $a(x^2 - 3) - 2(x^2 - 3)$

13. $(x - 2)(x + 5) + (x - 2)(x + 8)$

14. $(m - 4)(m + 3) + (m - 4)(m - 3)$

15. $a^2(x - y) + a^2(x - y)$

16. $3x^2(x - 6) + 3x^2(x - 6)$

17. $ac + ad + bc + bd$

18. $xy + xz + wy + wz$

19. $b^3 - b^2 + 2b - 2$

20. $y^3 - y^2 + 3y - 3$

21. $y^2 - 8y - y + 8$

22. $t^2 + 6t - 2t - 12$

Factor. Remember to look first for a common factor.

23. $x^2 - 16$

24. $y^2 - 9$

25. $6x^2 - 6y^2$

26. $8x^2 - 8y^2$

27. $4xy^4 - 4xz^4$

28. $25ab^4 - 25az^4$

29. $9a^4 - 25a^2b^4$ **30.** $16x^6 - 121x^2y^4$ **31.** $\frac{1}{25} - x^2$

32. $\frac{1}{16} - y^2$ **33.** $0.04x^2 - 0.09y^2$ **34.** $0.01x^2 - 0.04y^2$

☆

The number of diagonals of a polygon having n sides is given by the polynomial $\frac{1}{2}n^2 - \frac{3}{2}n$.

35. Find the number of diagonals of a polygon with 5 sides.

36. Find the number of diagonals of a polygon with 7 sides.

4.6 FACTORING TRINOMIALS

TRINOMIAL SQUARES

Some trinomials are squares of binomials; for example,

$$x^2 + 6x + 9 = (x + 3)^2.$$

Trinomials like this can be factored. We must first learn to recognize trinomial squares.

> **a)** Two of the terms must be squares, such as A^2 and B^2.
> **b)** There must be no minus sign before A^2 or B^2.
> **c)** If we multiply A and B (the square roots of these expressions) and double the result, we get the remaining term, $2AB$, or its additive inverse, $-2AB$.

Examples   Determine whether these are trinomial squares.

  1. $x^2 + 10x + 25$

    a) Two terms are squares: x^2 and 25.
    b) There is no minus sign before either x^2 or 25.
    c) If we multiply the square roots, x and 5, and double, we get $10x$, the remaining term.

Thus the trinomial is a square.

  2. $3x^2 + 4x + 16$

    a) Only one term, 16, is a square ($3x^2$ is not a square). Therefore the trinomial is not a square.

To factor trinomial squares, we use these equations.

$$A^2 + 2AB + B^2 = (A + B)^2$$
$$A^2 - 2AB + B^2 = (A - B)^2$$

Example 3 Factor: $x^2 - 10x + 25$.

$x^2 - 10x + 25 = (x - 5)^2$ We find the square terms and write their square roots with a minus sign between them.

Example 4 Factor: $16y^2 + 49 + 56y$.

$16y^2 + 49 + 56y = (4y + 7)^2$ We find the square terms and write their square roots with a plus sign between them.

Example 5 Factor: $-20xy + 4y^2 + 25x^2$.

$-20xy + 4y^2 + 25x^2 = (2y - 5x)^2$

Example 6 Factor: $2x^2 - 12x + 18$.

$2x^2 - 12x + 18 = 2(x^2 - 6x + 9)$ Removing the common factor

$= 2(x - 3)^2$ Factoring the trinomial square

Always remember to look first for a common factor.

Example 7 Factor: $-4y^2 - 144y^8 + 48y^5$.

$-4y^2 - 144y^8 + 48y^5 = -4y^2(1 + 36y^6 - 12y^3)$ Removing the common factor

$= -4y^2(1 - 12y^3 + 36y^6)$ Changing order

$= -4y^2(1 - 6y^3)^2$ Factoring the trinomial square

FACTORING TRINOMIALS OF THE TYPE $x^2 + ax + b$

Consider this product:

$$\text{F} \quad \text{O} \quad \text{I} \quad \text{L}$$
$$(x + 3)(x + 5) = x^2 + 5x + 3x + 15$$
$$= x^2 + 8x + 15.$$

Note that the coefficient 8 is the sum of 3 and 5, and the 15 is the product of 3 and 5. In general, $(x + a)(x + b) = x^2 + (a + b)x + ab$.

To factor we can use the preceding equation in reverse, as follows:

$$x^2 + (a + b)x + ab = (x + a)(x + b).$$

Finding a and b is a trial-and-error process. Practice helps.

Example 8   Factor: $x^2 - 3x - 10$.

We look for pairs of integers whose product is -10 and whose sum is -3.

Pairs of factors	Sum of factors
$-2,\quad 5$	3
$2, -5$	-3
$10, -1$	9
$-10,\quad 1$	-9

Thus the desired integers are 2 and -5. Then

$$x^2 - 3x - 10 = (x + 2)(x - 5).$$

We can check by multiplying.

TRINOMIALS OF THE TYPE $ax^2 + bx + c$

Consider a multiplication.

$$
\begin{array}{cccc}
 & \text{F} & \text{O} \quad \text{I} & \text{L} \\
(2x + 3)(5x + 4) = & 10x^2 + & 8x \; + \; 15x & + \; 12 \\
 & 10x^2 + & 23x & + \; 12 \\
 & \uparrow & \uparrow & \uparrow \\
 & \text{F} & \text{O} + \text{I} & \text{L} \\
 & 2 \cdot 5 & 2 \cdot 4 + 3 \cdot 5 & 3 \cdot 4
\end{array}
$$

To factor $ax^2 + bx + c$ we look for two binomials

$$(\underline{\quad}x + \underline{\quad})(\underline{\quad}x + \underline{\quad})$$

where products of numbers in the blanks are as follows.

1. The numbers in the *first* blanks have product a.

2. The *outside* product and the *inside* product add up to b.

3. The numbers in the *last* blanks have product c.

Example 9   Factor: $5x^2 - 9x - 2$.

We first look for a common factor. There is none other than 1.

We look for numbers whose product is 5. These are 1, 5 and $-1, -5$.

We have these possibilities:

$$(x + \quad)(5x + \quad) \quad \text{or} \quad (-x + \quad)(-5x + \quad).$$

Now we look for numbers whose product is -2. These are 1, -2 and $-1, 2$.

We have these as some of the possibilities for factorization.

We multiply each. We must get $5x^2 - 9x - 2$.

a) $(x + 1)(5x - 2)$ **a)** $5x^2 + 3x - 2$

b) $(x - 2)(5x + 1)$ **b)** $5x^2 - 9x - 2$

c) $(-x + 1)(-5x - 2)$ **c)** $5x^2 - 3x - 2$

d) $(-x + 2)(-5x - 1)$ **d)** $5x^2 - 9x - 2$

We see that (b) and (d) are both factorizations. We prefer to have the first coefficients positive when that is possible. Thus the factorization we prefer is $(x - 2)(5x + 1)$.

Example 10   Factor: $12x^2 + 34x + 14$.

We first look for a common factor. The number 2 is a common factor, so we factor it out: $2(6x^2 + 17x + 7)$. Now we consider $6x^2 + 17x + 7$. We look for numbers whose product is 6. These are 6, 1 and 2, 3. (From Example 9 we found that we needed to consider only positive factors of the first term.) We then have these possibilities:

$$(6x + \quad)(x + \quad) \quad \text{and} \quad (2x + \quad)(3x + \quad).$$

Next we look for pairs of numbers whose product is 7. They are

$$7, 1 \quad \text{and} \quad -7, -1. \qquad \text{Both positive or both negative}$$

By multiplying, we find that the answer is $2(2x + 1)(3x + 7)$.

Example 11   Factor: $x^2y^2 + 5xy + 4$.

In this case, we can treat xy as if it were a single variable.

$$x^2y^2 + 5xy + 4 = (xy)^2 + 5xy + 4$$
$$= (xy + 4)(xy + 1)$$

Optional

Another way to factor $ax^2 + bx + c$ is as follows:

a) First look for a common factor.

b) Multiply the first and last coefficient, a and c.

c) Try to factor the product ac so that the sum of the factors is b.

d) Write a sum equivalent to the middle term, bx.

e) Factor by grouping.

Example 12   Factor: $2x^2 - 3x - 35$.

a) First look for a common factor. There is none (other than 1).

b) Multiply the first and last coefficients, 2 and -35:

$$2(-35) = -70.$$

c) Try to factor -70 so that the sum of the factors is -3.

Some pairs of factors	Sums of factors
-2, 35	33
2, -35	-33
-14, 5	-9
7, -10	-3

The desired factors are 7 and -10.

d) Write a sum equivalent to $-3x$ using the results of (c):

$$-3x = -10x + 7x.$$

e) Factor by grouping:

$$\begin{aligned}
2x^2 - 3x - 35 &= 2x^2 - 10x + 7x - 35 \\
&= (2x^2 - 10x) + (7x - 35) \\
&= 2x(x - 5) + 7(x - 5) \\
&= (2x + 7)(x - 5).
\end{aligned}$$

EXERCISE SET 4.6

Factor. Remember to look first for a common factor.

1. $y^2 - 6y + 9$

2. $x^2 - 8x + 16$

3. $x^2 + 14x + 49$

4. $x^2 + 16x + 64$

5. $-18y^2 + y^3 + 81y$

6. $24a^2 + a^3 + 144a$

7. $12a^2 + 36a + 27$

8. $20y^2 + 100y + 125$

9. $2x^2 - 40x + 200$

10. $32x^2 + 48x + 18$

11. $0.25x^2 + 0.30x + 0.09$

12. $0.04x^2 - 0.28x + 0.49$

13. $x^2 + 9x + 20$

15. $2y^2 - 16y + 32$

16. $2a^2 - 20a + 50$

17. $x^2 - 27 - 6x$

18. $t^2 - 15 - 2t$

19. $32y + 4y^2 - y^3$ **20.** $56x + x^2 - x^3$ **21.** $15 + t^2 + 8t$

22. $27 + y^2 + 12y$ **23.** $x^4 + 11x^2 - 80$ **24.** $y^4 + 5y^2 - 84$

25. $3x^2 - 16x - 12$ **26.** $6x^2 - 5x - 25$ **27.** $6x^3 - 15x - x^2$

28. $10y^3 - 12y - 7y^2$ **29.** $3a^2 - 10a + 8$ **30.** $12a^2 - 7a + 1$

31. $35y^2 + 34y + 8$ **32.** $9a^2 + 18a + 8$ **33.** $4t + 10t^2 - 6$

34. $8x + 30x^2 - 6$ **35.** $8x^2 - 16 - 28x$ **36.** $18x^2 - 24 - 6x$

☆

37. The area of a ring, as shown, is $\pi R^2 - \pi r^2$. Factoring this polynomial, we get $\pi(R + r)(R - r)$. Find the area of a ring, where $R = 10$ cm and $r = 4$ cm. Use 3.14 for π.

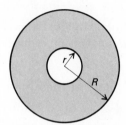

4.7 COMPLETING THE SQUARE

FACTORING DIFFERENCES OF SQUARES

A difference of squares can have more than two terms. For example, one of the squares may be a trinomial. We can factor by a kind of grouping.

Example 1 Factor: $x^2 + 6x + 9 - y^2$.

$$x^2 + 6x + 9 - y^2 = (x^2 + 6x + 9) - y^2$$

Grouping as a trinomial minus y^2 to show a difference of squares

$$= (\,x + 3\,)^2 - y^2$$

$$= (\,x + 3\, + y)(\,x + 3\, - y)$$

COMPLETING THE SQUARE

A trinomial such as $x^2 + 10x + 25$ may be the square of a binomial. This one is because $x^2 + 10x + 25 = (x + 5)^2$. Given the first two terms of a trinomial, we can find the third term that will make it a square as follows.

Example 2   What must be added to $x^2 + 12x$ to make it a trinomial square?

We take half the coefficient of x and square it.

$$x^2 + 12x$$

$\longrightarrow$ Half of 12 is 6, and $6^2 = 36$. We add 36.

$x^2 + 12x + 36$ is a trinomial square. It is equivalent to $(x + 6)^2$.

The process illustrated in Example 2 is called *completing the square*. Completing the square can be used in factoring and is used in a number of other ways.

Example 3   Complete the square on $x^2 - 8ax$.

We take half of $-8a$ (the coefficient of x) and get $-4a$.

We square $-4a$: $(-4a)^2 = 16a^2$.

We add the result to obtain $x^2 - 8ax + 16a^2$.

This is a trinomial square because it is equivalent to $(x - 4a)^2$.

Example 4   Complete the square on $y^2 + \frac{3}{4}y$.

We take half of $\frac{3}{4}$ (the coefficient of y) and get $\frac{1}{2} \cdot \frac{3}{4} = \frac{3}{8}$.

We square $\frac{3}{8}$: $\left(\frac{3}{8}\right)^2 = \frac{9}{64}$.

We add the result to obtain $y^2 + \frac{3}{4}y + \frac{9}{64}$.

FACTORING BY COMPLETING THE SQUARE

Completing the square can be used to factor trinomials.

Example 5   Factor $x^2 - 22x + 112$ by completing the square.

We first complete the square on $x^2 - 22x$. We take half of -22, which is -11.

We square -11: $(-11)^2 = 121$.

Now we add 0 to the original trinomial, naming it $121 - 121$.

$$x^2 - 22x + 112 = x^2 - 22x + 112 + (121 - 121) \qquad \text{Adding } 121 - 121$$

$$= (x^2 - 22x + 121) + 112 - 121 \qquad \text{Grouping}$$
$$= (x^2 - 22x + 121) - 9$$
$$= (x - 11)^2 - 3^2$$
$$= (x - 11 + 3)(x - 11 - 3), \text{ or}$$
$$(x - 8)(x - 14)$$

Example 6   Factor $2x^2 + 80x + 768$ by completing the square.

$$2x^2 + 80x + 768 = 2(x^2 + 40x + 384) \qquad \text{Removing the common factor 2}$$

To complete the square we take half of 40 to obtain 20.

We square 20 to get 400.

We add 0, naming it $400 - 400$.

$$2(x^2 + 40x + 384) = 2(x^2 + 40x + 400 - 400 + 384) \qquad \text{Adding } 400 - 400$$

$$= 2(x^2 + 40x + 400 - 16) \qquad \text{Grouping}$$
$$= 2[(x + 20)^2 - 4^2]$$
$$= 2(x + 20 + 4)(x + 20 - 4), \text{ or}$$
$$2(x + 24)(x + 16)$$

Example 7   Factor $x^2 - 6.2x + 8.61$ by completing the square.

To complete the square we take half of -6.2, obtaining -3.1.

We square -3.1: $(-3.1)^2 = 9.61$.

We then add 0, naming it $9.61 - 9.61$.

$$x^2 - 6.2x + 8.61 = x^2 - 6.2x + 8.61 + 9.61 - 9.61 \qquad \text{Adding } 9.61 - 9.61$$

$$= (x^2 - 6.2x + 9.61) + 8.61 - 9.61 \qquad \text{Grouping}$$
$$= (x - 3.1)^2 - 1$$
$$= (x - 3.1 + 1)(x - 3.1 - 1), \text{ or}$$
$$(x - 2.1)(x - 4.1)$$

EXERCISE SET 4.7

Factor.

1. $a^2 + 2ab + b^2 - 9$

2. $x^2 - 2xy + y^2 - 25$

3. $r^2 - 2r + 1 - 4s^2$

4. $c^2 + 4cd + 4d^2 - 9p^2$

5. $2m^2 + 4mn + 2n^2 - 50b^2$

6. $12x^2 + 12x + 3 - 3y^2$

7. $9 - (a^2 + 2ab + b^2)$

8. $16 - (x^2 - 2xy + y^2)$

Complete the square.

9. $x^2 + 16x$

10. $y^2 - 24y$

11. $x^2 - 4.2x$

12. $y^2 + 3.6y$

13. $x^2 + \dfrac{2}{3}bx$

14. $y^2 - \dfrac{3}{4}ay$

Factor by completing the square.

15. $x^2 + 24x + 119$

16. $x^2 + 30x + 176$

17. $x^2 - 26x + 105$

18. $x^2 - 32x + 192$

19. $2x^2 + 56x + 342$

20. $3x^2 + 54x + 231$

21. $2x^2 - 32ax + 96a^2$

22. $3x^2 - 60bx + 192b^2$

23. $x^2 + 2.6x + 0.69$

24. $x^2 - 3.2x + 1.56$

25. $x^2 + \dfrac{3}{5}x + \dfrac{8}{100}$

26. $x^2 - \dfrac{7}{2}x + \dfrac{45}{16}$

☆

Factor by completing the square.

27. ▦ $x^2 + 4.482x - 7.403544$

28. ▦ $5.72x^2 + 35.464x - 1319.2608$

4.8 SUMS OR DIFFERENCES OF TWO CUBES

To succeed with this section, you will need to review and memorize the cubes of integers from -10 to 10.

Next note the following:

$$(a + b)(a^2 - ab + b^2) = a(a^2 - ab + b^2) + b(a^2 - ab + b^2)$$
$$= a^3 - a^2b + ab^2 + a^2b - ab^2 + b^3$$
$$= a^3 + b^3$$

and

$$(a - b)(a^2 + ab + b^2) = a(a^2 + ab + b^2) - b(a^2 + ab + b^2)$$
$$= a^3 + a^2b + ab^2 - a^2b - ab^2 - b^3$$
$$= a^3 - b^3.$$

The above equations (reversed) show how we can factor a sum or a difference of two cubes.

$$A^3 + B^3 = (A + B)(A^2 - AB + B^2)$$
$$A^3 - B^3 = (A - B)(A^2 + AB + B^2)$$

You will need to memorize these patterns and remember to use them as you factor.

Example 1 Factor: $x^3 - 27$.

$$x^3 - 27 = x^3 - 3^3$$

In one set of parentheses we write the cube root of the first term, x. Then we write the cube root of the second term, -3. This gives us the expression $x - 3$.

$$(x - 3)(\qquad)$$

To get the next factor we think of $x - 3$ and do the following.

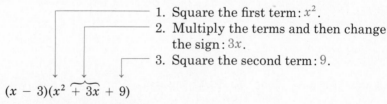

1. Square the first term: x^2.
2. Multiply the terms and then change the sign: $3x$.
3. Square the second term: 9.

$$(x - 3)(x^2 + 3x + 9)$$

Note: We cannot factor $x^2 + 3x + 9$. (It is not a trinomial square.)

Example 2 Factor: $125x^3 + y^3$.

$$125x^3 + y^3 = (5x)^3 + y^3$$

In one set of parentheses we write the cube root of the first term, then a plus sign, and then the cube root of the second term.

$$(5x + y)(\qquad)$$

To get the next factor, we think of $5x + y$ and do the following.

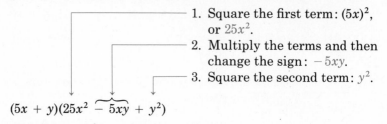

1. Square the first term: $(5x)^2$, or $25x^2$.
2. Multiply the terms and then change the sign: $-5xy$.
3. Square the second term: y^2.

$$(5x + y)(25x^2 - 5xy + y^2)$$

Example 3 Factor: $128y^7 - 250x^6y$.

We first look for a common factor.

$$2y(64y^6 - 125x^6) = 2y[(4y^2)^3 - (5x^2)^3]$$
$$= 2y(4y^2 - 5x^2)(16y^4 + 20x^2y^2 + 25x^4)$$

Example 4 Factor: $64a^6 - 729b^6$.

$$(8a^3 - 27b^3)(8a^3 + 27b^3)$$ Factoring a difference of squares

Each factor is a sum or difference of cubes. We factor.

$$(2a - 3b)(4a^2 + 6ab + 9b^2)(2a + 3b)(4a^2 - 6ab + 9b^2)$$

Example 5 Factor: $y^3 + 0.008$.

We note that $0.008 = (0.2)^3$. Thus we have a sum of two cubes, and we factor as follows.

$$(y + 0.2)(y^2 - 0.2y + 0.04)$$

> Remember the following about factoring sums or differences of squares and cubes:
>
> | | |
> |---|---|
> | Sum of cubes: | $A^3 + B^3 = (A + B)(A^2 - AB + B^2)$ |
> | Difference of cubes: | $A^3 - B^3 = (A - B)(A^2 + AB + B^2)$ |
> | Difference of squares: | $A^2 - B^2 = (A + B)(A - B)$ |
> | Sum of squares: | $A^2 + B^2$ **cannot be factored** |

EXERCISE SET 4.8

Factor.

1. $x^3 + 8$
2. $c^3 + 27$
3. $2y^3 - 128$
4. $3z^3 - 3$
5. $24a^3 + 3$
6. $54x^3 + 2$
7. $8 - 27b^3$
8. $64 - 125x^3$

9. $8x^3 + 27$

10. $27y^3 + 64$

11. $a^3 + \frac{1}{8}$

12. $b^3 + \frac{1}{27}$

$[Hint: \frac{1}{8} = (\frac{1}{2})^3.]$

13. $rs^3 + 64r$

14. $ab^3 + 125a$

15. $5x^3 - 40z^3$

16. $2y^3 - 54z^3$

17. $x^3 + 0.001$

18. $y^3 + 0.125$

19. $64x^6 - 8t^6$

20. $125c^6 - 8d^6$

☆

21. A 4-ft by 4-ft sandbox is placed on a square lawn x ft on a side. Express the area left over as a polynomial.

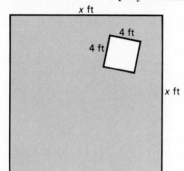

22. Express the colored area as a polynomial.

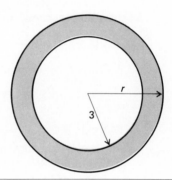

4.9 FACTORING: A GENERAL STRATEGY

Here is a general strategy for factoring.

A. **Always look first for a common factor.**

B. **Then look at the number of terms.**

Two terms: Try factoring as a difference of squares first. Next, try factoring as a sum or difference of cubes. Do *not* try to factor a *sum* of squares.

Three terms: Determine whether the trinomial is a square. If so, you know how to factor. If not, try trial and error.

Four or more terms: Try factoring by grouping and removing a common binomial factor. Next, try grouping into a difference of squares, one of which is a trinomial.

C. **Always *factor completely*. If a factor with more than one term can be factored, you should factor it.**

Example 1   Factor: $10a^2x - 40b^2x$.

A. We look first for a common factor:

$10x(a^2 - 4b^2)$.   Factoring out the largest common factor

B. The factor $a^2 - 4b^2$ has only two terms. It is a difference of squares.
We factor it:

$10x(a + 2b)(a - 2b)$.   It is a common error to forget to write the common factor in this step.

C. Have we factored completely? Yes, because no factor with more
than one term can be factored further.

Example 2   Factor: $x^6 - y^6$.

A. We look for a common factor. There isn't one (other than 1 or -1).

B. There are only two terms. It is a difference of squares: $(x^3)^2 - (y^3)^2$.
We factor it:

$(x^3 + y^3)(x^3 - y^3)$.

One factor is a sum of two cubes, and the other factor is a difference
of two cubes. We factor them:

$(x + y)(x^2 - xy + y^2)(x - y)(x^2 + xy + y^2)$.

C. We have factored completely because no factors can be factored
further.

In Example 2, had we thought of factoring first as a difference of two
cubes, we would have had

$$(x^2)^3 - (y^2)^3 = (x^2 - y^2)(x^4 + x^2y^2 + y^4)$$
$$= (x + y)(x - y)(x^4 + x^2y^2 + y^4).$$

In this case, we might have missed some factors; $x^4 + x^2y^2 + y^4$ can
be factored, but we might not have noticed.

Example 3   Factor: $10x^6 + 40y^2$.

A. We remove the largest common factor: $10(x^6 + 4y^2)$.

B. In the parentheses there are two terms, a sum of squares, which
cannot be factored.

Example 4   Factor: $2x^2 + 50a^2 - 20ax$.

A. We remove the largest common factor: $2(x^2 + 25a^2 - 10ax)$.

B. In the parentheses there are three terms. The trinomial is a square.
We factor it: $2(x - 5a)^2$

C. No factor with more than one term can be factored further.

Example 5 Factor: $6x^2 - 20x - 16$.

A. We remove the largest common factor: $2(3x^2 - 10x - 8)$.

B. In the parentheses there are three terms. The trinomial is not a square. We factor by trial: $2(x - 4)(3x + 2)$.

C. We cannot factor further.

Example 6 Factor: $3x + 12 + ax^2 + 4ax$.

A. There is no common factor (other than 1 or -1).

B. There are four terms. We try grouping to remove a common binomial factor.

$\quad\quad 3(x + 4) + ax(x + 4)$ Factoring two grouped binomials

$\quad\quad (x + 4)(3 + ax)$ Removing the common binomial factor

C. No factor with more than one term can be factored further.

Example 7 Factor: $y^2 - 9a^2 + 12y + 36$.

A. There is no common factor (other than 1 or -1).

B. There are four terms. We try grouping to remove a common binomial factor, but that is not possible.

We try grouping as a difference of squares.

$\quad\quad (y^2 + 12y + 36) - 9a^2$ Grouping

$\quad\quad (y + 6 + 3a)(y + 6 - 3a)$ Factoring the difference of squares

C. No factor with more than one term can be factored further.

Example 8 Factor: $x^3 - xy^2 + x^2y - y^3$.

A. There is no common factor (other than 1 or -1).

B. There are four terms. We try grouping to remove a common binomial factor.

$\quad\quad x(x^2 - y^2) + y(x^2 - y^2)$ Factoring two grouped binomials

$\quad\quad (x^2 - y^2)(x + y)$ Removing the common binomial factor

C. A factor can be factored further.

$\quad\quad (x + y)(x - y)(x + y)$ Factoring a difference of squares

No factor with more than one term can be factored further, so we have factored completely.

EXERCISE SET 4.9

Factor completely.

1. $x^2 - 144$

2. $y^2 - 81$

3. $2x^2 + 11x + 12$

4. $8a^2 + 18a - 5$

5. $3x^4 - 12$

6. $2xy^2 - 50x$

7. $a^2 + 25 + 10a$

8. $p^2 + 64 + 16p$

9. $2x^2 - 10x - 132$

10. $3y^2 - 15y - 252$

11. $9x^2 - 25y^2$

12. $16a^2 - 81b^2$

13. $4c^2 - 4cd + d^2$

14. $70b^2 - 3ab - a^2$

15. $-7x^2 + 2x^3 + 4x - 14$

16. $9m^2 + 3m^3 + 8m + 24$

17. $250x^3 - 128y^3$

18. $27a^3 - 343b^3$

19. $8m^3 + m^6 - 20$

20. $-37x^2 + x^4 + 36$

21. $ac + cd - ab - bd$

22. $xw - yw + xz - yz$

23. $m^6 - 1$

24. $64t^6 - 1$

25. $x^2 + 6x - y^2 + 9$

26. $t^2 + 10t - p^2 + 25$

27. $36y^2 - 35 + 12y$

28. $2b - 28a^2b + 10ab$

29. $a^8 - b^8$

30. $2x^4 - 32$

31. $a^3b - 16ab^3$

32. $x^3y - 25xy^3$

33. $2x^3 + 6x^2 - 8x - 24$

34. $3x^3 + 6x^2 - 27x - 54$

35. $16x^3 + 54y^3$

36. $250a^3 + 54b^3$

☆

Factor. Assume that variables in exponents represent positive integers.

37. $20x^2 - 4x - 39$

38. $30y^4 - 97xy^2 + 60x^2$

39. $72a^2 + 284a - 160$

40. $3x^2y^2z + 25xyz^2 + 28z^3$

41. $7x^3 - \dfrac{7}{8}$

42. $-16 + 17(5 - y^2) - (5 - y^2)^2$

43. $[(c - d)^3 - d^3]^2$

44. $(x - p)^2 - p^2$

45. $x^4 - 50x^2 + 49$

46. $(y - 1)^4 - (y - 1)^2$

47. $s^6 - 729t^6$

48. $x^6 - 2x^5 + x^4 - x^2 + 2x - 1$

49. $27x^{6s} + 64y^{3t}$

50. $5c^{100} - 80d^{100}$

51. $4x^2 + 4xy + y^2 - r^2 + 6rs - 9s^2$

52. $(y + x)^3 - x^3$

53. $c^4d^4 - a^{16}$

54. $(1 - x)^3 + (x - 1)^6$

55. $c^{2w+1} + 2c^{w+1} + c$

56. $24x^{2a} - 6$

57. $y^9 - y$

58. $1 - \dfrac{x^{27}}{1000}$

59. $3a^2 + 3b^2 - 3c^2 - 3d^2 + 6ab - 6cd$

60. $3(x + 1)^2 + 9(x + 1) - 12$

61. $(a + 2)^3 - (a - 2)^3$

62. $8(a - 3)^2 - 64(a - 3) + 128$

63. Let $\left(x + \dfrac{2}{x}\right)^2 = 6$. Find $x^3 + \dfrac{8}{x^3}$.

4.10 SOLVING EQUATIONS BY FACTORING

In Chapter 2 we solved equations using the *principle of zero products*. To use this principle in solving equations we make sure that there is a 0 on one side; then we factor the other side and set each factor equal to 0. (See p. 50.)

Example 1 Solve: $x^2 - 3x - 28 = 0$.

We first factor the polynomial. Then we use the principle of zero products.

$$x^2 - 3x - 28 = 0$$
$$(x - 7)(x + 4) = 0 \qquad \text{Factoring}$$

The expression $x^2 - 3x - 28$ and $(x - 7)(x + 4)$ name the same number for any replacement. Hence the equations have the same solutions.

$$x - 7 = 0 \quad \text{or} \quad x + 4 = 0 \qquad \text{Using the principle of zero products}$$
$$x = 7 \quad \text{or} \qquad x = -4$$

Check:

For 7:
$$\begin{array}{r} x^2 - 3x - 28 = 0 \\ \hline 7^2 - 3(7) - 28 \ \big|\ 0 \\ 49 - 21 - 28 \\ 0 \end{array}$$

For -4:
$$\begin{array}{r} x^2 - 3x - 28 = 0 \\ \hline (-4)^2 - 3(-4) - 28 \ \big|\ 0 \\ 16 + 12 - 28 \\ 0 \end{array}$$

The solutions are 7 and -4.

When we use the principle of zero products, a check is not necessary except to detect errors.

Example 2 Solve: $7y + 3y^2 = -2.$

$$3y^2 + 7y + 2 = 0 \qquad \text{Adding 2 to get 0 on one side and arranging in descending order}$$

$$(3y + 1)(y + 2) = 0 \qquad \text{Factoring}$$

$$3y + 1 = 0 \quad \text{or} \quad y + 2 = 0 \qquad \text{Using the principle of zero products}$$

$$3y = -1 \quad \text{or} \qquad y = -2$$

$$y = -\tfrac{1}{3} \quad \text{or} \qquad y = -2$$

The solutions are $-\tfrac{1}{3}$ and -2.

Important: There *must* be a 0 on one side of the equation. It is a common error to forget this.

Example 3 Solve: $5b^2 - 10b = 0.$

$$5b(b - 2) = 0 \qquad \text{Factoring}$$

$$5b = 0 \quad \text{or} \quad b - 2 = 0 \qquad \text{Using the principle of zero products}$$

$$b = 0 \quad \text{or} \qquad b = 2$$

The solutions are 0 and 2.

SOLVING PROBLEMS

To solve problems, we first translate the problem situation to an equation and then solve. We check to see whether the solution(s) of the equation(s) satisfies the conditions of the problem.

Example 4 The square of a number minus the number is 20. Find the number.

$$\underbrace{\text{The square of a number}}_{x^2} \quad \underbrace{\text{minus}}_{-} \quad \underbrace{\text{the number}}_{x} \quad \underbrace{\text{is}}_{=} \; \underbrace{20}_{20} \qquad \text{Translating}$$

We solve the equation.

$$x^2 - x = 20$$

$$x^2 - x - 20 = 0 \qquad \text{Adding } -20 \text{ to get 0 on one side}$$

$$(x - 5)(x + 4) = 0 \qquad \text{Factoring}$$

$$x - 5 = 0 \quad \text{or} \quad x + 4 = 0 \qquad \text{Using the principle of zero products}$$

$$x = 5 \quad \text{or} \qquad x = -4$$

The numbers 5 and -4 both check. They are the answers to the problem.

Example 5 The width of a rectangle is 2 m less than the length. The area is 15 m². Find the dimensions.

We first make a drawing and label the length and width.

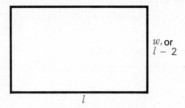

w, or
$l - 2$

l

Area is length times width, so we have the translation

$$l \cdot (l - 2) = 15.$$

We solve the equation.

> A common error here is to write $l = 0$ or $l - 2 = 0$, using the principle of zero products. That is incorrect because there is not a 0 on the other side of the equation.

$$l \cdot (l - 2) = 15$$
$$l^2 - 2l = 15 \qquad \text{Multiplying}$$
$$l^2 - 2l - 15 = 0 \qquad \text{Adding } -15 \text{ to get 0 on one side}$$
$$(l - 5)(l + 3) = 0 \qquad \text{Factoring}$$
$$l - 5 = 0 \quad \text{or} \quad l + 3 = 0 \qquad \text{Using the principle of zero products}$$
$$l = 5 \quad \text{or} \qquad l = -3$$

The solutions of the equation are 5 and -3. Now we check in the problem. The length of a rectangle cannot be negative. Thus the length is 5 m. Since the width is 2 m less than the length, the width is 3 m.

EXERCISE SET 4.10

Solve.

1. $x^2 + 3x - 28 = 0$

2. $y^2 - 4y - 45 = 0$

3. $y^2 - 8y + 16 = 0$

4. $r^2 - 2r + 1 = 0$

5. $9x + x^2 + 20 = 0$

6. $8y + y^2 + 15 = 0$

7. $x^2 + 8x = 0$

8. $t^2 + 9t = 0$

9. $z^2 = 36$

10. $y^2 = 81$

11. $x^2 + 14x + 45 = 0$

12. $y^2 + 12y + 32 = 0$

13. $p^2 - 11p = -28$

14. $x^2 - 14x = -45$

15. $2x^2 - 15x = -7$

16. $x^2 - 9x = -8$

17. $14 = x(x - 5)$

18. $24 = x(x - 5)$

19. Four times the square of a number is 21 more than eight times the number. What is the number?

20. The square of a number plus the number is 132. What is the number?

21. The length of the top of a table is 5 in. more than the width. Find the length and width if the area is 84 in².

22. Sam Sylow has a garden 25 feet longer than it is wide. The garden has an area of 7500 ft². What are its dimensions?

23. The sum of the squares of two consecutive odd positive integers is 202. Find the integers.

24. The sum of the squares of two consecutive odd positive integers is 394. Find the integers.

25. The square of a number plus the number is 156. What is the number?

26. Four times the square of a number is 45 more than eight times the number. What is the number?

27. The length of the top of a workbench is 4 cm greater than the width. The area is 96 cm². Find the length and the width.

28. A flower bed is to be 3 m longer than it is wide. The flower bed will have an area of 108 m². What will its dimensions be?

29. The perimeter of a square is 4 more than its area. Find the length of a side.

30. The area of a square is 12 more than its perimeter. Find the length of a side.

31. The base of a triangle is 9 cm greater than the height. The area is 56 cm². Find the height and base.

32. The base of a triangle is 5 cm less than the height. The area is 18 cm². Find the height and base.

33. Three consecutive even integers are such that the square of the first plus the square of the third is 136. Find the three integers.

34. Three consecutive even integers are such that the square of the third is 76 more than the square of the second. Find the three integers.

☆ ──────────────────────────────────

35. A rectangular piece of tin is twice as long as it is wide. Squares 2 cm on a side are cut out of each corner and the ends are turned up to make a box whose volume is 480 cm³. What are the dimensions of the piece of tin?

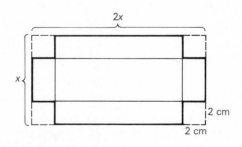

36. The sum of two numbers is 17, and the sum of their squares is 205. Find the numbers.

37. The top and base of a fish tank are rectangles in which the length is 10 inches more than the width. If the depth and the width of the tank total 50 inches and the combined area of the top and base is 400 square inches less than the total area of the four sides, what are the dimensions of the tank?

38. A rectangular swimming pool with dimensions 11 meters and 8 meters is built in a rectangular backyard. The area of the backyard is 1120 square meters. If the strip of yard surrounding the pool is of uniform width, how wide is the strip?

39. The hypotenuse of a right triangle is 3 cm more than one of its legs and 6 cm more than the other leg What is the area of the triangle?

4.11 EQUATIONS AND FORMULAS

EQUATIONS WITH LETTERS FOR CONSTANTS

If a letter is used to represent one specific number (although it may be unknown) we call it a constant. Usually, but not always, we use letters at the beginning of the alphabet (a, b, c) for constants and letters near the end (x, y, z) for variables.

To solve an equation with unknown constants, we treat them as if they were known. The laws of numbers apply in any case.

Example 1   Solve for y: $ay + c^2 = 5$.

$$ay + c^2 = 5$$

$$ay = 5 - c^2 \qquad \text{Adding } -c^2$$

$$y = \frac{5 - c^2}{a} \qquad \text{Multiplying by } \frac{1}{a}$$

In Example 1, we treated all constants—a, c^2, and 5—in the same way.

Usually when we solve an equation, we try to get the variable alone on one side. If the variable appears in several terms, we get those terms on one side and factor out the variable.

Examples   Solve for the variable.

2. $5cdx + 7 = -3cdx$

$5cdx + 3cdx = -7$ Adding -7 and $3cdx$ to get all the x-terms on one side

$(5cd + 3cd)x = -7$ Factoring out an x

$8cdx = -7$

$x = \dfrac{-7}{8cd}$ Multiplying by $\dfrac{1}{8cd}$

3. $(x + b)(x - b) = x(x - 2)$

$x^2 - b^2 = x^2 - 2x$ Multiplying to eliminate parentheses

$-b^2 = -2x$ Adding $-x^2$

$b^2 = 2x$ Multiplying by -1

$\dfrac{b^2}{2} = x$ Multiplying by $\dfrac{1}{2}$

FORMULAS

Formulas are important in business, science, etc. It is important to be able to solve a formula for a given letter. It is impossible to merely look at a formula and tell which letters are constants. In fact, a letter can be a constant in one situation and a variable in another.

Examples

4. The formula

$$N = \frac{kP_1 P_2}{d^2}$$

is used to find the number N of telephone calls between two cities of populations P_1 and P_2 at a distance d from each other. Solve it for P_1.

Note that P_1 and P_2 are different variables (or constants). So, don't forget to write the subscripts, 1 and 2. We first clear of fractions.

$d^2 N = kP_1 P_2$ Multiplying by d^2

Now there is no fractional expression in the equation. Next,

$\dfrac{d^2 N}{kP_2} = P_1.$ Multiplying by $\dfrac{1}{kP_2}$ to get P_1 alone

5. Solve $P = \dfrac{k}{4}(a + b)$ for k.

$$4P = k(a + b) \qquad \text{Multiplying by 4}$$

$$\frac{4P}{a + b} = k \qquad \text{Multiplying by } \frac{1}{a + b}$$

Example 6   Solve the following formula for R^2:

$$V = \frac{1}{3}\pi h(R^2 + r^2).$$

Note carefully that R and r are different. Be careful how you write them.

$$3V = \pi h(R^2 + r^2) \qquad \text{Multiplying by 3}$$

$$\frac{3V}{\pi h} = R^2 + r^2 \qquad \text{Multiplying by } \frac{1}{\pi h}$$

$$\frac{3V}{\pi h} - r^2 = R^2 \qquad \text{Adding } -r^2$$

In solving, you could multiply $(R^2 + r^2)$ by πh before using the addition principle. If so, you would get

$$\frac{3V - \pi h r^2}{\pi h} = R^2.$$

That answer is also correct, but it is not as simply arrived at as the one we obtained. There are also other correct answers.

EXERCISE SET 4.11

Solve for x, y, or z.

1. $15 - ay - b = 3y$

2. $bx - a = b - ax$

3. $ab - bz = cz - ac$

4. $bx - a + cx = a + b$

5. $10(a + x) = 8(a - x)$

6. $y(b + c) - a = d$

7. $3(2x - 4) + 2(x - 3) = 5(x + 3)$

8. $4y(2 + b) - b(3y + 1) = 5b$

9. $a^2 - ax = b^2 - bx$

10. $a^2 = b^2 - b^3y + a^2by$

11. Solve $V = \pi r^2 h$ for r^2.
(A volume formula)

12. Solve $V = \dfrac{1}{3}Bh$ for B.
(A volume formula)

13. Solve $s = \dfrac{1}{2}gt^2$ for t^2.
(A motion formula)

14. Solve $F = \dfrac{mv^2}{r}$ for m.
(A motion formula)

15. Solve $A = \pi r^2 + 2\pi rh$ for h.
(An area formula)

16. Solve $A = \pi rs + \pi r^2$ for s.
(An area formula)

17. Solve $A = P(1 + rt)$ for r.
(An interest formula)

18. Solve $V = \dfrac{1}{3}\pi h^2(3R - h)$ for R.
(A volume formula)

☆

Solve for x.

19. $(x + 6)(x - 3) + b = (x - 5)(x - 2) - 3b$ **20.** $x^2 - 3(x - b) + c = (x - b)^2$

4.12 FUNCTIONS

A *function* is a certain kind of correspondence, from one set to another.

Examples

1. To each person in a math class   there corresponds   his or her age.

2. To each item in a store   there corresponds   its price.

3. To each positive real number   there corresponds   its positive square root.

In each example, the first set is called the *domain*. The second set is called the *range*. Given a member of the domain, there is *just one* member of the range to which it corresponds. This kind of correspondence is called a *function*.

> A *function* is a correspondence between a first set, called the *domain*, and a second set, called the *range*, such that to any member of the domain, there corresponds *exactly one* member of the range.

Examples   Which of the following are functions?

Domain	Correspondence	Range
4. A family	Each person's weight	A set of positive numbers

This correspondence is a function, because each person has *only one* weight.

Domain	Correspondence	Range
5. The natural numbers	Square of each number	A set of natural numbers

This correspondence is a function, because each natural number has *only one* square.

Domain	Correspondence	Range
6. The set of positive real numbers	Square roots of each number	The set of all real numbers

This correspondence is not a function, because every positive real number has two square roots.*

NOTATION FOR FUNCTIONS

To understand function notation, it helps to think of a *function machine*. Think of putting a member of the domain (an *input*) into the machine. The machine knows the correspondence and gives you a member of the range (the *output*).

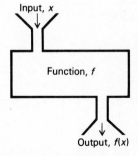

---

* In a function, two different members of the domain can correspond to the same member of the range. In Example 4, for instance, two different family members could have the same weight.

The function has been named f. We call the input x, and its output is named $f(x)$. This is read "f of x," or "f at x," or "the value of f at x." Note that $f(x)$ does *not* mean "f times x."

Some functions can be described by formulas or equations. For example, $f(x) = 2x + 3$ describes the function that takes an input x, multiplies it by 2, and then adds 3.

$$f(x) \;=\; \underset{\text{Double}}{2x} \;\; \underset{\text{Add 3}}{+\;3}$$

To find the output $f(4)$, we take the input, 4, double it, and add 3 to get 11. That is, we substitute 4 into the formula for $f(x)$:

$$f(4) = 2 \cdot 4 + 3$$
$$= 11.$$

Examples   Find these function values.

7. $f(x) = 3x + 2$; find $f(5)$.

    $f(5) = 3 \cdot 5 + 2 = 17$

> Remember, $f(x)$ does *not* mean f times x. Rather, $f(x)$ is the output when the input is x.

8. $g(x) = 5x^2 - 4$; find $g(3)$.

    $g(3) = 5(3)^2 - 4 = 41$

9. $A(x) = 3x^2 + 2x$; find $A(-2)$.

    $A(-2) = 3(-2)^2 + 2(-2) = 8$

POLYNOMIAL FUNCTIONS

Any function that can be described by a polynomial such as

$$f(x) = 5x^7 + 3x^5 - 4x^2 - 5,$$

where the right-hand side is a polynomial, is called a *polynomial function*.

GRAPHS OF FUNCTIONS

Consider the function $f(x) = 3x + 2$. If you think of this as $y = 3x + 2$, you already know how to draw a graph. The members of the domain are shown along the horizontal axis, and the members of the range along the vertical axis. Given an x (an input), the output, which is y or $f(x)$, is found by going up to the line and then over to the $f(x)$ axis.

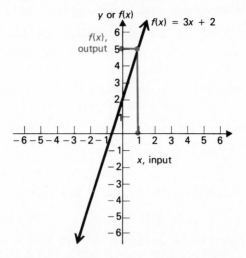

Some graphs are graphs of functions and some are not. If a vertical line drawn anywhere on the graph intersects the curve in more than one point, then there will be more than one number in the range corresponding to a member of the domain. The graph is not the graph of a function.

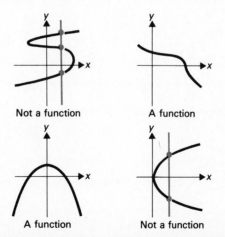

The vertical line test: **If it is possible for a vertical line to intersect a graph more than once, the graph is not the graph of a function.**

EXERCISE SET 4.12

Which of the following are functions?

Domain	*Correspondence*	*Range*
1. A math class	Each person's seat number	A set of numbers
2. A set of numbers	Square each number and then add 4	A set of numbers

Given the function g described by $g(x) = -2x - 4$, find:

3. $g(-3)$ **4.** $g\left(-\dfrac{1}{2}\right)$ **5.** $g\left(\dfrac{3}{2}\right)$ **6.** $g(1.2)$

Given the function h described by $h(x) = 3x^2$, find:

7. $h(-1)$ **8.** $h(1)$ **9.** $h(2)$ **10.** $h(-2)$

A polynomial function P is described by $P(x) = x^3 - x$. Find:

11. $P(2)$ **12.** $P(-2)$

A polynomial function Q is described by $Q(x) = x^4 - 2x^3 + x^2 - x + 2$. Find:

13. $Q(2)$ **14.** $Q(-2)$

The function A described by $A(s) = s^2\sqrt[2]{3}/4$ gives the area of an equilateral triangle with side s. Find:

15. $A(4)$ **16.** $A(6)$

The function V described by $V(r) = \frac{4}{3}\pi r^3$ gives the volume of a sphere with radius r. Find:

17. $V(3)$ **18.** $V(6)$

Which of the following are graphs of functions?

19.

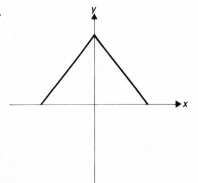

20.

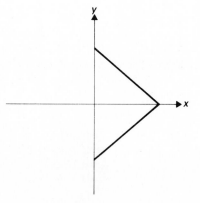

21.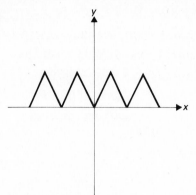

22.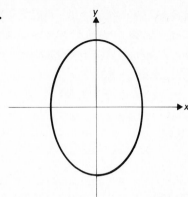

☆
Which of the following are functions?

Domain	*Correspondence*	*Range*
23. A set of pairs of numbers (a, b)	Take the larger of a and b.	A set of numbers
24. A set of ordered pairs of numbers (a, b)	Subtract a from b. Then take the larger of a and $b - a$.	A set of numbers

25. Given the functions f, g, and h described by $f(x) = (x + 1)^3$, $g(x) = (1 - x)^2$, and $h(x) = -2x + 1$, find:

   a) $3h(x) - g(x) + f(x)$ b) $f(x) \cdot g(x)$ c) $[g(x) - h(x)]^3$

   d) $3h(10) - g(10) + f(10)$ e) $f(-1) \cdot g(-1)$ f) $[g(2) - h(2)]^3$

   g) $h(5) - 2g(-3)$ h) $f(0) \cdot g(-1) \cdot h(1)$ i) $\left[g\left(\frac{1}{2}\right) - h\left(\frac{1}{2}\right) \right]^2$

TEST OR REVIEW—CHAPTER 4

1. Given the polynomial $3xy^3 - 4x^2y + 5x^5y^4 - 2x^4y$:
   a) Arrange in descending powers of x.
   b) What is the degree of the polynomial?

2. Collect like terms: $5xy - 2xy^2 - 2xy + 5xy^2$.

3. Add: $3x^2y^4 - 2x^2y^2 - 5x$ and
       $-3xy^2 + 2x^2y + 4$.

4. Subtract: $(6y^2 - 2y - 5y^3) - (4y^2 - 7y - 6y^3)$.

5. Multiply: $4x^3 - 2x + 3$ and $-2x^2 - 3x + 5$.

6. Multiply: $(3x - 4y)^2$.   **7.** Multiply: $(3x + 5)(3x - 5)$.

8. Factor: $6a^2b - 21a^3b^2 + 3a^2b^3$.   **9.** Factor: $y^2 + 5y - 4y - 20$.

10. Factor: $20a^2 - 5b^2$.   **11.** Factor: $9x^2 - 30x + 25$.

12. Factor: $24x^2 - 46x + 10$.   **13.** Factor: $y^2 + 8y + 16 - 100t^2$.

14. Complete the square: $x^2 - 5x$.   **15.** Factor by completing the square: $x^2 + 4.4x + 0.84$

16. Factor: $16a^7b + 54ab^7$.   **17.** Solve: $2x^2 = 3 - 5x$.

18. A photograph is 3 cm longer than it is wide. Its area is 40 cm². Find its length and width.   **19.** Solve for r: $F = \dfrac{mv^2}{r}$.

20. Given the function f described by $f(x) = x^2 - 3x + 5$, find $f(-2)$.

5

FRACTIONAL EXPRESSIONS AND EQUATIONS

5.1 MULTIPLYING AND DIVIDING

The following are fractional expressions. They are quotients of polynomials.

$$\frac{7}{8}, \quad \frac{8}{y + 5}, \quad \frac{x^2 + 7xy - 4}{x^3 - y^3}$$

Fractional expressions indicate division.

$$\frac{7}{8} \quad \text{means} \quad 7 \div 8 \quad \text{and} \quad \frac{2x + y}{x - y} \quad \text{means} \quad (2x + y) \div (x - y).$$

Certain substitutions are not sensible in fractional expressions. Since division by 0 is not defined, any number that makes a denominator 0 is not a sensible replacement. In $8/(y + 5)$, -5 is not a sensible replacement.

Most of the calculations we do with fractional expressions are very much like calculations with fractional notation in arithmetic.

MULTIPLYING

> **To multiply two fractional expressions, multiply numerators and multiply denominators.**

Example 1   Multiply.

$$\frac{x + 3}{y - 4} \cdot \frac{x^3}{y + 5} = \frac{(x + 3)x^3}{(y - 4)(y + 5)} \qquad \text{Multiplying numerators and multiplying denominators}$$

$$= \frac{x^4 + 3x^3}{y^2 + y - 20} \qquad \text{Simplifying}$$

In algebra we do the same thing as in arithmetic. For example,

$$\frac{2}{3} \cdot \frac{5}{7} = \frac{2 \cdot 5}{3 \cdot 7},$$

multiplying numerators and denominators. We could multiply out to get $\frac{10}{21}$, but we shall not do it yet for fractional expressions because we will learn to simplify first.

MULTIPLYING BY 1

Any number multiplied by 1 is that same number. Any fractional expression with the same numerator and denominator names the number 1.

$$\frac{y + 5}{y + 5}, \quad \frac{4x^2 - 5}{4x^2 - 5}, \quad \frac{-1}{-1}$$ All name the number 1 for all sensible replacements

We can multiply by 1 to get equivalent expressions. For example, let us multiply $(x + y)/5$ by 1.

$$\frac{x + y}{5} \cdot \boxed{\frac{x - y}{x - y}} = \frac{(x + y)(x - y)}{5(x - y)}$$ Multiplying by $\boxed{\dfrac{x - y}{x - y}}$, which is 1

We know that

$$\frac{x + y}{5}$$

and

$$\frac{(x + y)(x - y)}{5(x - y)}$$

are equivalent. That means they will name the same number for all replacements that do not make a denominator 0.

Examples   Multiply to obtain equivalent expressions.

2. $\dfrac{x^2 + 3}{x - 1} \cdot \boxed{\dfrac{x + 1}{x + 1}} = \dfrac{(x^2 + 3)(x + 1)}{(x - 1)(x + 1)}$

3. $\boxed{\dfrac{-1}{-1}} \cdot \dfrac{x - 4}{x - y} = \dfrac{-1 \cdot (x - 4)}{-1 \cdot (x - y)}$

SIMPLIFYING FRACTIONAL EXPRESSIONS

We can simplify fractional expressions by "removing" factors that are equal to 1. We first factor numerator and denominator, then factor the fractional expression, so that a factor is equal to 1.

Examples   Simplify by removing factors equal to 1.

4. $\dfrac{5x^2}{x} = \dfrac{5x \cdot x}{1 \cdot x}$ Factoring numerator and denominator

$\quad\quad = \dfrac{5x}{1} \cdot \boxed{\dfrac{x}{x}}$ Factoring the fractional expression

$\quad\quad = 5x \cdot 1 \quad\quad \dfrac{x}{x} = 1$ for all sensible replacements

$\quad\quad = 5x \quad\quad$ "Removing" a factor of 1

In this example we supplied a 1 in the denominator. This can always be done, if necessary.

5. $\dfrac{4a + 8}{2} = \dfrac{2(2a + 4)}{2 \cdot 1}$ Factoring numerator and denominator

$\quad\quad = \boxed{\dfrac{2}{2}} \cdot \dfrac{2a + 4}{1}$ Factoring the fractional expression

$\quad\quad = \dfrac{2a + 4}{1} \quad\quad$ "Removing" a factor of 1

$\quad\quad = 2a + 4$

> A common error here is to get $2a + 8$. This happens because of taking an incorrect short-cut. You should go through the *entire* process of removing a factor of 1.

6. $\dfrac{2x^2 + 4x}{6x^2 + 2x} = \dfrac{2x(x + 2)}{2x(3x + 1)}$ Factoring numerator and denominator

$\quad\quad = \boxed{\dfrac{2x}{2x}} \cdot \dfrac{x + 2}{3x + 1}$ Factoring the fractional expression

$\quad\quad = \dfrac{x + 2}{3x + 1} \quad\quad$ "Removing" a factor of 1

7. $\dfrac{x^2 - 1}{2x^2 - x - 1} = \dfrac{(x - 1)(x + 1)}{(2x + 1)(x - 1)}$ Factoring numerator and denominator

$\quad\quad = \boxed{\dfrac{x - 1}{x - 1}} \cdot \dfrac{x + 1}{2x + 1}$ Factoring the fractional expression

$\quad\quad = \dfrac{x + 1}{2x + 1} \quad\quad$ "Removing" a factor of 1

8. $\dfrac{9x^2 + 6xy - 3y^2}{12x^2 - 12y^2} = \dfrac{3(x + y)(3x - y)}{12(x + y)(x - y)}$ Factoring numerator and denominator

$$= \dfrac{3(x + y)}{3(x + y)} \cdot \dfrac{3x - y}{4(x - y)}$$ Factoring the fractional expression

$$= \dfrac{3x - y}{4(x - y)}$$ "Removing" a factor of 1

After removing all possible factors of 1, we usually multiply out the numerator and the denominator. The final answer is then

$$\dfrac{3x - y}{4x - 4y}.$$

MULTIPLYING AND SIMPLIFYING

After multiplying you should usually simplify, if possible.

Examples Multiply. Then simplify by removing factors of 1.

9. $\dfrac{x + 2}{x - 3} \cdot \dfrac{x^2 - 4}{x^2 + x - 2} = \dfrac{(x + 2)(x^2 - 4)}{(x - 3)(x^2 + x - 2)}$ Multiplying numerators and also denominators

$$= \dfrac{(x + 2)(x - 2)(x + 2)}{(x - 3)(x + 2)(x - 1)}$$ Factoring numerators and denominators

$$= \dfrac{x + 2}{x + 2} \cdot \dfrac{(x + 2)(x - 2)}{(x - 3)(x - 1)}$$ Factoring the fractional expression

$$= \dfrac{(x + 2)(x - 2)}{(x - 3)(x - 1)}$$ "Removing" a factor of 1

$$= \dfrac{x^2 - 4}{x^2 - 4x + 3}$$ Multiplying out numerator and denominator

10. $\dfrac{a^3 - b^3}{a^2 - b^2} \cdot \dfrac{a^2 + 2ab + b^2}{a^2 + ab + b^2} = \dfrac{(a^3 - b^3)(a^2 + 2ab + b^2)}{(a^2 - b^2)(a^2 + ab + b^2)}$

$$= \dfrac{(a - b)(a^2 + ab + b^2)(a + b)(a + b)}{(a - b)(a + b)(a^2 + ab + b^2)}$$

$$= \dfrac{(a - b)(a^2 + ab + b^2)(a + b)}{(a - b)(a^2 + ab + b^2)(a + b)} \cdot \dfrac{a + b}{1} = a + b$$

DIVIDING AND SIMPLIFYING

Two expressions are reciprocals of each other if their product is 1. To find the reciprocal of a fractional expression we interchange numerator and denominator.

The reciprocal of $\dfrac{x + 2y}{x + y - 1}$ is $\dfrac{x + y - 1}{x + 2y}$.

The reciprocal of $y - 8$ is $\dfrac{1}{y - 8}$.

> **We divide with fractional expressions the way we did in arithmetic. We multiply by the reciprocal of the divisor.**

We sometimes say "invert the divisor and multiply."

Examples   Divide. Simplify by removing a factor of 1 if possible. Then multiply out the numerator and denominator.

11. $\dfrac{x - 2}{x + 1} \div \dfrac{x + 5}{x - 3} = \dfrac{x - 2}{x + 1} \cdot \dfrac{x - 3}{x + 5}$ Multiplying by the reciprocal

$\qquad\qquad\qquad = \dfrac{x^2 - 5x + 6}{x^2 + 6x + 5}$ Multiplying numerators and denominators

12. $\dfrac{a^2 - 1}{a - 1} \div \dfrac{a^2 - 2a + 1}{a + 1}$

$\qquad = \dfrac{a^2 - 1}{a - 1} \cdot \dfrac{a + 1}{a^2 - 2a + 1}$ Multiplying by the reciprocal

$\qquad = \dfrac{(a^2 - 1)(a + 1)}{(a - 1)(a^2 - 2a + 1)}$ Multiplying numerators and denominators

$\qquad = \dfrac{(a + 1)(a - 1)(a + 1)}{(a - 1)(a - 1)(a - 1)}$ Factoring numerator and denominator

$\qquad = \dfrac{a - 1}{a - 1} \cdot \dfrac{(a + 1)(a + 1)}{(a - 1)(a - 1)}$ Factoring the fractional expression

$\qquad = \dfrac{(a + 1)(a + 1)}{(a - 1)(a - 1)}$ "Removing" a factor of 1

$\qquad = \dfrac{a^2 + 2a + 1}{a^2 - 2a + 1}$ Multiplying out numerator and denominator

EXERCISE SET 5.1

Multiply to obtain equivalent expressions. Do not simplify.

1. $\dfrac{3x}{3x} \cdot \dfrac{x+1}{x+3}$

2. $\dfrac{4-y^2}{6-y} \cdot \dfrac{-1}{-1}$

3. $\dfrac{t-3}{t+2} \cdot \dfrac{t+3}{t+3}$

4. $\dfrac{p-4}{p-5} \cdot \dfrac{p+5}{p+5}$

Simplify by removing factors of 1.

5. $\dfrac{9y^2}{15y}$

6. $\dfrac{6x^3}{18x^2}$

7. $\dfrac{2a-6}{2}$

8. $\dfrac{3a-6}{3}$

9. $\dfrac{4y-12}{4y+12}$

10. $\dfrac{8x+16}{8x-16}$

11. $\dfrac{t^2-16}{t^2-8t+16}$

12. $\dfrac{p^2-25}{p^2+10p+25}$

Multiply. Simplify by removing a factor of 1 if possible. Then multiply out the numerator and denominator.

13. $\dfrac{x^2-16}{x^2} \cdot \dfrac{x^2-4x}{x^2-x-12}$

14. $\dfrac{y^2+10y+25}{y^2-9} \cdot \dfrac{y+3}{y+5}$

15. $\dfrac{y^2-16}{2y+6} \cdot \dfrac{y+3}{y-4}$

16. $\dfrac{m^2-n^2}{4m+4n} \cdot \dfrac{m+n}{m-n}$

17. $\dfrac{x^2-2x-35}{2x^3-3x^2} \cdot \dfrac{4x^3-9x}{7x-49}$

18. $\dfrac{y^2-10y+9}{y^2-1} \cdot \dfrac{y+4}{y^2-5y-36}$

19. $\dfrac{c^3+8}{c^2-4} \cdot \dfrac{c^2-4c+4}{c^2-2c+4}$

20. $\dfrac{x^3-27}{x^2-9} \cdot \dfrac{x^2-6x+9}{x^2+3x+9}$

21. $\dfrac{x^2-y^2}{x^3-y^3} \cdot \dfrac{x^2+xy+y^2}{x^2+2xy+y^2}$

22. $\dfrac{4x^2-9y^2}{8x^3-27y^3} \cdot \dfrac{4x^2+6xy+9y^2}{4x^2+12xy+9y^2}$

Divide. Simplify by removing a factor of 1 if possible. Then multiply out the numerator and denominator.

23. $\dfrac{3y+15}{y} \div \dfrac{y+5}{y}$

24. $\dfrac{6x+12}{x} \div \dfrac{x+2}{x^3}$

25. $\dfrac{y^2-9}{y} \div \dfrac{y+3}{y+2}$

26. $\dfrac{x^2-4}{x} \div \dfrac{x-2}{x+4}$

27. $\dfrac{4a^2-1}{a^2-4} \div \dfrac{2a-1}{a-2}$

28. $\dfrac{25x^2-4}{x^2-9} \div \dfrac{5x-2}{x+3}$

29. $\dfrac{x^2-16}{x^2-10x+25} \div \dfrac{3x-12}{x^2-3x-10}$

30. $\dfrac{y^2-36}{y^2-8y+16} \div \dfrac{3y-18}{y^2-y+12}$

31. $\dfrac{y^3 + 3y}{y^2 - 9} \div \dfrac{y^2 + 5y - 14}{y^2 + 4y - 21}$

32. $\dfrac{a^3 + 4a}{a^2 - 16} \div \dfrac{a^2 + 8a + 15}{a^2 + a - 20}$

33. $\dfrac{x^3 - 64}{x^3 + 64} \div \dfrac{x^2 - 16}{x^2 - 4x + 16}$

34. $\dfrac{8y^3 + 27}{64y^3 - 1} \div \dfrac{4y^2 - 9}{16y^2 + 4y + 1}$

Perform the indicated operations and simplify.

35. $\left[\dfrac{r^2 - 4s^2}{r + 2s} \div (r + 2s) \right] \cdot \dfrac{2s}{r - 2s}$

36. $\left[\dfrac{d^2 - d}{d^2 - 6d + 8} \cdot \dfrac{d - 2}{d^2 + 5d} \right] \div \dfrac{5d}{d^2 - 9d + 20}$

☆
Multiply or divide.

37. ▦ $\dfrac{834x}{y - 427.2} \cdot \dfrac{26.3x}{y + 427.2}$

38. ▦ $\dfrac{527}{x + 93.87} \div \dfrac{x - 93.87}{468}$

Simplify.

39. $\dfrac{x(x + 1) - 2(x + 3)}{(x + 1)(x + 2)(x + 3)}$

40. $\dfrac{2x - 5(x + 2) - (x - 2)}{x^2 - 4}$

41. $\dfrac{m^2 - t^2}{m^2 + t^2 + m + t + 2mt}$

42. $\dfrac{a^3 - 2a^2 + 2a - 4}{a^3 - 2a^2 - 3a + 6}$

43. $\dfrac{x^3 + x^2 - y^3 - y^2}{x^2 - 2xy + y^2}$

44. $\dfrac{u^6 + v^6 + 2u^3v^3}{u^3 - v^3 + u^2v - uv^2}$

45. $\dfrac{x^5 - x^3 + x^2 - 1 - (x^3 - 1)(x + 1)^2}{(x^2 - 1)^2}$

5.2 LEAST COMMON MULTIPLES

FINDING LCM'S BY FACTORING

To add fractional expressions when denominators are different, we first find a common denominator. Let us review the procedure in arithmetic first. To do the addition

$$\frac{5}{42} + \frac{7}{12}$$

we first find a common denominator. We look for a number that is a multiple of both 42 and 12 (a common multiple). Usually we try to get the smallest such number, or the *least common multiple* (LCM). To find the LCM we factor both numbers completely (into primes).

$42 = 2 \cdot 3 \cdot 7$ ← Any multiple of 42 has these factors.

$12 = 2 \cdot 2 \cdot 3$ ← Any multiple of 12 has these factors.

The LCM is the number that has 2 as a factor twice, 3 as a factor once, and 7 as a factor once. The LCM is $2 \cdot 2 \cdot 3 \cdot 7$, or 84.

> **To obtain the LCM we use each factor the greatest number of times it occurs in any one factorization.**

Example 1   Find the LCM of 18 and 24.

$18 = 3 \cdot 3 \cdot 2$

$24 = 2 \cdot 2 \cdot 2 \cdot 3$ The LCM is $3 \cdot 3 \cdot 2 \cdot 2 \cdot 2$, or 72.

ADDING

Now let us return to adding $\frac{5}{42}$ and $\frac{7}{12}$.

$$\frac{5}{42} + \frac{7}{12} = \frac{5}{2 \cdot 3 \cdot 7} + \frac{7}{2 \cdot 2 \cdot 3} \qquad \text{Factoring denominators}$$

The LCM is $2 \cdot 2 \cdot 3 \cdot 7$. To get this LCM in the first denominator we need a 2. In the second denominator we need a 7. We multiply by 1, as follows:

$$\frac{5}{2 \cdot 3 \cdot 7} \cdot \frac{2}{2} + \frac{7}{2 \cdot 2 \cdot 3} \cdot \frac{7}{7} = \frac{10}{2 \cdot 2 \cdot 3 \cdot 7} + \frac{49}{2 \cdot 2 \cdot 3 \cdot 7}$$

$$= \frac{59}{2 \cdot 2 \cdot 3 \cdot 7}$$

$$= \frac{59}{84}$$

Multiplying the first fraction by $\frac{2}{2}$ gave us a denominator that is the LCM. Multiplying the second fraction by $\frac{7}{7}$ also gave us a denominator that is the LCM. When we have a common denominator, we add the numerators.

LCM'S OF ALGEBRAIC EXPRESSIONS

Examples

2. Find the LCM of $12xy^2$ and $15x^3y$.

We factor each expression completely.

$$12xy^2 = 2 \cdot 2 \cdot 3 \cdot x \cdot y \cdot y$$

$$15x^3y = 3 \cdot 5 \cdot x \cdot x \cdot x \cdot y$$

Remember: To find the LCM, use each factor the greatest number of times it occurs in any one factorization.

$$\text{LCM} = 2 \cdot 2 \cdot 3 \cdot 5 \cdot x \cdot x \cdot x \cdot y \cdot y = 60x^3y^2$$

3. Find the LCM of $x^2 + 2x + 1, 5x^2 - 5x$, and $x^2 - 1$.

$$x^2 + 2x + 1 = (x + 1)(x + 1)$$

$$5x^2 - 5x = 5x(x - 1)$$

$$x^2 - 1 = (x + 1)(x - 1)$$

Factoring

$$\text{LCM} = 5x(x + 1)(x + 1)(x - 1)$$

4. Find the LCM of $x^2 - y^2, x^3 + y^3$, and $x^2 + 2xy + y^2$.

$$x^2 - y^2 = (x - y)(x + y)$$

$$x^3 + y^3 = (x + y)(x^2 - xy + y^2)$$

$$x^2 + 2xy + y^2 = (x + y)(x + y)$$

$$\text{LCM} = (x - y)(x + y)(x + y)(x^2 - xy + y^2)$$

The additive inverse of an LCM is also an LCM. For example if $(x + 2)(x - 3)$ is an LCM, then $-(x + 2)(x - 3)$ is also an LCM. We can name the latter

$$(x + 2)(-1)(x - 3), \quad \text{or} \quad (x + 2)(3 - x).$$

If, when finding LCMs, factors that are additive inverses occur, we do not use them both. For example if $(a - b)$ occurs in one factorization and $(b - a)$ occurs in another, we do not use them both since they are additive inverses.

Example 5 Find the LCM of $x^2 - y^2$ and $3y - 3x$.

$$x^2 - y^2 = \boxed{(x + y) (x - y)} \longleftarrow$$

We can use $(x - y)$ or $(y - x)$ but we do not use both.

$$3y - 3x = \boxed{3(y - x)} \longleftarrow$$

$$\text{LCM} = 3(x + y)(x - y), \quad \text{or} \quad 3(x + y)(y - x)$$

EXERCISE SET 5.2

Find the LCM by factoring.

1. 12, 18   **2.** 15, 20   **3.** 18, 48

4. 45, 54   **5.** 24, 36   **6.** 30, 75

7. 9, 15, 5   **8.** 27, 35, 63

Add, first finding the LCM of the denominators.

9. $\dfrac{5}{6} + \dfrac{4}{15}$ **10.** $\dfrac{5}{36} + \dfrac{5}{24}$ **11.** $\dfrac{7}{12} + \dfrac{11}{18}$

12. $\dfrac{11}{30} + \dfrac{19}{75}$ **13.** $\dfrac{3}{4} + \dfrac{2}{5} + \dfrac{1}{6}$ **14.** $\dfrac{5}{8} + \dfrac{7}{12} + \dfrac{11}{40}$

Find the LCM.

15. $12x^2y, \quad 4xy$ **16.** $18r^2s, \quad 12rs^3$

17. $y^2 - 9, \quad 3y + 9$ **18.** $a^2 - b^2, \quad ab + b^2$

19. $15ab^2, \quad 3ab, \quad 10a^3b$ **20.** $6x^2y^2, \quad 9x^3y, \quad 15y^3$

21. $5y - 15, \quad y^2 - 6y + 9$ **22.** $x^2 + 10x + 25, \quad x^2 + 2x - 15$

23. $x^2 - 4, \quad 2 - x$ **24.** $y^2 - 9, \quad 3 - y$

25. $2r^2 - 5r - 12, \quad 3r^2 - 13r + 4, \quad r^2 - 16$

26. $2x^2 - 5x - 3, \quad 2x^2 - x - 1, \quad x^2 - 6x + 9$

☆

Find the LCM.

27. 18, 42, 82, 120, 300, 700   **28.** $x^8 - x^4, x^5 - x^2, x^5 - x^3, x^5 + x^2$

29. The LCM of two expressions is $8a^4b^7$. One of the expressions is $2a^3b^7$. List all possibilities for the other expression.

5.3 ADDITION AND SUBTRACTION

WHEN DENOMINATORS ARE THE SAME

> When denominators are the same, we add or subtract the numerators and keep the same denominator.

Example 1   Add.

$$\frac{3 + x}{x} + \frac{4}{x} = \frac{7 + x}{x}$$ ← This expression does *not* simplify to 7.

Example 1 shows that

$$\frac{3 + x}{x} + \frac{4}{x} \quad \text{and} \quad \frac{7 + x}{x}$$

are equivalent expressions.

They name the same number for all replacements except 0.

Example 2   Add.

$$\frac{4x^2 - 5xy}{x^2 - y^2} + \frac{2xy - y^2}{x^2 - y^2} = \frac{4x^2 - 3xy - y^2}{x^2 - y^2} \qquad \text{Adding numerators}$$

$$= \frac{(4x + y)(x - y)}{(x + y)(x - y)} \qquad \begin{array}{l}\text{Factoring numerator}\\\text{and denominator}\end{array}$$

$$= \frac{x - y}{x - y} \cdot \frac{4x + y}{x + y} \qquad \begin{array}{l}\text{Factoring the frac-}\\\text{tional expression}\end{array}$$

$$= \frac{4x + y}{x + y} \qquad \text{``Removing'' a factor of 1}$$

Example 3   Subtract.

$$\frac{4x + 5}{x + 3} - \frac{x - 2}{x + 3} = \frac{4x + 5 - (x - 2)}{x + 3} \qquad \text{Subtracting numerators}$$

$$= \frac{4x + 5 - x + 2}{x + 3}$$

$$= \frac{3x + 7}{x + 3}$$

A common error: Forgetting these parentheses. If you forget them, you will be subtracting only *part* of the numerator, $x - 2$.

WHEN DENOMINATORS ARE ADDITIVE INVERSES OF EACH OTHER

When one denominator is the additive inverse of the other, we multiply one expression by $-1/-1$. This gives a common denominator.

Example 4   Add.

$$\frac{a}{2a} + \frac{a^3}{-2a} = \frac{a}{2a} + \frac{-1}{-1} \cdot \frac{a^3}{-2a} \qquad \text{Multiplying by } \frac{-1}{-1} \leftarrow$$

This is equal to 1 (not -1).

$$= \frac{a}{2a} + \frac{-a^3}{2a}$$

$$= \frac{a - a^3}{2a} \qquad \text{Adding numerators}$$

$$= \frac{a(1 - a^2)}{2a} \qquad \text{Factoring numerator and denominator}$$

$$= \frac{a}{a} \cdot \frac{1 - a^2}{2} \qquad \text{Factoring the expression}$$

$$= \frac{1 - a^2}{2} \qquad \text{"Removing" a factor of 1}$$

Example 5   Subtract.

$$\frac{5x}{x - 2y} - \frac{3y - 7}{2y - x} = \frac{5x}{x - 2y} - \frac{-1}{-1} \cdot \frac{3y - 7}{2y - x}$$

Remember: $-1(2y - x) = -(2y - x) = (x - 2y)$.

$$= \frac{5x}{x - 2y} - \frac{7 - 3y}{x - 2y}$$

$$= \frac{5x - (7 - 3y)}{x - 2y} \qquad \text{Subtracting numerators}$$

Don't forget these parentheses.

$$= \frac{5x - 7 + 3y}{x - 2y}$$

WHEN DENOMINATORS ARE DIFFERENT

When denominators are different, but not additive inverses of each other, we first find a common denominator, using the LCM, and then add or subtract numerators.

Example 6   Add: $\dfrac{2a}{5} + \dfrac{3b}{2a}$.

First find the LCM of the denominators.

$$\begin{matrix} 5 \\ 2a \end{matrix} \qquad \text{The LCM is } 5 \cdot 2a \quad \text{or} \quad 10a.$$

Now we multiply each expression by 1. We choose whatever symbol for 1 will give us the LCM in each denominator. In this case we use $2a/2a$ and $5/5$.

$$\frac{2a}{5} \cdot \boxed{\frac{2a}{2a}} + \frac{3b}{2a} \cdot \boxed{\frac{5}{5}} = \frac{4a^2}{10a} + \frac{15b}{10a} = \frac{4a^2 + 15b}{10a}$$

Multiplying by $2a/2a$ in the first term gave us a denominator of $10a$. Multiplying by $\frac{5}{5}$ in the second term also gave us a denominator of $10a$.

Example 7   Add: $\dfrac{3x^2 + 3xy}{x^2 - y^2} + \dfrac{2 - 3x}{x - y}$.

We first find the LCM of the denominators.

$$\left. \begin{matrix} x^2 - y^2 = (x + y)(x - y) \\ x - y = x - y \end{matrix} \right\} \qquad \text{The LCM is } (x + y)(x - y)$$

We now multiply by 1 to get the LCM in the second expression. Then we add and simplify if possible.

$$\frac{3x^2 + 3xy}{(x + y)(x - y)} + \frac{2 - 3x}{x - y} \cdot \boxed{\frac{x + y}{x + y}} \qquad \text{Multiplying by 1 to get LCM}$$

$$= \frac{3x^2 + 3xy}{(x + y)(x - y)} + \frac{(2 - 3x)(x + y)}{(x - y)(x + y)}$$

$$= \frac{3x^2 + 3xy}{(x + y)(x - y)} + \frac{2x + 2y - 3x^2 - 3xy}{(x - y)(x + y)} \qquad \text{Multiplying out in numerator}$$

$$= \frac{3x^2 + 3xy + 2x + 2y - 3x^2 - 3xy}{(x + y)(x - y)} \qquad \text{Adding numerators}$$

$$= \frac{2x + 2y}{(x + y)(x - y)} = \frac{2(x + y)}{(x + y)(x - y)} \qquad \text{Combining like terms and then factoring numerator}$$

$$= \boxed{\frac{x + y}{x + y}} \cdot \frac{2}{x - y} = \frac{2}{x - y} \qquad \text{Simplifying by ''removing'' a factor of 1}$$

Example 8 Subtract.

$$\frac{2y + 1}{y - 7y + 6} - \frac{y + 3}{y - 5y - 6}$$

$$= \frac{2y + 1}{(y - 6)(y - 1)} - \frac{y + 3}{(y - 6)(y + 1)} \qquad \begin{array}{l} \text{The LCM is} \\ (y - 6)(y - 1)(y + 1). \end{array}$$

$$= \frac{2y + 1}{(y - 6)(y - 1)} \cdot \frac{y + 1}{y + 1} - \frac{y + 3}{(y - 6)(y + 1)} \cdot \frac{y - 1}{y - 1}$$

$$= \frac{(2y + 1)(y + 1) - (y + 3)(y - 1)}{(y - 6)(y - 1)(y + 1)}$$

$$= \frac{2y^2 + 3y + 1 - (y^2 + 2y - 3)}{(y - 6)(y - 1)(y + 1)}$$

$$= \frac{2y^2 + 3y + 1 - y^2 - 2y + 3}{(y - 6)(y - 1)(y + 1)} = \frac{y^2 + y + 4}{(y - 6)(y - 1)(y + 1)}$$

If a denominator has 3 or more factors, we usually do not multiply it out.

COMBINED ADDITIONS AND SUBTRACTIONS

Example 9 Do this calculation.

$$\frac{2x}{x^2 - 4} + \frac{5}{2 - x} - \frac{1}{2 + x} = \frac{2x}{(x - 2)(x + 2)} + \frac{5}{2 - x} - \frac{1}{2 + x}$$

$$= \frac{2x}{(x - 2)(x + 2)} + \frac{-1}{-1} \cdot \frac{5}{(2 - x)} - \frac{1}{x + 2}$$

$$= \frac{2x}{(x - 2)(x + 2)} + \frac{-5}{x - 2} - \frac{1}{x + 2} \qquad \begin{array}{l} \text{The LCM is} \\ (x - 2)(x + 2). \end{array}$$

$$= \frac{2x}{(x - 2)(x + 2)} + \frac{-5}{x - 2} \cdot \frac{x + 2}{x + 2} - \frac{1}{x + 2} \cdot \frac{x - 2}{x - 2}$$

$$= \frac{2x - 5(x + 2) - (x - 2)}{(x - 2)(x + 2)} = \frac{2x - 5x - 10 - x + 2}{(x - 2)(x + 2)}$$

$$= \frac{-4x - 8}{(x - 2)(x + 2)} = \frac{-4(x + 2)}{(x - 2)(x + 2)} = \frac{-4}{x - 2} \cdot \frac{x + 2}{x + 2} = \frac{-4}{x - 2}$$

EXERCISE SET 5.3

Add or subtract. Simplify by removing a factor of 1 when possible.

1. $\dfrac{a - 3b}{a + b} + \dfrac{a + 5b}{a + b}$

2. $\dfrac{x - 5y}{x + y} + \dfrac{x + 7y}{x + y}$

3. $\dfrac{4y + 3}{y - 2} - \dfrac{y - 2}{y - 2}$

4. $\dfrac{3t + 2}{t - 4} - \dfrac{t - 4}{t - 4}$

5. $\dfrac{a^2}{a - b} + \dfrac{b^2}{b - a}$

6. $\dfrac{r^2}{r - s} + \dfrac{s^2}{s - r}$

7. $\dfrac{3}{x} - \dfrac{8}{-x}$

8. $\dfrac{2}{a} - \dfrac{5}{-a}$

9. $\dfrac{2x - 10}{x^2 - 25} - \dfrac{5 - x}{25 - x^2}$

10. $\dfrac{y - 9}{y^2 - 16} - \dfrac{7 - y}{16 - y^2}$

Add or subtract. Simplify by removing a factor of 1 when possible. If a denominator has three or more factors (other than monomials), leave it factored.

11. $\dfrac{y - 2}{y + 4} + \dfrac{y + 3}{y - 5}$

12. $\dfrac{x - 2}{x + 3} + \dfrac{x + 2}{x - 4}$

13. $\dfrac{4xy}{x^2 - y^2} + \dfrac{x - y}{x + y}$

14. $\dfrac{5ab}{a^2 - b^2} + \dfrac{a + b}{a - b}$

15. $\dfrac{9x + 2}{3x^2 - 2x - 8} + \dfrac{7}{3x^2 + x - 4}$

16. $\dfrac{3y + 2}{2y^2 - y - 10} + \dfrac{8}{2y^2 - 7y + 5}$

17. $\dfrac{4}{x + 1} + \dfrac{x + 2}{x^2 - 1} + \dfrac{3}{x - 1}$

18. $\dfrac{-2}{y + 2} + \dfrac{5}{y - 2} + \dfrac{y + 3}{y^2 - 4}$

19. $\dfrac{x - 1}{3x + 15} - \dfrac{x + 3}{5x + 25}$

20. $\dfrac{y - 2}{4y + 8} - \dfrac{y + 6}{5y + 10}$

21. $\dfrac{5ab}{a^2 - b^2} - \dfrac{a - b}{a + b}$

22. $\dfrac{6xy}{x^2 - y^2} - \dfrac{x + y}{x - y}$

23. $\dfrac{3y}{y^2 - 7y + 10} - \dfrac{2y}{y^2 - 8y + 15}$

24. $\dfrac{5x}{x^2 - 6x + 8} - \dfrac{3x}{x^2 - x - 12}$

25. $\dfrac{y}{y^2 - y - 20} + \dfrac{2}{y + 4}$

26. $\dfrac{6}{y^2 + 6y + 9} + \dfrac{5}{y^2 - 9}$

27. $\dfrac{3y + 2}{y^2 + 5y - 24} + \dfrac{7}{y^2 + 4y - 32}$

28. $\dfrac{3y + 2}{y^2 - 7y + 10} + \dfrac{2y}{y^2 - 8y + 15}$

29. $\dfrac{3x - 1}{x^2 + 2x - 3} - \dfrac{x + 4}{x^2 - 9}$

30. $\dfrac{3p - 2}{p^2 + 2p - 24} - \dfrac{p - 3}{p^2 - 16}$

Do these calculations.

31. $\dfrac{1}{x + 1} - \dfrac{x}{x - 2} + \dfrac{x^2 + 2}{x^2 - x - 2}$

32. $\dfrac{2}{y + 3} - \dfrac{y}{y - 1} + \dfrac{y^2 + 2}{y^2 + 2y - 3}$

33. $\dfrac{x - 1}{x - 2} - \dfrac{x + 1}{x + 2} + \dfrac{x - 6}{x^2 - 4}$

34. $\dfrac{y - 3}{y - 4} - \dfrac{y + 2}{y + 4} + \dfrac{y - 7}{y^2 - 16}$

35. $\dfrac{4x}{x^2 - 1} + \dfrac{3x}{1 - x} - \dfrac{4}{x - 1}$

36. $\dfrac{5y}{1 - 2y} - \dfrac{2y}{2y + 1} + \dfrac{3}{4y^2 - 1}$

☆

Do these calculations.

37. $2x^{-2} + 3x^{-2}y^{-2} - 7xy^{-1}$

38. $2(a^2 - b^2)^{-1} + 2(a^2 + b^2)^{-1} + 4a^2(a^4 - b^4)^{-1}$

39. $4(y - 1)(2y - 5)^{-1} + 5(2y + 3)(5 - 2y)^{-1} + (y - 4)(2y - 5)^{-1}$

5.4 COMPLEX FRACTIONAL EXPRESSIONS

> A *complex fractional expression* is one that has a fractional expression in its numerator or its denominator, or both.

These are complex fractional expressions:

$$\frac{x}{x - \dfrac{1}{3}}, \qquad \frac{2x - \dfrac{4x}{3y}}{\dfrac{5x^2 + 2x}{6y^2}},$$

$$\frac{\dfrac{1}{a} + \dfrac{1}{b}}{\dfrac{1}{a} - \dfrac{1}{b}}, \qquad \frac{\dfrac{5}{x}}{\dfrac{x}{y}}.$$

> **To simplify a complex fractional expression:**
>
> 1. **Add or subtract, as necessary, to get a single fractional expression in the numerator.**
> 2. **Add or subtract, as necessary, to get a single fractional expression in the denominator.**
> 3. **Divide the numerator by the denominator.**
> 4. **If possible, simplify by removing a factor of 1.**

Example 1   Simplify: $\dfrac{x + \dfrac{1}{5}}{x - \dfrac{1}{3}}.$

$$\frac{x + \dfrac{1}{5}}{x - \dfrac{1}{3}} = \frac{x \cdot \dfrac{5}{5} + \dfrac{1}{5}}{x - \dfrac{1}{3}}$$

$$= \frac{\dfrac{5x + 1}{5}}{x - \dfrac{1}{3}}$$

Finding LCM in numerator, multiplying by 1 and adding

$$= \frac{\dfrac{5x + 1}{5}}{x \cdot \dfrac{3}{3} - \dfrac{1}{3}}$$

$$= \frac{\dfrac{5x + 1}{5}}{\dfrac{3x - 1}{3}}$$

Finding LCM in denominator, multiplying by 1, and subtracting

$$= \frac{5x + 1}{5} \cdot \frac{3}{3x - 1}$$

Multiplying by the reciprocal of the denominator

$$= \frac{15x + 3}{15x - 5}$$

Multiplying numerators and denominators

Simplifying by removing a factor of 1 is not possible in this case.

If you feel more comfortable doing so, you can always write denominators of 1 where there are no denominators. For example, you could start out by writing

$$\frac{\dfrac{x}{1} + \dfrac{1}{5}}{\dfrac{x}{1} - \dfrac{1}{3}}.$$

Example 2   Simplify: $\dfrac{1 + \dfrac{1}{x}}{1 - \dfrac{1}{x^2}}.$

$$\dfrac{1 + \dfrac{1}{x}}{1 - \dfrac{1}{x^2}} = \dfrac{\dfrac{x}{x} + \dfrac{1}{x}}{\dfrac{x^2}{x^2} - \dfrac{1}{x^2}} \qquad \text{Finding the LCM and multiplying by 1}$$

$$= \dfrac{\dfrac{x + 1}{x}}{\dfrac{x^2 - 1}{x^2}} \qquad \begin{array}{l}\text{Adding in the numerator and} \\ \text{subtracting in the denominator}\end{array}$$

$$= \dfrac{x + 1}{x} \cdot \dfrac{x^2}{x^2 - 1} \qquad \begin{array}{l}\text{Multiplying by the reciprocal} \\ \text{of the denominator}\end{array}$$

$$= \dfrac{(x + 1) \cdot x^2}{x(x^2 - 1)}$$

$$= \dfrac{x(x + 1)}{x(x + 1)} \cdot \dfrac{x}{x - 1} \qquad \begin{array}{l}\text{Factoring the fractional} \\ \text{expression}\end{array}$$

$$= \dfrac{x}{x - 1} \qquad \text{Removing a factor of 1}$$

Example 2 shows that the expressions

$$\dfrac{1 + \dfrac{1}{x}}{1 - \dfrac{1}{x^2}}$$

and

$$\dfrac{x}{x - 1}$$

are equivalent.

They name the same number for all replacements except 0, 1, and -1.

Example 3 Simplify.

$$\frac{\dfrac{1}{a}+\dfrac{1}{b}}{\dfrac{1}{a^3}+\dfrac{1}{b^3}} = \frac{\dfrac{1}{a}\cdot\dfrac{b}{b}+\dfrac{1}{b}\cdot\dfrac{a}{a}}{\dfrac{1}{a^3}\cdot\dfrac{b^3}{b^3}+\dfrac{1}{b^3}\cdot\dfrac{a^3}{a^3}} = \frac{\dfrac{b}{ab}+\dfrac{a}{ab}}{\dfrac{b^3}{a^3b^3}+\dfrac{a^3}{a^3b^3}}$$

$$= \frac{\dfrac{b+a}{ab}}{\dfrac{b^3+a^3}{a^3b^3}} \qquad \text{Adding in the numerator and denominator}$$

$$= \frac{b+a}{ab}\cdot\frac{a^3b^3}{b^3+a^3} \qquad \begin{array}{l}\text{Multiplying by the reciprocal}\\ \text{of the denominator}\end{array}$$

$$= \frac{(b+a)a^3b^3}{ab(b^3+a^3)} = \frac{(b+a)\cdot ab\cdot a^2b^2}{ab(b+a)(b^2-ab+a^2)}$$

$$= \frac{(b+a)ab}{(b+a)ab}\cdot\frac{a^2b^2}{b^2-ab+a^2}$$

$$= \frac{a^2b^2}{b^2-ab+a^2}$$

EXERCISE SET 5.4

Simplify.

1. $\dfrac{\dfrac{1}{x}+4}{\dfrac{1}{x}-3}$

2. $\dfrac{\dfrac{1}{y}+7}{\dfrac{1}{y}-5}$

3. $\dfrac{x-\dfrac{1}{x}}{x+\dfrac{1}{x}}$

4. $\dfrac{y+\dfrac{1}{y}}{y-\dfrac{1}{y}}$

5. $\dfrac{\dfrac{3}{x}+\dfrac{4}{y}}{\dfrac{4}{x}-\dfrac{3}{y}}$

6. $\dfrac{\dfrac{2}{y}+\dfrac{5}{z}}{\dfrac{1}{y}-\dfrac{4}{z}}$

7. $\dfrac{\dfrac{x^2 - y^2}{xy}}{\dfrac{x - y}{y}}$

8. $\dfrac{\dfrac{a^2 - b^2}{ab}}{\dfrac{a - b}{b}}$

9. $\dfrac{a - \dfrac{3a}{b}}{b - \dfrac{b}{a}}$

10. $\dfrac{1 - \dfrac{2}{3x}}{x - \dfrac{4}{9x}}$

11. $\dfrac{\dfrac{1}{a} + \dfrac{1}{b}}{\dfrac{a^2 - b^2}{ab}}$

12. $\dfrac{\dfrac{1}{x} + \dfrac{1}{y}}{\dfrac{x^2 - y^2}{xy}}$

13. $\dfrac{\dfrac{1}{x + h} - \dfrac{1}{x}}{h}$ It may help you to write $\dfrac{h}{1}$.

14. $\dfrac{\dfrac{1}{a - h} - \dfrac{1}{a}}{h}$

15. $\dfrac{\dfrac{y^2 - y - 6}{y^2 - 5y - 14}}{\dfrac{y^2 + 6y + 5}{y^2 - 6y - 7}}$

16. $\dfrac{\dfrac{x^2 - x - 12}{x^2 - 2x - 15}}{\dfrac{x^2 + 8x + 12}{x^2 - 5x - 14}}$

☆ _____

Simplify.

17. $\dfrac{5x^{-1} - 5y^{-1} + 10x^{-1}y^{-1}}{6x^{-1} - 6y^{-1} + 12x^{-1}y^{-1}}$

18. $\dfrac{\dfrac{4a}{2a^2 - a - 1} - \dfrac{4}{a - 1}}{\dfrac{1}{a - 1} + \dfrac{6}{2a + 1}}$

19. $2 + \dfrac{2}{2 + \dfrac{2}{2 + \dfrac{2}{2 + \dfrac{2}{x}}}}$

20. $\left[\dfrac{\dfrac{x + 3}{x - 3} + 1}{\dfrac{x + 3}{x - 3} - 1} \right]^4$

21. $(a^2 - ab + b^2)^{-1}(a^2b^{-1} + b^2a^{-1})(a^{-2} - b^{-2})(a^{-2} + 2a^{-1}b^{-1} + b^{-2})^{-1}$

Find the reciprocal of each of the following and simplify.

22. $x^2 - \dfrac{1}{x}$

23. $\dfrac{1 - \dfrac{1}{a}}{a - 1}$

24. $\dfrac{a^3 + b^3}{a + b}$

5.5 DIVISION OF POLYNOMIALS

DIVISOR A MONOMIAL

Remember that fractional expressions indicate division. Division by a monomial can be done by first writing a fractional expression.

Example 1   Divide $12x^3 + 8x^2 + x + 4$ by $4x$.

$$\frac{12x^3 + 8x^2 + x + 4}{4x} \qquad \text{Writing a fractional expression}$$

$$= \frac{12x^3}{4x} + \frac{8x^2}{4x} + \frac{x}{4x} + \frac{4}{4x} \qquad \text{Doing the reverse of adding*}$$

$$= 3x^2 + 2x + \frac{1}{4} + \frac{1}{x} \qquad \text{Doing the four indicated divisions}$$

Example 2   Divide: $(8x^4 - 3x^3 + 5x^2) \div x^2$.

$$\frac{8x^4 - 3x^3 + 5x^2}{x^2} = \frac{8x^4}{x^2} - \frac{3x^3}{x^2} + \frac{5x^2}{x^2}$$

$$= 8x^2 - 3x + 5$$

You should try to write only the answer.

> **To divide a polynomial by a monomial we can divide each term by the monomial.**

DIVISOR NOT A MONOMIAL

When the divisor is not a monomial, we use a procedure very much like long division in arithmetic.

---

* To see this, do the addition

$$\frac{12x^3}{4x} + \frac{8x^2}{4x} + \frac{x}{4x} + \frac{4}{4x}$$

and you will get the expression

$$\frac{12x^3 + 8x^2 + x + 4}{4x}.$$

Example 3 Divide $x^2 + 5x + 8$ by $x + 3$.

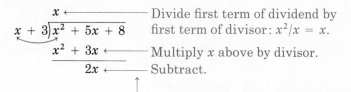

Divide first term of dividend by
first term of divisor: $x^2/x = x$.

Multiply x above by divisor.

Subtract.

> The subtraction we do here is $(x^2 + 5x) -$ $(x^2 + 3x)$. Remember, to subtract, add the inverse (change the sign of every term, then add).

We now "bring down" the other terms of the dividend—in this case, 8.

$$
\begin{array}{r}
x \;\; + 2 \\
x + 3 \overline{)\, x^2 + 5x + 8} \\
x^2 + 3x \\
\hline
2x + 8 \\
2x + 6 \\
\hline
2
\end{array}
$$

Divide first term by first term: $2x/x = 2$.

The 8 has been "brought down."

Multiply 2 by divisor.

Subtract: $(2x + 8) - (2x + 6)$.

Answer: Quotient $x + 2$, remainder 2.

To check, we multiply quotient by divisor and add the remainder, to see if we get the dividend.

Divisor Quotient   Remainder
$(x + 3)$ $(x + 2)$ $+$ 2 $= (x^2 + 5x + 6) + 2$
$$= \underline{x^2 + 5x + 8}$$
$$\text{Dividend}$$

The answer checks.

Always remember the following.

1. Arrange polynomials in descending order.

2. If there are missing terms in the dividend, either write them with 0 coefficients or leave space for them.

Example 4   Divide: $(125y^3 - 8) \div (5y - 2)$.

a)

$$\begin{array}{r} 25y^2 + 10y \;\; + \;\; 4 \\ 5y - 2 \overline{\smash{)}125y^3 + \;\; 0y^2 + \;\; 0y - 8} \\ 125y^3 - 50y^2 \\ \hline 50y^2 + \;\; 0y - 8 \\ 50y^2 - 20y \\ \hline 20y - 8 \\ 20y - 8 \\ \hline 0 \end{array}$$

When there are missing terms, we can write them in as in (a), or leave space for them as in (b).

This subtraction is $125y^3 - (125y^3 - 50y^2)$. We get $50y^2$.

b)

$$\begin{array}{r} 25y^2 + 10y \;\; + \;\; 4 \\ 5y - 2 \overline{\smash{)}125y^3 \qquad\qquad - 8} \\ 125y^3 - 50y^2 \\ \hline 50y^2 \qquad - 8 \\ 50y^2 - 20y \\ \hline 20y - 8 \\ 20y - 8 \\ \hline 0 \end{array}$$

This subtraction is $50y^2 - (50y^2 - 20y)$. We get $20y$.

Example 5   Divide: $(x^4 - 9x^2 - 5) \div (x - 2)$.

$$\begin{array}{r} x^3 + 2x^2 - 5x \;\; - 10 \\ x - 2 \overline{\smash{)}x^4 \qquad\quad - 9x^2 \qquad\quad - 5} \\ x^4 - 2x^3 \\ \hline 2x^3 - 9x^2 \qquad\quad - 5 \\ 2x^3 - 4x^2 \\ \hline - 5x^2 \qquad\quad - 5 \\ - 5x^2 + 10x \\ \hline - 10x - 5 \\ - 10x + 20 \\ \hline - 25 \end{array}$$

The first subtraction is $x^4 - (x^4 - 2x^3)$.

The second subtraction is $(2x^3 - 9x^2) - (2x^3 - 4x^2)$.

The answer is $x^3 + 2x^2 - 5x - 10$, with $R = -25$, or

$$x^3 + 2x^2 - 5x - 10 + \frac{-25}{x - 2}.$$

When dividing, we may "come out even" (have a remainder of 0), or we may not. If not, how long should we keep working? We continue until the degree of the remainder is less than the degree of the divisor, as in the next example. The answer can be given by writing the quotient and remainder, or by using a fractional expression, as you saw in Example 5 and as you will see in the next example.

Example 6 Divide: $(x^3 + 9x^2 - 5) \div (x^2 - 1)$.

$$
\begin{array}{r}
x + 9 \\
x^2 - 1 \overline{) x^3 + 9x^2 + 0x - 5} \\
x^3 - x \\
\hline
9x^2 + x - 5 \\
9x^2 - 9 \\
\hline
x + 4
\end{array}
$$

Again we have a missing term, so we can write it in.

The degree of the remainder is less than the degree of the divisor, so we are finished.

The answer is $x + 9$, with $R = x + 4$, or

$$x + 9 + \frac{x + 4}{x^2 - 1}.$$

This expression is the remainder over the divisor.

EXERCISE SET 5.5

Divide.

1. $\dfrac{30x^8 - 15x^6 + 40x^4}{5x^4}$

2. $\dfrac{24y^6 + 18y^5 - 36y^2}{6y^2}$

3. $(9y^4 - 18y^3 + 27y^2) \div 9y$

4. $(24a^3 + 28a^2 - 20a) \div 2a$

5. $(x^2 + 10x + 21) \div (x + 3)$

6. $(y^2 - 8y + 16) \div (y - 4)$

7. $(a^2 - 8a - 16) \div (a + 4)$

8. $(y^2 - 10y - 25) \div (y - 5)$

9. $(x^2 + 7x + 14) \div (x + 5)$

10. $(t^2 - 7t - 9) \div (t - 3)$

11. $(4y^3 + 6y^2 + 14) \div (2y + 4)$

12. $(6x^3 - x^2 - 10) \div (3x + 4)$

13. $(10y^3 + 6y^2 - 9y + 10) \div (5y - 2)$

14. $(6x^3 - 11x^2 + 11x - 2) \div (2x - 3)$

15. $(2x^4 - x^3 - 5x^2 + x - 6) \div (x^2 + 2)$

16. $(3x^4 + 2x^3 - 11x^2 - 2x + 5) \div (x^2 - 2)$

☆ ──

Divide.

17. $(2x^6 + 5x^4 - x^3 + 1) \div (x^2 + x + 1)$ **18.** $(2y^5 + y^3 - 2y - 3) \div (y^2 - 3y + 1)$

19. $(x^4 - x^3y + x^2y^2 + 2x^2y - 2xy^2 + 2y^3) \div (x^2 - xy + y^2)$

20. $(4a^3b + 5a^2b^2 + a^4 + 2ab^3) \div (a^2 + 2b^2 + 3ab)$

21. $(x^4 - y^4) \div (x - y)$ **22.** $(a^7 + b^7) \div (a + b)$

23. Find k so that when $x^3 - kx^2 + 3x + 7k$ is divided by $x + 2$ the remainder will be 0.

──

5.6 SYNTHETIC DIVISION

To divide a polynomial by a binomial of the type $x - a$, we can stream-line the usual procedure, by a process called *synthetic division*.

Compare the following. In **A** we perform a division. In **B** we do not write the variables.

A.
$$\begin{array}{r} 4x^2 + 5x + 11 \\ x - 2\overline{)4x^3 - 3x^2 + x + 7} \\ 4x^3 - 8x^2 \\ \hline 5x^2 + x \\ 5x^2 - 10x \\ \hline 11x + 7 \\ 11x - 22 \\ \hline 29 \end{array}$$

B.
$$\begin{array}{r} 4 + 5 + 11 \\ 1 - 2\overline{)4 - 3 + 1 + 7} \\ 4 - 8 \\ \hline 5 + 1 \\ 5 - 10 \\ \hline 11 + 7 \\ 11 - 22 \\ \hline 29 \end{array}$$

In **B** there is still some duplication of writing. Also, since we can subtract by adding an inverse, we can use 2 instead of -2 and then add instead of subtracting.

C. *Synthetic Division*

a) $\underline{2}|\ 4 - 3 + 1 + 7$ Write the 2 of $x - 2$ and the
coefficients of the dividend.

$\overline{\qquad\qquad\qquad}$
$\ \ 4$ Bring down the first coefficient.

b) $\underline{2}|\ 4 - 3 + 1 + 7$
$\qquad\qquad 8$ Multiply 4 by 2, to get 8. Add 8 and -3.
$\overline{\ \ 4 \ \ \ 5}$

c) $\underline{2}|\ 4 - 3 + 1 + 7$
$\qquad\quad 8 \ \ \ 10$ Multiply 5 by 2 to get 10. Add 10 and 1.
$\overline{\ \ 4 \ \ \ 5 \ \ \ 11}$

d)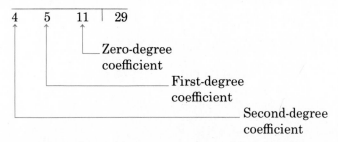

Multiply 11 by 2 to get 22. Add 22 and 7.

The last number, 29, is the remainder. The other numbers are the coefficients of the quotient, with that of the term of highest degree first, as follows.

$$4 \qquad 5 \qquad 11 \quad | \quad 29$$

Zero-degree coefficient

First-degree coefficient

Second-degree coefficient

The quotient is $4x^2 + 5x + 11$. The remainder is 29.

> It is important to remember that for this method to work, the divisor must be of the form $x - a$, that is, a variable minus a constant. The coefficient of the variable must be 1.

Example 1 Use synthetic division $(x^3 + 6x^2 - x - 30) \div (x - 2)$.

$$\underline{2 \,|\, 1 \quad 6 - 1 - 30}$$
$$\quad\quad 2 \quad 16 \quad 30$$
$$\overline{\,1 \quad 8 \quad 15|\quad 0}$$

The quotient is $x^2 + 8x + 15$. The remainder is 0.

When there are missing terms, be sure to write 0's.

Examples Use synthetic division to find the quotient and remainder.

2. $(2x^3 + 7x^2 - 5) \div (x + 3)$

There is no x-term, so we must write a 0 for the coefficient. Note that $x + 3 = x - (-3)$.

$$\underline{-3 \,|\, 2 \quad 7 \quad 0 - 5}$$
$$\quad\quad -6 - 3 \quad 9$$
$$\overline{\,2 \quad 1 - 3|\quad 4}$$

The quotient is $2x^2 + x - 3$. The remainder is 4.

3. $(x^3 + 4x^2 - x - 4) \div (x + 4)$

$$\underline{-4} \, \lfloor \, 1 \quad 4 \, - \, 1 \, - \, 4$$
$$\quad\quad -4 \quad 0 \quad 4$$
$$\overline{1 \quad 0 \, - \, 1 \, \lfloor \quad 0}$$

The quotient is $x^2 - 1$. The remainder is 0.

4. $(x^4 - 1) \div (x - 1)$

$$\underline{1} \, \lfloor \, 1 \quad 0 \quad 0 \quad 0 \, - \, 1$$
$$\quad\quad 1 \quad 1 \quad 1 \quad 1$$
$$\overline{1 \quad 1 \quad 1 \quad 1 \, \lfloor \quad 0}$$

The quotient is $x^3 + x^2 + x + 1$. The remainder is 0.

5. $(8x^5 - 6x^3 + x - 8) \div (x + 2)$

$$\underline{-2} \, \lfloor \, 8 \quad 0 \, - \, 6 \quad 0 \quad 1 \, - \, 8$$
$$\quad\quad\quad -16 \quad 32 \, - \, 52 \quad 104 \, - \, 210$$
$$\overline{8 \, - \, 16 \quad 26 \, - \, 52 \quad 105 \, \lfloor \, - \, 218}$$

The quotient is $8x^4 - 16x^3 + 26x^2 - 52x + 105$. The remainder is -218.

EXERCISE SET 5.6

Use synthetic division to find the quotient and remainder.

1. $(x^3 - 2x^2 + 2x - 5) \div (x - 1)$ **2.** $(x^3 - 2x^2 + 2x - 5) \div (x + 1)$

3. $(a^2 + 11a - 19) \div (a + 4)$ **4.** $(a^2 + 11a - 19) \div (a - 4)$

5. $(x^3 - 7x^2 - 13x + 3) \div (x - 2)$ **6.** $(x^3 - 7x^2 - 13x + 3) \div (x + 2)$

7. $(3x^3 + 7x^2 - 4x + 3) \div (x + 3)$ **8.** $(3x^3 + 7x^2 - 4x + 3) \div (x - 3)$

9. $(y^3 - 3y + 10) \div (y - 2)$ **10.** $(x^3 - 2x^2 + 8) \div (x + 2)$

11. $(3x^4 - 25x^2 - 18) \div (x - 3)$ **12.** $(6y^4 + 15y^3 + 28y + 6) \div (y + 3)$

13. $(x^3 - 27) \div (x - 3)$ **14.** $(y^3 + 27) \div (y + 3)$

15. $(y^4 - 1) \div (y - 1)$ **16.** $(x^5 - 32) \div (x - 2)$

☆
Divide, using synthetic division.

17. ▦ $(3.41x^4 - 24.25x^2 - 13.47) \div (x - 2.41)$

18. ▦ $(5.032x^5 + 11.414x^3 + 217.3) \div (x + 17.07)$

5.7 SOLVING FRACTIONAL EQUATIONS

A fractional equation is an equation that contains one or more fractional expressions. These are fractional equations:

$$\frac{2}{3} + \frac{5}{6} = \frac{1}{x}, \qquad \frac{x-1}{x-5} = \frac{4}{x-5}.$$

> **To solve a fractional equation we multiply on both sides by the LCM of all the denominators. This is called** *clearing of fractions.*

Example 1 Solve: $\dfrac{2}{3} - \dfrac{5}{6} = \dfrac{1}{x}$.

The LCM of all denominators is $6x$, or $2 \cdot 3 \cdot x$. We multiply on both sides of the equation by the LCM.

$$(2 \cdot 3 \cdot x) \cdot \left(\frac{2}{3} - \frac{5}{6}\right) = (2 \cdot 3 \cdot x) \cdot \frac{1}{x} \qquad \text{Multiplying by LCM}$$

$$2 \cdot 3 \cdot x \cdot \frac{2}{3} - 2 \cdot 3 \cdot x \cdot \frac{5}{6} = 2 \cdot 3 \cdot x \cdot \frac{1}{x} \qquad \begin{array}{l}\text{Multiplying to remove}\\ \text{parentheses}\end{array}$$

$$\frac{2 \cdot 3 \cdot x \cdot 2}{3} - \frac{2 \cdot 3 \cdot x \cdot 5}{6} = \frac{2 \cdot 3 \cdot x}{x}$$

$$4x - 5x = 6 \qquad \text{Simplifying}$$

$$-x = 6$$

$$-1 \cdot x = 6$$

$$x = -6$$

> When clearing of fractions, be sure to multiply *every* term in the equation by the LCM.

Check:

$$\frac{2}{3} - \frac{5}{6} = \frac{1}{x}$$

$\dfrac{2}{3} - \dfrac{5}{6}$	$\dfrac{1}{-6}$
$\dfrac{4}{6} - \dfrac{5}{6}$	$-\dfrac{1}{6}$
$-\dfrac{1}{6}$	

Note that when we *clear of fractions* all the denominators disappear. Thus we have an equation without fractional expressions, which we know how to solve.

> *Note:* We have introduced a new use of the LCM here. Before, you used the LCM in adding or subtracting fractional expressions. *Now* we have equations. There are = signs. We clear of fractions by multiplying on both sides of the equation by the LCM. This makes the denominators disappear. *Do not* make the mistake of trying to "clear of fractions" when you do not have an equation!

Example 2 Solve: $\dfrac{x+1}{2} - \dfrac{x-3}{3} = 3$.

The LCM of all the denominators is $2 \cdot 3$, or 6. We multiply on both sides of the equation by the LCM.

$$2 \cdot 3 \left(\frac{x+1}{2} - \frac{x-3}{3} \right) = 2 \cdot 3 \cdot 3 \qquad \text{Multiplying on both sides by the LCM}$$

$$2 \cdot 3 \cdot \frac{x+1}{2} - 2 \cdot 3 \cdot \frac{x-3}{3} = 2 \cdot 3 \cdot 3 \qquad \text{Multiplying to remove parentheses}$$

$$\frac{2 \cdot 3(x+1)}{2} - \frac{2 \cdot 3(x-3)}{3} = 18$$

$$3(x+1) - 2(x-3) = 18$$

$$\left. \begin{array}{r} 3x + 3 - 2x + 6 = 18 \\ x + 9 = 18 \end{array} \right\} \qquad \text{Multiplying and collecting like terms}$$

$$x = 9$$

Check:

$$\frac{x+1}{2} - \frac{x-3}{3} = 3$$

$$\begin{array}{c|c} \dfrac{9+1}{2} - \dfrac{9-3}{3} & 3 \\ 5 - 2 & \\ 3 & \end{array}$$

When we multiply by an expression with a variable, we may not get equivalent equations. Thus we must *always* check possible solutions in the original equation.

Example 3   Solve: $\dfrac{x-1}{x-5} = \dfrac{4}{x-5}$.

The LCM of the denominators is $x - 5$. We multiply by $x - 5$.

$$(x-5) \cdot \frac{x-1}{x-5} = (x-5) \cdot \frac{4}{x-5}$$

$$x - 1 = 4$$

$$x = 5$$

Check:
$$\frac{x-1}{x-5} = \frac{4}{x-5}$$

$$
\begin{array}{c|c}
\dfrac{5-1}{5-5} & \dfrac{4}{5-5} \\[2ex]
\dfrac{4}{0} & \dfrac{4}{0}
\end{array}
$$

5 is not a solution of the original equation because it results in division by 0. In fact, the equation has no solution.

> **When in solving an equation we multiply by an expression containing a variable, we may get an equation having solutions that are not solutions of the original equation. In such a case you *must* check.**

Example 4   Solve: $\dfrac{x^2}{x-2} = \dfrac{4}{x-2}$.

The LCM of the denominators is $x - 2$. We multiply by $x - 2$.

$$(x-2) \cdot \frac{x^2}{x-2} = (x-2) \cdot \frac{4}{x-2}$$

$$x^2 = 4$$

$$x^2 - 4 = 0$$

$$(x+2)(x-2) = 0$$

$$x = -2 \quad \text{or} \quad x = 2 \qquad \text{Using the principle of zero products}$$

Check: For 2: $\dfrac{x^2}{x-2} = \dfrac{4}{x-2}$ For -2: $\dfrac{x^2}{x-2} = \dfrac{4}{x-2}$

$$\dfrac{\dfrac{2^2}{2-2}}{} \;\bigg|\; \dfrac{4}{2-2} \qquad\qquad \dfrac{(-2)^2}{-2-2} \;\bigg|\; \dfrac{4}{-2-2}$$

$$\dfrac{4}{-4} \;\bigg|\; \dfrac{4}{-4}$$

The number -2 is a solution, but 2 is not (it results in division by 0).

Example 5   Solve: $x + \dfrac{6}{x} = 5.$

The LCM of the denominators is x. We multiply on both sides by x.

$$x\left(x + \dfrac{6}{x}\right) = 5 \cdot x \qquad \text{Multiplying on both sides by } x$$

$$x^2 + x \cdot \dfrac{6}{x} = 5x$$

$$x^2 + 6 = 5x \qquad \text{Simplifying}$$
$$x^2 - 5x + 6 = 0 \qquad \text{Getting 0 on one side}$$
$$(x - 3)(x - 2) = 0 \qquad \text{Factoring}$$
$$x = 3 \;\text{ or }\; x = 2 \qquad \text{Using the principle of zero products}$$

Check: For 3: $x + \dfrac{6}{x} = 5$ For 2: $x + \dfrac{6}{x} = 5$

$$3 + \dfrac{6}{3} \;\bigg|\; 5 \qquad\qquad 2 + \dfrac{6}{2} \;\bigg|\; 5$$

$$3 + 2 \qquad\qquad\quad 2 + 3$$

$$5 \qquad\qquad\qquad\quad 5$$

The solutions are 2 and 3.

Example 6   Solve: $\dfrac{2}{x-1} = \dfrac{3}{x+1}.$

The LCM of the denominators is $(x - 1)(x + 1)$.

$$(x-1)(x+1) \cdot \dfrac{2}{(x-1)} = (x-1)(x+1) \cdot \dfrac{3}{x+1} \qquad \text{Multiplying}$$

$$2(x + 1) = 3(x - 1) \qquad \text{Simplifying}$$
$$2x + 2 = 3x - 3$$
$$5 = x$$

Check:

$$\frac{2}{x-1} = \frac{3}{x+1}$$

$\dfrac{2}{5-1}$	$\dfrac{3}{5+1}$
$\dfrac{2}{4}$	$\dfrac{3}{6}$
$\dfrac{1}{2}$	$\dfrac{1}{2}$

The number 5 is the solution.

Example 7 Solve: $\dfrac{2}{x+5} + \dfrac{1}{x-5} = \dfrac{16}{x^2-25}.$

The LCM is $(x+5)(x-5)$. We multiply by $(x+5)(x-5)$.

$$(x+5)(x-5) \cdot \left[\frac{2}{x+5} + \frac{1}{x-5} \right] = (x+5)(x-5) \cdot \frac{16}{x^2-25}$$

$$(x+5)(x-5) \cdot \frac{2}{x+5} + (x+5)(x-5) \cdot \frac{1}{x-5}$$

$$= (x+5)(x-5) \cdot \frac{16}{x^2-25}$$

$$2(x-5) + (x+5) = 16$$
$$2x - 10 + x + 5 = 16$$
$$3x - 5 = 16$$
$$3x = 21$$
$$x = 7$$

Check:

$$\frac{2}{x+5} + \frac{1}{x-5} = \frac{16}{x^2-25}$$

$\dfrac{2}{7+5} + \dfrac{1}{7-5}$	$\dfrac{16}{7^2-25}$
$\dfrac{2}{12} + \dfrac{1}{2}$	$\dfrac{16}{49-25}$
$\dfrac{8}{12}$	$\dfrac{16}{24}$
$\dfrac{2}{3}$	$\dfrac{2}{3}$

The solution is 7.

EXERCISE SET 5.7

Solve.

1. $\dfrac{2}{5} + \dfrac{7}{8} = \dfrac{y}{20}$

2. $\dfrac{4}{5} + \dfrac{1}{3} = \dfrac{t}{9}$

3. $\dfrac{1}{3} - \dfrac{5}{6} = \dfrac{1}{x}$

4. $\dfrac{5}{8} - \dfrac{2}{5} = \dfrac{1}{y}$

5. $\dfrac{x}{3} - \dfrac{x}{4} = 12$

6. $\dfrac{y}{5} - \dfrac{y}{3} = 15$

7. $y + \dfrac{5}{y} = -6$

8. $x + \dfrac{4}{x} = -5$

9. $\dfrac{4}{z} + \dfrac{2}{z} = 3$

10. $\dfrac{4}{3y} - \dfrac{3}{y} = \dfrac{10}{3}$

11. $\dfrac{x-3}{x+2} = \dfrac{1}{5}$

12. $\dfrac{y-5}{y+1} = \dfrac{3}{5}$

13. $\dfrac{3}{y+1} = \dfrac{2}{y-3}$

14. $\dfrac{4}{x-1} = \dfrac{3}{x+2}$

15. $\dfrac{y-1}{y-3} = \dfrac{2}{y-3}$

16. $\dfrac{x-2}{x-4} = \dfrac{2}{x-4}$

17. $\dfrac{x+1}{x} = \dfrac{3}{2}$

18. $\dfrac{y+2}{y} = \dfrac{5}{3}$

19. $\dfrac{2}{x} - \dfrac{3}{x} + \dfrac{4}{x} = 5$

20. $\dfrac{4}{y} - \dfrac{6}{y} + \dfrac{8}{y} = 8$

21. $\dfrac{1}{2} - \dfrac{4}{9x} = \dfrac{4}{9} - \dfrac{1}{6x}$

22. $-\dfrac{1}{3} - \dfrac{5}{4y} = \dfrac{3}{4} - \dfrac{1}{6y}$

23. $\dfrac{60}{x} - \dfrac{60}{x-5} = \dfrac{2}{x}$

24. $\dfrac{50}{y} - \dfrac{50}{y-2} = \dfrac{4}{y}$

25. $\dfrac{7}{5x-2} = \dfrac{5}{4x}$

26. $\dfrac{5}{y+4} = \dfrac{3}{y-2}$

27. $\dfrac{x}{x-2} + \dfrac{x}{x^2-4} = \dfrac{x+3}{x+2}$

28. $\dfrac{3}{y-2} + \dfrac{2y}{4-y^2} = \dfrac{5}{y+2}$

29. $\dfrac{a}{2a-6} - \dfrac{3}{a^2-6a+9} = \dfrac{a-2}{3a-9}$

30. $\dfrac{2}{x+4} + \dfrac{2x-1}{x^2+2x-8} = \dfrac{1}{x-2}$

31. $\dfrac{2x+3}{x-1} = \dfrac{10}{x^2-1} + \dfrac{2x-3}{x+1}$

32. $\dfrac{y}{y+1} + \dfrac{3y+5}{y^2+4y+3} = \dfrac{2}{y+3}$

☆ ───
Solve.

33. $\left(\dfrac{1}{1+x} + \dfrac{x}{1-x} \right) \div \left(\dfrac{x}{1+x} - \dfrac{1}{1-x} \right) = -1$

34. $\dfrac{x+3}{x+2} - \dfrac{x+4}{x+3} = \dfrac{x+5}{x+4} - \dfrac{x+6}{x+5}$

5.8 SOLVING PROBLEMS

WORK PROBLEMS

Suppose a machine can do a certain job in 5 hours. Then in 1 hour it can do $\frac{1}{5}$ of the job. In 3 hours it can do $\frac{3}{5}$ of the job, and so on. This reasoning gives us a principle for solving certain problems.

> **If a job can be done in t hours (or days), then $1/t$ of it can be done in 1 hour (or day).**

> Don't forget the principles that hold in general for solving problems. You should review them. Go to Section 2.3, p. 53.

Example 1 Lon Moore can mow a lawn in 4 hours. Penny Push can mow the same lawn in 5 hours. How long would it take both of them, working together, to mow the lawn?

> Use some common sense here. Don't just start manipulating numbers. The answer is going to be less than 4 hours because Lon alone can do the job in 4 hours.

Lon can mow the lawn in 4 hours, so he can mow $\frac{1}{4}$ of it in 1 hour. Penny can mow the lawn in 5 hours, so she can mow $\frac{1}{5}$ of it in 1 hour. Thus they can mow $\frac{1}{4} + \frac{1}{5}$ of it in 1 hour working together.

Let t represent the time it takes them, working together. Then they mow $1/t$ of it in 1 hour. We now have an equation:

$$\frac{1}{4} + \frac{1}{5} = \frac{1}{t}. \qquad \text{This is the translation to mathematical language.}$$

We solve the equation:

$$20t \left(\frac{1}{4} + \frac{1}{5} \right) = 20t \cdot \frac{1}{t}$$

$$\frac{20t}{4} + \frac{20t}{5} = \frac{20t}{t}$$

Multiplying on both sides by the LCM of denominators and clearing of fractions

$$5t + 4t = 20$$

$$9t = 20$$

$$t = \frac{20}{9}, \quad \text{or} \quad 2\frac{2}{9}.$$

Now, of course, we check in the original problem to see if $\frac{20}{9}$ is the right answer.

Lon does $\frac{1}{4}$ of the lawn in 1 hour. In $\frac{20}{9}$ hr he does $\frac{1}{4} \cdot \frac{20}{9}$, or $\frac{5}{9}$ of it. Penny does $\frac{1}{5}$ of the job in 1 hour. In $\frac{20}{9}$ hr she does $\frac{1}{5} \cdot \frac{20}{9}$, or $\frac{4}{9}$ of it. Altogether they do $\frac{5}{9} + \frac{4}{9}$ of the job, or all of it. The number checks, so the answer is that it takes them $\frac{20}{9}$ hr to do the job working together.

Example 2   At a factory, smokestack A pollutes the air twice as fast as smokestack B. When the stacks operate together they yield a certain amount of pollution in 15 hours. Find the time it would take each to yield that same amount of pollution operating alone.

Let x represent the amount of time it takes A to yield the pollution. Then $2x$ is the time it takes B to yield the same amount of pollution. Then

$$\frac{1}{x}$$ is the fraction of the pollution by A in 1 hour,

and

$$\frac{1}{2x}$$ is the fraction of the pollution by B in 1 hour.

Common sense will help with this translation, too. The slower smokestack will take *longer* to generate pollution than the faster stack. Since B is slower, the time required is longer. That way we get $2x$ for the time required for B.

Together, the stacks yield $1/x + 1/2x$ of the pollution in 1 hour. They also yield $\frac{1}{15}$ of it in one hour. We now have an equation:

$$\frac{1}{x} + \frac{1}{2x} = \frac{1}{15}. \qquad \text{This is the translation.}$$

We solve the equation:

$$30x\left(\frac{1}{x} + \frac{1}{2x}\right) = 30x \cdot \frac{1}{15} \qquad \begin{array}{l}\text{Multiplying on both sides by the}\\ \text{LCM of denominators and clearing}\\ \text{of fractions}\end{array}$$

$$\frac{30x}{x} + \frac{30x}{2x} = \frac{30x}{15}$$

$$30 + 15 = 2x$$

$$45 = 2x$$

$$x = \frac{45}{2}, \quad \text{or} \quad 22\frac{1}{2}.$$

Thus if the numbers check, the answer will be A takes x, or $\frac{45}{2}$, hours and B takes $2x$, or 45, hours. Part of the check is obvious already. B will take twice as long, so A pollutes twice as fast.

Now if A does $1/x$ of the total pollution in 1 hour, or $1/\frac{45}{2}$, which is $\frac{2}{45}$ of the total pollution, then in $\frac{45}{2}$ hours it will do $\frac{2}{45} \cdot \frac{45}{2}$, or all of it. This checks.

If B does $1/2x$, or $\frac{1}{45}$, of the total pollution in 1 hour, it will do $\frac{1}{45} \cdot 45$ of it in 45 hours. This is the entire amount. This checks.

MOTION PROBLEMS

Problems dealing with speed, distance, and time are called *motion* problems. To translate them we use the definition of speed.

$$\textbf{Speed} = \frac{\textbf{Distance}}{\textbf{Time}}$$

$$r = \frac{d}{t}, \quad \text{or} \quad d = rt.$$

From the equation $r = d/t$ we can easily get $d = rt$ or $t = d/r$.

Example 3 A train leaves Sioux City traveling east at 30 mph. Two hours later another train leaves Sioux City in the same direction on a parallel track at 45 mph. How far from Sioux City will the faster train catch the slower one?

We first make a drawing.

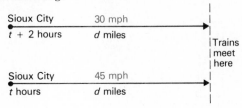

From the drawing we see that the distances are the same. Let's call the distance d. We don't know the times. Let t represent the time for the faster train. Then the time for the slower one will be $t + 2$. We can organize the information in a table.

	Distance	Speed	Time
Slow train	d	30	$t + 2$
Fast train	d	45	t

Using $d = rt$ in each row of the table, we get an equation. Thus we get a system of two equations:

$$d = 30(t + 2) \quad \text{and} \quad d = 45t. \qquad \text{This is the translation.}$$

We solve:

$$30(t + 2) = 45t \qquad \text{Using substitution}$$
$$30t + 60 = 45t$$
$$60 = 15t$$
$$t = 4.$$

Thus the time for the faster train should be 4 hours and the time for the slower train 6 hours. At 45 mph the faster train would go $45 \cdot 4$, or 180 miles in 4 hours. At 30 mph the slower train would go $30 \cdot 6$, or 180 miles in 6 hours. We have a check, and the answer is 180 miles.

Example 4 An airplane flies 1062 miles with the wind. In the same amount of time it can fly 738 miles against the wind. The speed of the plane in still air is 200 mph. Find the speed of the wind.

We first make a drawing. We let r represent the speed of the wind, and organize the facts in a chart.

$$\overset{\text{1062 mi}}{\underset{200\,+\,r}{\longrightarrow}}\qquad \text{t hours}$$
(The wind increases the speed over the ground.)

$$\overset{\text{738 mi}}{\underset{200\,-\,r}{\longleftarrow}}\qquad \text{t hours}$$
(The wind decreases the speed over the ground.)

	Distance	Speed	Time
With wind	1062	$200 + r$	t
Against wind	738	$200 - r$	t

Using $t = d/r$ with each row of the table, we get a system of equations:

$$t = \frac{1062}{200 + r} \quad \text{and} \quad t = \frac{738}{200 - r}. \qquad \text{This is the translation.}$$

We solve:

$$\frac{1062}{200 + r} = \frac{738}{200 - r} \qquad \text{Using substitution}$$

$$(200 + r)(200 - r)\left(\frac{1062}{200 + r}\right) = (200 + r)(200 - r)\left(\frac{738}{200 - r}\right)$$

Multiplying by LCM to clear of fractions

$$(200 - r)1062 = (200 + r)738$$
$$212{,}400 - 1062r = 147{,}600 + 738r$$
$$64{,}800 = 1800r$$
$$r = 36.$$

Thus, if it checks, the speed of the wind is 36 mph. With the wind, the speed of the plane would be 236 mph. Dividing the distance, 1062 miles, by the speed, 236 mph, we get 4.5 hours for the time. Against the wind, the speed of the plane would be 164 mph. Dividing the distance, 738 miles, by the speed, 164 mph, we get 4.5 hours. The answer checks, so the speed of the wind is 36 mph.

EXERCISE SET 5.8

Solve.

1. Sam Strong, an experienced shipping clerk, can fill a certain order in 5 hours. Willy Weak, a new clerk, needs 9 hours to do the same job. Working together, how long would it take them to fill the order?

2. Paul Putty can paint a room in 4 hours. Sally Spackle can paint the same room in 3 hours. Working together, how long would it take them to paint the room?

3. A swimming pool can be filled in 12 hours if water enters through a pipe alone, or in 30 hours if water enters through a hose alone. If water is entering through both the pipe and the hose, how long will it take to fill the pool?

4. A tank can be filled in 18 hours by pipe A alone and in 22 hours by pipe B alone. How long would it take to fill the tank if both pipes were working?

5. Bill Dozer can clear a lot in 5.5 hours. His partner can do the same job in 7.5 hours. How long would it take them to clear the lot working together?

6. One of Ty Psetter's printing presses can print an order of booklets in 4.5 hours. Another press can do the same job in 5.5 hours. How long would it take if both presses are used?

Hint: You may find that multiplying on both sides of your equation by $\frac{1}{10}$ will clear of decimals.

7. A can paint the neighbor's house 4 times as fast as B. The year they worked together it took them 8 days. How long would it take each to paint the house alone?

8. A can deliver papers 3 times as fast as B. If they work together, it takes them 1 hour. How long would it take each to deliver the papers alone?

9. The speed of a stream is 3 mph. A boat travels 4 mi upstream in the same time it takes to travel 10 mi downstream. What is the speed of the boat in still water?

10. The speed of a stream is 4 mph. A boat travels 6 mi upstream in the same time it takes to travel 12 mi downstream. What is the speed of the boat in still water?

11. A train leaves a station and travels north at 75 km/h. Two hours later a second train leaves on a parallel track traveling north at 125 km/h. How far from the station will the second train overtake the first train?

12. A private airplane leaves an airport and flies due east at 180 km/h. Two hours later a jet leaves the same airport and flies due east at 900 km/h. How far from the airport will the jet overtake the private plane?

13. The speed of train A is 12 mph slower than the speed of train B. Train A travels 230 mi in the same time it takes train B to travel 290 mi. Find the speed of each train.

14. The speed of a passenger train is 14 mph faster than the speed of a freight train. The passenger train travels 400 mi in the same time it takes the freight train to travel 330 mi. Find the speed of each train.

15. Suzie Skiff has a boat that can move at a speed of 15 km/h in still water. She rides 140 km downstream in a river in the same time it takes to ride 35 km upstream. What is the speed of the river?

16. A paddleboat can move at a speed of 2 km/h in still water. The boat is paddled 4 km downstream in a river in the same time it takes to go 1 km upstream. What is the speed of the river?

☆

17. At what time after 4:00 will the minute hand and the hour hand of a clock first be in the same position?

18. At what time after 10:30 will the hands of a clock first be perpendicular?

19. A boat travels 96 km downstream in 4 hr. It travels 28 km upstream in 7 hr. Find the speed of the boat and the speed of the stream.

20. An airplane carries enough fuel for 6 hours of flight time, and its speed in still air is 240 mph. It leaves an airport against a wind of 40 mph and returns to the same airport with a wind of 40 mph. How far can it fly under those conditions without refueling?

21. A motor boat travels 3 times as fast as the current. A trip up the river and back takes 10 hours, and the total distance of the trip is 100 km. Find the speed of the current.

22. An employee drives to work at 50 mph and arrives one minute late. The employee drives to work at 60 mph and arrives 5 minutes early. How far does the employee live from work?

Average speed is defined as *total distance divided by total time.*

23. A driver went 200 km. For the first half of the trip the driver traveled at a speed of 40 km/h. For the second half of the trip the driver traveled at a speed of 60 km/h. What was the average speed for the entire trip? (It is *not* 50 km/h.)

24. For half of a trip a driver travels at 40 mph. What speed would the driver have to travel for the last half of the trip so that the average speed for the entire trip would be 45 mph?

25. Three trucks A, B, and C, working together, can move a load of sand in t hours. When working alone, it takes A, 1 extra hour to move the sand; B, 6 extra hours; and C, t extra hours. Find t.

5.9 FORMULAS

As we mentioned earlier, solving formulas is important in business, technology, science, and engineering.

Example 1   The formula $A = 9R/I$ tells how to calculate a pitcher's earned run average (ERA). In this formula, A is the ERA, R is the number of earned runs, and I is the number of innings pitched. Solve this formula for I.

We multiply by the LCM, which is I.

$$I \cdot A = I \cdot \frac{9R}{I} \qquad I \cdot A = 9R$$

Next we multiply by $\frac{1}{A}$.

$$I \cdot A \cdot \frac{1}{A} = 9R \cdot \frac{1}{A} \qquad I = \frac{9R}{A}$$

> In solving formulas we proceed as if we were solving equations with numbers. In this case we have a fractional equation, so the first thing we do is clear of fractions.

From the formula solved for I, you can calculate the number of innings a pitcher pitched if you know the earned run average and the number of earned runs given up.

Example 2   The formula

$$\frac{1}{R} = \frac{1}{r_1} + \frac{1}{r_2}$$

gives the resistance R of two resistors r_1 and r_2 connected in parallel.* Solve it for r_1.

We multiply by the LCM, which is Rr_1r_2:

$$Rr_1r_2 \cdot \frac{1}{R} = Rr_1r_2 \cdot \left[\frac{1}{r_1} + \frac{1}{r_2}\right] \qquad \text{Multiplying by LCM}$$

$$Rr_1r_2 \cdot \frac{1}{R} = Rr_1r_2 \cdot \frac{1}{r_1} + Rr_1r_2 \cdot \frac{1}{r_2} \qquad \text{Multiplying to remove parentheses}$$

$$\left.\begin{array}{l} \dfrac{R}{R} \cdot r_1r_2 = \dfrac{r_1}{r_1} \cdot Rr_2 + \dfrac{r_2}{r_2} \cdot Rr_1 \\[2mm] r_1r_2 = Rr_2 + Rr_1. \end{array}\right\} \quad \begin{array}{l}\text{Simplifying by}\\ \text{removing factors}\\ \text{of 1}\end{array}$$

---

* Note that R, r_1, and r_2 are all different variables. It is common to use subscripts, as in r_1 and r_2, to distinguish variables. Often this helps to keep ideas straight.

One might be tempted at this point to multiply by $1/r_2$ to get r_1 alone on the left, BUT note that there is an r_1 on the right. We must get all the terms involving r_1 on the *same side* of the equation.

$$r_1 r_2 - R r_1 = R r_2 \qquad \text{Adding} - R r_1$$
$$r_1(r_2 - R) = R r_2 \qquad \text{Factoring out } r_1$$
$$r_1 = \frac{R r_2}{r_2 - R} \qquad \text{Multiplying by } \frac{1}{r_2 - R} \text{ to get } r_1 \text{ alone}$$

To solve a formula we do the same things we would do to solve an equation with numbers. For fractional formulas the procedure is as follows.

To solve a fractional formula:

1. **Clear of fractions.**
2. **Multiply if necessary to remove parentheses.**
3. **Get all terms with the unknown alone on one side of the equation.**
4. **Factor out the unknown.**
5. **Use the multiplication principle to get the unknown alone on one side.**

Example 3 Solve the formula $I = \dfrac{pT}{M + pn}$ for p.

We first clear of fractions by multiplying by the LCM, $M + pn$.

$$I(M + pn) = (M + pn)\frac{pT}{M + pn}$$

$$IM + Ipn = pT$$

We must now get all terms containing p alone on one side.

$$IM = pT - Ipn \qquad \text{Adding} - Ipn$$
$$IM = p(T - In) \qquad \text{Factoring out the unknown } p$$

$$\frac{IM}{T - In} = p \qquad \text{Multiplying by } \frac{1}{T - In}$$

We now have p alone on one side, and p does *not* appear on the other side, so we have solved the formula for p.

EXERCISE SET 5.9

1. Solve $\dfrac{W_1}{W_2} = \dfrac{d_1}{d_2}$ for d_1.

2. Solve $\dfrac{W_1}{W_2} = \dfrac{d_1}{d_2}$ for W_2.

3. Solve $s = \dfrac{(v_1 + v_2)t}{2}$ for t.

4. Solve $s = \dfrac{(v_1 + v_2)t}{2}$ for v_1.

5. Solve $\dfrac{1}{R} = \dfrac{1}{r_1} + \dfrac{1}{r_2}$ for r_2.

6. Solve $\dfrac{1}{R} = \dfrac{1}{r_1} + \dfrac{1}{r_2}$ for R.

7. Solve $R = \dfrac{gs}{g + s}$ for s.

8. Solve $I = \dfrac{2V}{R + 2r}$ for r.

9. Solve $\dfrac{1}{p} + \dfrac{1}{q} = \dfrac{1}{f}$ for p.

10. Solve $\dfrac{1}{p} + \dfrac{1}{q} = \dfrac{1}{f}$ for f.

11. Solve $I = \dfrac{nE}{R + nr}$ for r.

12. Solve $I = \dfrac{nE}{R + nr}$ for n.

13. Solve $S = \dfrac{H}{m(t_1 - t_2)}$ for H.

14. Solve $S = \dfrac{H}{m(t_1 - t_2)}$ for t_1.

15. Solve $\dfrac{E}{e} = \dfrac{R + r}{r}$ for e.

16. Solve $\dfrac{E}{e} = \dfrac{R + r}{r}$ for r.

5.10 VARIATION

DIRECT VARIATION

A plumber earns \$9 per hour. In 1 hour, \$9 is earned. In 2 hours, \$18 is earned. In 3 hours, \$27 is earned, and so on. This gives rise to a set of ordered pairs of numbers, all having the same ratio:

$(1, 9), \quad (2, 18), \quad (3, 27), \quad (4, 36), \quad$ and so on.

The ratio of earnings to time is $\frac{9}{1}$ in every case.

Whenever a situation gives rise to pairs of numbers in which the ratio is constant, we say that there is *direct variation*. Here the earnings *vary directly* as the time:

$$\frac{E}{t} = 9 \text{ (a constant)}, \quad \text{or} \quad E = 9t.$$

Note that the constant, 9, is positive. Note also that when one variable increases, so does the other. When one decreases, so does the other.

> **Whenever a situation gives rise to a relation among variables**
> $y = kx$, **where** k **is a positive constant, we say that there is**
> *direct variation*, **or that** y *varies directly* **as** x. **The number** k **is**
> **called the** *variation constant*.

Example 1   Find the variation constant and an equation of variation,
where y varies directly as x, and where $y = 32$ when $x = 2$.

We know that $(2, 32)$ is a solution of $y = kx$. Thus,

   $32 = k \cdot 2$ Substituting

   $\dfrac{32}{2} = k,$ or $k = 16.$ Solving for k

The variation constant is 16.

The equation of variation is $y = 16x$.

PROPORTIONS

Suppose that y varies directly as x. Then we have $y = kx$. Suppose
also that (x_1, y_1) and (x_2, y_2) are solutions of the equation. Then,
substituting, we have

   $y_1 = kx_1$ and $y_2 = kx_2.$

Dividing, we get

   $$\frac{y_1}{y_2} = \frac{kx_1}{kx_2} = \frac{k}{k} \cdot \frac{x_1}{x_2} = \frac{x_1}{x_2}.$$

The equation

   $$\frac{y_1}{y_2} = \frac{x_1}{x_2}$$

is called a *proportion*. Such equations are helpful in solving problems.

DIRECT-VARIATION PROBLEMS

Example 2   The number of centimeters W of water, produced from
melting snow, varies directly as S, the number of centimeters of snow.
Meteorologists have found that 150 cm of snow will melt to 16.8 cm
of water. To how many cm of water will 200 cm of snow melt?

Method 1.   First find the variation constant using the data and then find an equation of variation.

$$W = kS$$
$$16.8 = k \cdot 150 \qquad \text{Substituting}$$
$$\frac{16.8}{150} = k \qquad \text{Solving for } k$$
$$0.112 = k \qquad \text{This is the variation constant.}$$

The equation of variation is $W = 0.112S$.

Next, use the equation to find how many cm of water will result from melting 200 cm of snow.

$$W = 0.112S$$
$$W = 0.112(200) \qquad \text{Substituting}$$
$$W = 22.4$$

Thus 200 cm of snow will melt to 22.4 cm of water.

Method 2.   Use a proportion and solve for W without finding the variation constant.

$$\frac{W_1}{W_2} = \frac{S_1}{S_2} \qquad \text{Writing the proportion}$$
$$\frac{16.8}{W_2} = \frac{150}{200} \qquad \text{Substituting} \leftarrow \boxed{\text{Think of (16.8, 150) as } (W_1, S_1). \text{ For a second solution we have } (W_2, 200).}$$
$$\frac{200}{150} \cdot 16.8 = W_2 \qquad \text{Solving}$$
$$22.4 = W_2$$

INVERSE VARIATION

A bus is traveling a distance of 20 mi. At a speed of 20 mph, it will take 1 hour. At 40 mph it will take $\frac{1}{2}$ hour. At 60 mph it will take $\frac{1}{3}$ hour, and so on. This gives rise to a set of pairs of numbers, all having the same product:

$$(20, 1), \quad \left(40, \frac{1}{2}\right), \quad \left(60, \frac{1}{3}\right), \quad \left(80, \frac{1}{4}\right), \quad \text{and so on.}$$

Whenever a situation gives rise to pairs of numbers whose product is constant, we say that there is *inverse variation*. Here the time *varies inversely* as the speed:

$$rt = 20 \text{ (a constant)} \quad \text{or} \quad t = \frac{20}{r}.$$

Note that the constant, 20, is positive. Note also that when one variable increases, the other decreases.

> **Whenever a situation gives rise to a relation among variables $y = k/x$, where k is a positive constant, we say that there is** *inverse variation,* **or that** *y varies inversely as x.* **The number k is called the** *variation constant.*

Example 3 Find the variation constant and then an equation of variation where y varies inversely as x, and $y = 32$ when $x = 0.2$.

We know that $(0.2, 32)$ is a solution of $y = k/x$. We substitute

$$y = \frac{k}{x}$$

$$32 = \frac{k}{0.2} \qquad \text{Substituting}$$

$$(0.2)32 = k$$

$$6.4 = k. \qquad \text{Solving}$$

The variation constant is 6.4.

The equation of variation is $y = \dfrac{6.4}{x}$.

PROPORTIONS

Suppose that y varies inversely as x. Then we have $y = k/x$. Suppose also that (x_1, y_1) and (x_2, y_2) are solutions of the equation.

We substitute into $y = k/x$. Then we have

$$y_1 = \frac{k}{x_1} \quad \text{and} \quad y_2 = \frac{k}{x_2}.$$

Dividing, we get

$$\frac{y_1}{y_2} = \frac{\dfrac{k}{x_1}}{\dfrac{k}{x_2}} = \frac{k}{x_1} \cdot \frac{x_2}{k} = \frac{k}{k} \cdot \frac{x_2}{x_1} = \frac{x_2}{x_1}.$$

The equation

$$\frac{y_1}{y_2} = \frac{x_2}{x_1}$$

is called a *proportion*. Compare this with proportions obtained in direct variation. Note how the x-values are interchanged.

INVERSE-VARIATION PROBLEMS

Example 4   The time t required to do a certain job varies inversely as the number of people P who work on the job (assuming that all do the same amount of work). It takes 4 hours for 12 people to erect some football bleachers. How long would it take 3 people to do the same job?

Method 1.   First find the variation constant using the data and then find an equation of variation:

$$t = \frac{k}{P}$$

$$4 = \frac{k}{12} \qquad \text{Substituting}$$

$$48 = k. \qquad \text{Solving for } k, \text{ the variation constant}$$

The equation of variation is $t = \dfrac{48}{P}$.

Next, use the equation to find the time it would take 3 people to do the job:

$$t = \frac{48}{P}$$

$$t = \frac{48}{3} \qquad \text{Substituting}$$

$$t = 16.$$

It would take 16 hours.

Method 2. Use a proportion and solve for t without finding the variation constant.

> We think of $(4, 12)$ as (t_1, P_1). For a second solution we have $(t_2, 3)$.

$$\frac{t_1}{t_2} = \frac{P_2}{P_1} \qquad \text{Writing the proportion}$$

$$\frac{4}{t_2} = \frac{3}{12} \qquad \text{Substituting}$$

$$4 \cdot 12 = 3t_2 \qquad \text{Solving}$$

$$16 = t_2$$

EXERCISE SET 5.10

Find the variation constant and an equation of variation where y varies directly as x and the following are true.

1. $y = 24$ when $x = 3$

2. $y = 5$ when $x = 12$

3. $y = 3.6$ when $x = 1$

4. $y = 2$ when $x = 5$

5. $y = 0.8$ when $x = 0.5$
(*Hint:* After you have substituted, clear of decimals.)

6. $y = 0.6$ when $x = 0.4$

Solve.

7. The electric current I, in amperes, in a circuit varies directly as the voltage V. When 12 volts are applied, the current is 4 amperes. What is the current when 18 volts are applied?

8. Hooke's law states that the distance d that a spring is stretched by a hanging object varies directly as the weight w of the object. If the distance is 40 cm when the weight is 3 kg, what is the distance when the weight is 5 kg?

9. The number N of plastic straws produced by a machine varies directly as the amount of time t the machine is operating. If the machine produces 20,000 straws in 8 hours, how many straws can it produce in 50 hours?

10. The number N of aluminum cans used each year varies directly as the number of people using the cans. If 250 people use 60,000 cans in one year, how many cans are used each year in Dallas, which has a population of 850,000?

Find the variation constant and an equation of variation where y varies inversely as x and the following are true.

11. $y = 6$ when $x = 10$

12. $y = 16$ when $x = 4$

13. $y = 12$ when $x = 3$

14. $y = 9$ when $x = 5$

15. $y = 0.4$ when $x = 0.8$
(*Hint:* After you have substituted, clear of decimals.)

16. $y = 1.5$ when $x = 0.3$

Solve.

17. The current I in an electrical conductor varies inversely as the resistance R of the conductor. If the current is $\frac{1}{2}$ ampere when the resistance is 240 ohms, what is the current when the resistance is 540 ohms?

18. The time t required to empty a tank varies inversely as the rate r of pumping. If a pump can empty a tank in 45 minutes at the rate of 600 kL per minute, how long will it take the pump to empty the same tank at the rate of 1000 kL per minute?

19. The weight W that a horizontal beam can support varies inversely as the length L of the beam. Suppose an 8-meter beam can support 1200 kilograms. How many kilograms can a 14-meter beam support?

20. The wavelength W of a radio wave varies inversely as its frequency F. A wave with a frequency of 1200 kilohertz has a length of 300 meters. What is the length of a wave with a frequency of 800 kilohertz?

☆

21. A loan of $9000 earns simple interest of $1665. Find an equation of variation that describes the simple interest I in terms of the loan principal P.

22. A 96-fluid-ounce container of fabric softener costs $2.99. A 128-fluid-ounce container of the same softener costs $3.99. Which is the better value? Find a way to use proportions to solve this problem.

23. During the summer Steve sold 4 sweepers for every 3 sweepers sold by his sister, Stephanie. Together they sold 98 sweepers. How many did each sell?

24. In each of the following equations state whether y varies directly as x, inversely as x, or neither directly nor inversely as x.

a) $7xy = 14$ b) $x - 2y = 12$

c) $-2x + 3y = 0$ d) $x = \frac{3}{4}y$

e) $\frac{x}{y} = 2$

TEST OR REVIEW—CHAPTER 5

1. Multiply and simplify: $\dfrac{2x^2 + 20x + 50}{x^2 - 4} \cdot \dfrac{x + 2}{x + 5}$.

2. Divide and simplify: $\dfrac{y^2 - 16}{2y + 6} \div \dfrac{y - 4}{y + 3}$.

3. Find the LCM: $x^2 + x - 6,\ x^2 + 8x + 15$.

4. Add and simplify: $\dfrac{x^2}{x - y} + \dfrac{y^2}{y - x}$.

5. Calculate and simplify: $\dfrac{1}{x + 1} - \dfrac{x + 2}{x^2 - 1} + \dfrac{3}{x - 1}$.

6. Simplify: $\dfrac{1 - \dfrac{1}{x^2}}{1 - \dfrac{1}{x}}$.

7. Divide: $(x^2 - 10x - 25) \div (x - 3)$.

8. Divide. Use synthetic division.

$(6y^4 + 15y^3 - 28y + 6) \div (y + 2)$

9. Solve: $\dfrac{2}{x - 1} = \dfrac{3}{x + 3}$.

10. Two pipes are filling a tank. One of them can fill the tank alone in 6 hours. The other can fill the tank alone in 4 hours. How long will it take them together?

11. A boat can go upstream a certain distance in 6 hours. It can go that same distance downstream in 2 hours and 15 minutes. The speed of the stream is 5 km/h. What is the speed of the boat in still water?

12. Solve $m = \dfrac{2t}{r + t}$ for t.

13. Income varies directly as the time worked. If a job pays $275 for 40 hours, what does it pay for 72 hours?

14. The time required to drive a certain distance varies inversely as the speed. If you can make a trip in $3\frac{1}{2}$ hours at 45 mph, how long will it take you at 60 mph?

6

EXPONENTS, POWERS, AND ROOTS

6.1 SCIENTIFIC NOTATION

INTEGER EXPONENTS

Let us recall from Chapter 1 some of the properties of integer exponents.

$a^m a^n = a^{m+n}$ Add exponents when multiplying.

$\dfrac{a^m}{a^n} = a^{m-n}$ Subtract exponents when dividing.

$(a^m)^n = a^{mn}$ Multiply exponents when raising a power to a power.

$a^{-n} = \dfrac{1}{a^n}$, hence $\dfrac{1}{a^{-n}} = a^n$. a^n and a^{-n} are reciprocals.

$a^1 = a$

$a^0 = 1$ if $a \neq 0$

Examples   Simplify.

1. $(4xy)^1 = 4xy$

2. $(-8x^3y^7)^0 = 1$, assuming neither x nor y is 0

3. $4x^2x^{-5} = 4x^{2-5} = 4x^{-3}$, or $\dfrac{4}{x^3}$

4. $(-8x^5y^7)(4x^3y^{-2}) = -8 \cdot 4x^{5+3}y^{7-2} = -32x^8y^5$

5. $\dfrac{-10x^7}{2x^6} = \dfrac{-10}{2}x^{7-6} = -5x$

6. $\dfrac{42x^8y^3}{21x^4y^8} = \dfrac{42}{21}x^{8-4}y^{3-8} = 2x^4y^{-5}$, or $\dfrac{2x^4}{y^5}$

7. $(x^{-3})^5 = x^{-15}$, or $\dfrac{1}{x^{15}}$

8. $\dfrac{-8^6y^4z^{-3}}{2^{-1}y^{-2}z^2} = \dfrac{-(2^3)^6y^4z^{-3}}{2^{-1}y^{-2}z^2} = -2^{18}2^{-(-1)}y^4y^{-(-2)}z^{-3}z^{-2}$

$$= \dfrac{-2^{19}y^6}{z^5}$$

9. $(5x^{-2}y^4)^3 = 5^3x^{-6}y^{12} = 125x^{-6}y^{12}$, or $\dfrac{125y^{12}}{x^6}$

The following illustrates a new property of exponents.

$$\left(\frac{x}{y}\right)^3 = \frac{x}{y} \cdot \frac{x}{y} \cdot \frac{x}{y} = \frac{x \cdot x \cdot x}{y \cdot y \cdot y} = \frac{x^3}{y^3}$$

For any real numbers a and b, where $b \neq 0$, and any integer n,

$$\left(\frac{a}{b}\right)^n = \frac{a^n}{b^n}.$$

Examples Simplify.

To raise a fractional expression to a power we can raise both numerator and denominator to that power.

10. $\left(\dfrac{4x^2}{y}\right)^3 = \dfrac{(4x^2)^3}{y^3} = \dfrac{4^3(x^2)^3}{y^3} = \dfrac{64x^6}{y^3}$

11. $\left(\dfrac{x^{-5}}{y^4}\right)^{-3} = \dfrac{(x^{-5})^{-3}}{(y^4)^{-3}} = \dfrac{x^{15}}{y^{-12}}$, or $x^{15}y^{12}$

12. $\left(\dfrac{2x^{-2}}{y^3}\right)^4 = \dfrac{(2x^{-2})^4}{(y^3)^4} = \dfrac{2^4(x^{-2})^4}{(y^3)^4} = \dfrac{16x^{-8}}{y^{12}}$, or $\dfrac{16}{x^8 y^{12}}$

13. $\left(\dfrac{4x^3 y^{-1}}{3z^3}\right)^2 = \dfrac{(4x^3 y^{-1})^2}{(3z^3)^2} = \dfrac{4^2 x^6 y^{-2}}{3^2 z^6}$, or $\dfrac{16}{9}x^6 y^{-2}z^{-6}$

SCIENTIFIC NOTATION

The following are examples of scientific notation, which is very useful when naming or calculating with very large or very small numbers. It is also very helpful when estimating.

6.4×10^{13} means 64000000000000; 4.6×10^{-6} means 0.0000046.

Scientific notation **for a number consists of exponential notation for a power of 10, and, if needed, decimal notation for a number a between 1 and 10 and a multiplication sign. The following are both scientific notation: $a \times 10^b$; 10^b.**

We can convert to scientific notation by multiplying by 1, choosing a name like $10^b \cdot 10^{-b}$ for the number 1.

Example 14 Light travels about 9,460,000,000,000 kilometers in one year. Write scientific notation for the number.

We want to move the decimal point 12 places, between the 9 and the 4, so we choose $10^{-12} \times 10^{12}$ as a name for 1. Then we multiply.

$$9{,}460{,}000{,}000{,}000 \times 10^{-12} \times 10^{12} \qquad \text{Multiplying by 1}$$
$$= 9.46 \times 10^{12} \qquad \text{The } 10^{-12} \text{ moved the decimal point 12 places to the left and we have scientific notation.}$$

Example 15 Write scientific notation for 0.0000000000156.

We want to move the decimal point 11 places. We choose $10^{11} \times 10^{-11}$ as a name for 1, and then multiply.

$$0.0000000000156 \times 10^{11} \times 10^{-11} \qquad \text{Multiplying by 1}$$
$$= 1.56 \times 10^{-11} \qquad \text{The } 10^{11} \text{ moved the decimal point 11 places to the right and we have scientific notation.}$$

You should try to make conversions to scientific notation mentally as much as possible.

Examples Convert mentally to decimal notation.

16. $7.893 \times 10^5 = 789{,}300$ Moving decimal point 5 places to the right

17. $4.7 \times 10^{-8} = 0.000000047$ Moving decimal point 8 places to the left

> A positive exponent indicates a large number. A negative exponent indicates a small number.

MULTIPLYING AND DIVIDING

Multiplying and dividing with scientific notation is easy because we can use the properties of exponents.

Example 18 Multiply and write scientific notation for the answer: $(3.1 \times 10^5)(4.5 \times 10^{-3})$.

We apply the commutative and associative laws to get

$$(3.1 \times 4.5)(10^5 \times 10^{-3}) = 13.95 \times 10^2.$$

To find scientific notation for the result, we convert 13.95 to scientific notation and then simplify.

$$13.95 \times 10^2 = (1.395 \times 10^1) \times 10^2 = 1.395 \times 10^3$$

Example 19 Divide and write scientific notation for the answer: $\dfrac{6.4 \times 10^{-7}}{8.0 \times 10^6}$.

> This shows two divisions.

$$\frac{6.4 \times 10^{-7}}{8.0 \times 10^6} = \frac{6.4}{8.0} \times \frac{10^{-7}}{10^6} \qquad \text{Factoring}$$

$$= 0.8 \times 10^{-13} \qquad \text{Doing the divisions separately}$$

$$= (8.0 \times 10^{-1}) \times 10^{-13} \qquad \text{Converting 0.8 to scientific notation}$$

$$= 8.0 \times 10^{-14}$$

> The answer 0.8×10^{-13} is not incorrect, except that we have not converted it to scientific notation.

ESTIMATING AND APPROXIMATING

Scientific notation is helpful in estimating or approximating.

Example 20 Estimate: $\dfrac{780{,}000{,}000 \times 0.00071}{0.000005}$.

We first convert to scientific notation:

$$\frac{(7.8 \times 10^8)(7.1 \times 10^{-4})}{5 \times 10^{-6}}.$$

Next, we group as follows:

$$\frac{7.8 \times 7.1}{5} \times \frac{(10^8)(10^{-4})}{10^{-6}}.$$

We round the numbers that are between 1 and 10 and calculate:

$$\frac{8 \times 7}{5} = \frac{56}{5} \approx \frac{55}{5} = 11.$$

> This answer will vary, depending on how you round.

Next, we consider the powers of 10.

$$\frac{10^8 \times 10^{-4}}{10^{-6}} = \frac{10^4}{10^{-6}} = 10^{10}$$

> With some practice you will be able to estimate mentally.

We now have an approximation of 11×10^{10}, or 1.1×10^{11}.

Some examples using scientific notation

The distance from the planet Pluto to the sun is 3.664×10^9 miles.

The mass of an electron is 9.11×10^{-28} grams.

The population of the United States was 2.04×10^8.

The wavelength of a certain red light is 6.6×10^{-5} cm.

The gross national product (GNP) one year was $\$9.32 \times 10^{11}$.

An electron has a charge of 4.8×10^{-10} electrostatic units.

EXERCISE SET 6.1

Simplify.

1. $(-4m^2n^3)^0$

2. $(-9x^3y^5)^0$

3. $(x^2 + y)^1$

4. $(y^3 + z)^1$

5. $(-3r^2s^{-3})(5r^{-6}s^5)$

6. $(8s^{-4}t^5)(-2s^8t^{-7})$

7. $(-2m^{-2}n^3)^3$

8. $(-3x^{-3}y^5)^2$

9. $\dfrac{5x^{-4}}{25x^{-10}}$

10. $\dfrac{40z^{-8}}{5z^{-6}}$

11. $(-2a^2b^3)(6a^{-4}b^{-1})$

12. $(-5c^3d^{-5})(7d^4c^{-1})$

13. $\dfrac{a^2b^{-3}}{a^4b^{-2}}$

14. $\dfrac{x^3y^{-5}}{x^5y^{-7}}$

15. $\dfrac{-6^5y^4z^{-5}}{2^{-2}y^{-2}z^3}$

16. $\dfrac{9^{-2}x^{-4}y}{3^{-3}x^{-3}y^2}$

17. $(2x^{-4}y^{-2})^{-3}$

18. $(4a^{-4}b^{-5})^{-3}$

19. $\left(\dfrac{3a^{-2}b}{c^2}\right)^3$

20. $\left(\dfrac{2x^2y^{-2}}{3a^2}\right)^3$

Convert to scientific notation.

21. 47,000,000,000

22. 2,600,000,000,000

23. 863,000,000,000,000,000

24. 957,000,000,000,000,000

25. 0.000000016

26. 0.000000263

27. 0.00000000007

28. 0.00000000009

Convert to decimal notation.

29. 6.73×10^8

30. 9.24×10^7

31. 8.923×10^{-10}

32. 7.034×10^{-2}

Multiply and write scientific notation for the answer.

33. $(2.3 \times 10^6)(4.2 \times 10^{-11})$ **34.** $(6.5 \times 10^3)(5.2 \times 10^{-8})$

35. $(2.34 \times 10^{-8})(5.7 \times 10^{-4})$ **36.** $(3.26 \times 10^{-6})(8.2 \times 10^9)$

Divide and write scientific notation for the answer.

37. $\dfrac{8.5 \times 10^8}{3.4 \times 10^5}$ **38.** $\dfrac{5.1 \times 10^6}{3.4 \times 10^3}$ **39.** $\dfrac{4.0 \times 10^{-6}}{8.0 \times 10^{-3}}$ **40.** $\dfrac{7.5 \times 10^{-9}}{2.5 \times 10^{-4}}$

Estimate.

41. $\dfrac{(6.1 \times 10^4)(7.2 \times 10^{-6})}{9.8 \times 10^{-4}}$ **42.** $\dfrac{(8.05 \times 10^{-11})(5.9 \times 10^7)}{3.1 \times 10^{14}}$

☆ _____

43. The distance light travels in 100 years is approximately 5.8×10^{14} miles. How far does light travel in 13 weeks? Write scientific notation for your answer.

44. Compare $8 \cdot 10^{-90}$ and $9 \cdot 10^{-91}$. Which is the larger value? How much larger? Write scientific notation for the difference.

45. Find 8^{-x} when $8^{3x} = 27$.

46. Evaluate: $(4096)^{0.05}(4096)^{0.2}$

47. What is the unit's digit in $(513)^{127}$?

48. Given that $a = 4 - 3^b$ and $c = 2 - 3^{-b}$, find c in terms of a.

Simplify.

49. $[7y(7-8)^{-2} - 8y(8-7)^{-2}]^{-(-2)^2}$ **50.** $\dfrac{3^{q+3} - 3^2(3^q)}{3(3^{q+4})}$

51. $[3^{-(3s+1)} - 3^{-(3s-1)} + 3^{-3s}]^{-2^2}$

_____

6.2 RADICAL EXPRESSIONS

SQUARE ROOTS

A *square root* of a number a is a number c whose second power is a, that is, $c^2 = a$.

> 5 is a square root of 25 because $5^2 = 5 \cdot 5 = 25$.
>
> -5 is a square root of 25 because $(-5)^2 = (-5)(-5) = 25$.

> -4 does not have a real square root because there is no real number b such that $b^2 = -4$.

> **Every positive real number has two real-number square roots. The number 0 has just one square root, 0 itself. Negative numbers do not have real number roots.** *

Example 1   Find the two square roots of 64.

The square roots are 8 and -8 because $8^2 = 64$ and $(-8)^2 = 64$.

> **The *principal* square root of a nonnegative number is its nonnegative square root. The symbol $\sqrt{a}$ represents the principal square root of a. To name the negative square root of a we write $-\sqrt{a}$.**

Examples   Simplify.

2. $\sqrt{25} = 5$ $\boxed{\text{Remember: } \sqrt{} \text{ indicates the principal (nonnegative) square root.}}$

3. $\sqrt{\dfrac{25}{64}} = \dfrac{5}{8}$ **4.** $-\sqrt{64} = -8$

5. $\sqrt{0} = 0$ **6.** $\sqrt{0.0049} = 0.07$

7. $\sqrt{-49}$ Does not exist. Negative numbers do not have real number square roots.

> **The symbol $\sqrt{}$ is called a *radical*. An expression written with a radical is called a *radical expression*. The expression written under the radical is called the *radicand*.**

These are radical expressions:

$$\sqrt{5}, \qquad \sqrt{a}, \qquad -\sqrt{5x}, \qquad \frac{\sqrt{y^2 + 7}}{\sqrt{x}}.$$

Example 8   Identify the radicand in $\sqrt{x^2 - 9}$.

The radicand is $x^2 - 9$.

---

* Later, we will see that there is a number system other than the real-number system, in which negative numbers do have square roots.

FINDING $\sqrt{a^2}$

In the expression $\sqrt{a^2}$, the radicand is a perfect square.

Suppose $a = 5$. Then we have $\sqrt{5^2}$, which is $\sqrt{25}$, or 5.

Suppose $a = -5$. Then we have $\sqrt{(-5)^2}$, which is $\sqrt{25}$, or 5.

Suppose $a = 0$. Then we have $\sqrt{0^2}$, which is $\sqrt{0}$, or 0.

The symbol $\sqrt{a^2}$ does not represent a negative number. It represents the principal square root of a^2. Note that if a represents a positive number or 0, then $\sqrt{a^2}$ represents a. If a is negative, then $\sqrt{a^2}$ represents the additive inverse of a. In all cases the radical expression represents the absolute value of a.

> For any real number a, $\sqrt{a^2} = |a|$. The principal (nonnegative) square root of a^2 is the absolute value of a.

Examples Find the following. Assume that letters can represent any real numbers.

9. $\sqrt{(-16)^2} = |-16|$, or 16

10. $\sqrt{(3b)^2} = |3b|$, or $3|b|$

> $|3b|$ can be simplified to $3|b|$ because the absolute value of any product is the product of the absolute values. That is, $|a \cdot b| = |a| \cdot |b|$. In this case, $|3b| = |3| \cdot |b|$, or $3|b|$.

11. $\sqrt{(x - 1)^2} = |x - 1|$

12. $\sqrt{x^2 + 8x + 16} = \sqrt{(x + 4)^2}$

$= |x + 4|$

> $|x + 4|$ does *not* mean the same as $|x| + 4$.

CUBE ROOTS

The number c is the cube root of a if its third power is a, that is, $c^3 = a$. For example,

2 is the cube root of 8 because $2^3 = 2 \cdot 2 \cdot 2 = 8$;

-4 is the cube root of -64 because $(-4)^3 = (-4)(-4)(-4) = -64$.

We used the word "the" with cube roots because of the following.

> **Every real number has exactly one cube root in the system of real numbers.**

The symbol $\sqrt[3]{a}$ represents the cube root of a.

Examples Find the following.

13. $\sqrt[3]{8} = 2$

14. $\sqrt[3]{-27} = -3$

15. $\sqrt{-\dfrac{216}{125}} = -\dfrac{6}{5}$

16. $\sqrt[3]{x^3} = x$

17. $\sqrt[3]{-8y^3} = -2y$

No absolute value signs are needed when finding cube roots because a real number has just one cube root. The cube root of a positive number is positive. The cube root of a negative number is negative.

ODD AND EVEN kth ROOTS

The 5th root of a number a is the number c for which $c^5 = a$. There are also 7th roots, 9th roots, and so on. Whenever the number k in $\sqrt[k]{\ }$ is an odd number, we say that we are taking an *odd* root. The number k is called the *index*. When the index is 2 we do not write it.

Every number has just one odd root. If the number is positive, the root is positive. If the number is negative, the root is negative.

Examples Find the following.

18. $\sqrt[5]{32} = 2$

19. $\sqrt[5]{-32} = -2$

20. $-\sqrt[5]{32} = -2$

21. $-\sqrt[5]{-32} = -(-2) = 2$

22. $\sqrt[7]{x^7} = x$

23. $\sqrt[9]{(x-1)^9} = x - 1$

Absolute value signs are never needed when finding odd roots.

When the index k in $\sqrt[k]{\ }$ is an even number, we say that we are taking an *even* root. Every positive real number has two kth roots when k is even. One of those roots is positive and one is negative. Negative real numbers do not have kth roots when k is even. When finding even kth roots, absolute value signs are sometimes necessary, as they are with square roots.

Examples Find the following. Assume that letters can represent any real number.

24. $\sqrt[4]{16} = 2$　　　　　　　　　**25.** $-\sqrt[4]{16} = -2$

26. $\sqrt[4]{-16}$　　Does not exist　　**27.** $\sqrt[4]{81x^4} = 3|x|$

28. $\sqrt[6]{(y + 7)^6} = |y + 7|$

For any real number a:

a) $\sqrt[k]{a^k} = |a|$ when k is even. We use absolute value when k is even unless a is nonnegative.

b) $\sqrt[k]{a^k} = a$ when k is odd. We do not use absolute value when k is odd.

EXERCISE SET 6.2

Find the square roots of each number.

1. 16　　　　**2.** 225　　　　**3.** 144　　　　**4.** 9　　　　**5.** 400　　　　**6.** 81

Find the following.

7. $-\sqrt{\dfrac{49}{36}}$　　　**8.** $-\sqrt{\dfrac{361}{9}}$　　　**9.** $\sqrt{196}$　　　**10.** $\sqrt{441}$

11. $-\sqrt{\dfrac{16}{81}}$　　**12.** $-\sqrt{\dfrac{81}{144}}$　　**13.** $\sqrt{0.09}$　　**14.** $\sqrt{0.36}$

15. $-\sqrt{0.0049}$　　**16.** $\sqrt{0.0144}$

Identify the radicand in each expression.

17. $5\sqrt{p^2 + 4}$　　**18.** $-7\sqrt{y^2 - 8}$　　**19.** $x^2y^2\sqrt{\dfrac{x}{y + 4}}$　　**20.** $a^2b^3\sqrt{\dfrac{a}{a^2 - b}}$

Find the following. Assume that letters can represent any real number.

21. $\sqrt{16x^2}$　　　　　　　**22.** $\sqrt{25t^2}$　　　　　　　**23.** $\sqrt{(-7c)^2}$

24. $\sqrt{(-6b)^2}$　　　　　　**25.** $\sqrt{(a + 1)^2}$　　　　　**26.** $\sqrt{(5 - b)^2}$

27. $\sqrt{x^2 - 4x + 4}$　　　　　　**28.** $\sqrt{y^2 + 16y + 64}$

29. $\sqrt{4x^2 + 28x + 49}$　　　　　**30.** $\sqrt{9x^2 - 30x + 25}$

Simplify.

31. $\sqrt[3]{27}$ **32.** $-\sqrt[3]{64}$ **33.** $\sqrt[3]{-64x^3}$

34. $\sqrt[3]{-125y^3}$ **35.** $\sqrt[3]{-216}$ **36.** $-\sqrt[3]{-1000}$

37. $\sqrt[3]{0.343(x+1)^3}$ **38.** $\sqrt[3]{0.000008(y-2)^3}$

Find the following. Assume that letters can represent any real number.

39. $\sqrt[4]{625}$ **40.** $-\sqrt[4]{256}$ **41.** $\sqrt[5]{-1}$ **42.** $-\sqrt[5]{-32}$

43. $\sqrt[5]{-\dfrac{32}{243}}$ **44.** $\sqrt[5]{-\dfrac{1}{32}}$ **45.** $\sqrt[6]{x^6}$ **46.** $\sqrt[8]{y^8}$

47. $\sqrt[4]{(5a)^4}$ **48.** $\sqrt[4]{(7b)^4}$ **49.** $\sqrt[10]{(-6)^{10}}$ **50.** $\sqrt[12]{(-10)^{12}}$

51. $\sqrt[414]{(a+b)^{414}}$ **52.** $\sqrt[1976]{(2a+b)^{1976}}$ **53.** $\sqrt[7]{y^7}$ **54.** $\sqrt[3]{(-6)^3}$

55. $\sqrt[5]{(x-2)^5}$ **56.** $\sqrt[9]{(2xy)^9}$

☆

57. *Parking.* A parking lot has attendants to park the cars. The number N of temporary stalls needed for waiting cars before attendants can get to them is given by the formula $N = 2.5\sqrt{A}$, where A is the number of arrivals in peak hours. Find the number of spaces needed when the average number of arrivals in peak hours is (a) 25; (b) 36; (c) 49; (d) 64.

6.3 MULTIPLYING AND SIMPLIFYING WITH RADICALS

MULTIPLYING

Notice that $\sqrt{4}\sqrt{25} = 2 \cdot 5 = 10$.

Also $\sqrt{4 \cdot 25} = \sqrt{100} = 10$.

Likewise, $\sqrt[3]{27}\sqrt[3]{8} = 3 \cdot 2 = 6$ and $\sqrt[3]{27 \cdot 8} = \sqrt[3]{216} = 6$.

These examples suggest the following.

> **For any nonnegative real numbers a and b and any index k,**
> $\sqrt[k]{a} \cdot \sqrt[k]{b} = \sqrt[k]{a \cdot b}.$ **(To multiply, we multiply the radicands.)**

Examples   Multiply.

Note that the index k must be the same throughout.

1. $\sqrt{3} \cdot \sqrt{5} = \sqrt{3 \cdot 5} = \sqrt{15}$

2. $\sqrt{x+2}\sqrt{x-2} = \sqrt{(x+2)(x-2)} = \sqrt{x^2-4}$

3. $\sqrt[3]{4}\sqrt[3]{5} = \sqrt[3]{4 \cdot 5} = \sqrt[3]{20}$

4. $\sqrt[4]{\dfrac{y}{5}}\sqrt[4]{\dfrac{7}{x}} = \sqrt[4]{\dfrac{y}{5} \cdot \dfrac{7}{x}}$

$$= \sqrt[4]{\dfrac{7y}{5x}}$$

A common error is to omit the index in the answer.

SIMPLIFYING BY FACTORING

Reversing the above statement we have $\sqrt[k]{a \cdot b} = \sqrt[k]{a} \cdot \sqrt[k]{b}$. This shows a way to factor and thus simplify radical expressions. Consider $\sqrt{20}$. The number 20 has the factor 4, which is a perfect square. Therefore,

$$\sqrt{20} = \sqrt{4 \cdot 5} \qquad \text{Factoring the radicand (4 is a perfect square)}$$
$$= \sqrt{4} \cdot \sqrt{5} \qquad \text{Factoring into two radicals}$$
$$= 2\sqrt{5}. \qquad \text{Taking the square root of 4}$$

> **To simplify a radical expression by factoring, look for factors of the radicand that are perfect kth powers (where k is the index). Then take the kth root of those factors.**

Examples   Simplify by factoring.

5. $\sqrt{50} = \sqrt{25 \cdot 2} = \sqrt{25} \cdot \sqrt{2} = 5\sqrt{2}$

This factor is a perfect square.

6. $\sqrt[3]{32} = \sqrt[3]{8 \cdot 4} = \sqrt[3]{8} \cdot \sqrt[3]{4} = 2\sqrt[3]{4}$

This factor is a perfect cube (third power).

In many situations, expressions under radicals never represent negative numbers. In such cases, absolute value notation is not necessary.

Examples   Simplify by factoring. Assume that all expressions under radicals represent nonnegative numbers.

7. $\sqrt{5x^2} = \sqrt{x^2 \cdot 5} \qquad \text{Factoring the radicand}$
$$= \sqrt{x^2} \cdot \sqrt{5} \qquad \text{Factoring into two radicals}$$
$$= x \cdot \sqrt{5} \qquad \text{Taking the square root of } x^2$$

8. $\sqrt{2x^2 - 4x + 2} = \sqrt{2(x-1)^2}$

$\qquad\qquad\qquad = \sqrt{(x-1)^2} \cdot \sqrt{2}$

$\qquad\qquad\qquad = (x-1) \cdot \sqrt{2}$

> Absolute value is not needed because expressions are not negative.

9. $\sqrt{216x^5y^3} = \sqrt{36 \cdot 6 \cdot x^4 \cdot x \cdot y^2 \cdot y}$ $\Big\}$ Factoring the radicand

$\qquad\qquad\quad = \sqrt{36 \cdot x^4 \cdot y^2 \cdot 6 \cdot x \cdot y}$

$\qquad\qquad\quad = \sqrt{36}\sqrt{x^4}\sqrt{y^2}\sqrt{6xy}$ Factoring into several radicals

$\qquad\qquad\quad = 6x^2y\sqrt{6xy}$ Taking square roots

Note: Had we not seen that $216 = 36 \cdot 6$, where 36 is the largest square factor of 216, we could have found the prime factorization $2 \cdot 2 \cdot 2 \cdot 3 \cdot 3 \cdot 3$. Each pair of factors makes a square, so

$$\sqrt{2 \cdot 2 \cdot 2 \cdot 3 \cdot 3 \cdot 3} = \sqrt{2^2 \cdot 3^2 \cdot 2 \cdot 3} = 2 \cdot 3\sqrt{2 \cdot 3}.$$

MULTIPLYING AND SIMPLIFYING

Sometimes after we multiply we can then simplify by factoring.

Examples Multiply and then simplify by factoring. Assume that all expressions under radicals represent nonnegative numbers.

10. $\sqrt{15}\sqrt{6} = \sqrt{15 \cdot 6} = \sqrt{90} = \sqrt{9 \cdot 10} = 3\sqrt{10}$

11. $3\sqrt[3]{25} \cdot 2\sqrt[3]{5} = 6 \cdot \sqrt[3]{25 \cdot 5}$ $\Big\}$ Multiplying radicands

$\qquad\qquad\quad = 6 \cdot \sqrt[3]{125}$

$\qquad\qquad\quad = 6 \cdot 5,\ \text{or}\ 30$ Taking the cube root of 125

12. $\sqrt[3]{18y^3}\sqrt[3]{4x^2} = \sqrt[3]{18y^3 \cdot 4x^2} = \sqrt[3]{72y^3x^2}$ Multiplying radicands

$\qquad\qquad\qquad\quad = \sqrt[3]{8y^3 \cdot 9x^2}$ Factoring the radicand

$\qquad\qquad\qquad\quad = \sqrt[3]{8y^3}\sqrt[3]{9x^2}$ Factoring into two radicals

$\qquad\qquad\qquad\quad = 2y\sqrt[3]{9x^2}$ Taking the cube root

APPROXIMATING SQUARE ROOTS

Table 1 in the back of the book contains approximate square roots for the natural numbers 1 through 100. If a radicand is not listed in the table we can factor the radical expression, find exact or approximate square roots of the factors, and then find the product of those square roots.

Example 13 Approximate to the nearest tenth.

$$\sqrt{275} = \sqrt{25 \cdot 11} \qquad \text{Factoring the radicand}$$
$$= \sqrt{25} \cdot \sqrt{11} \qquad \text{Factoring into two radicals}$$
$$\approx 5 \times 3.317 \approx 16.6 \qquad \text{From Table 1, } \sqrt{11} \approx 3.317$$

Example 14 Approximate to the nearest tenth.

$$\frac{16 - \sqrt{640}}{4} = \frac{16 - \sqrt{64 \cdot 10}}{4} \qquad \text{Factoring the radicand}$$

$$= \frac{16 - \sqrt{64} \cdot \sqrt{10}}{4} \qquad \text{Factoring into two radicals}$$

$$= \frac{16 - 8\sqrt{10}}{4} \qquad \longleftarrow \boxed{\begin{array}{l}\text{A common error here is to "cancel"}\\ \text{and get } 4 - 8\sqrt{10}. \text{ Always remem-}\\ \text{ber to } remove\ a\ factor\ of\ 1.\end{array}}$$

$$= \frac{4(4 - 2\sqrt{10})}{4} = 4 - 2\sqrt{10} \qquad \text{Removing a factor of 1}$$

$$\approx 4 - 2 \times 3.162 \qquad \text{Using Table 1}$$
$$\approx -2.324, \text{ or } -2.3 \text{ to nearest tenth}$$

RATIONAL AND IRRATIONAL NUMBERS

The real numbers consist of the rational numbers and the irrational numbers. We can describe these kinds of numbers in several ways.

> **Rational numbers are:**
> 1. **Those numbers that can be named with fractional notation** a/b, **where** a **and** b **are integers. (Definition from p. 2)**
> 2. **Those numbers for which decimal notation either ends or repeats.**

Examples

15. $\dfrac{5}{16} = 0.3125$ \qquad Ending (terminating) decimal

16. $\dfrac{8}{7} = 1.142857142857\ldots = 1.\overline{142857}$ \qquad Repeating decimal

17. $\dfrac{3}{11} = 0.2727\ldots = 0.\overline{27}$ Repeating decimal

> **Irrational numbers are:**
>
> 1. **Those real numbers that are not rational.**
> 2. **Those real numbers that cannot be named with fractional notation** a/b**, where** a **and** b **are integers.**
> 3. **Those real numbers for which decimal notation does not end and does not repeat.**

There are many irrational numbers. For example, $\sqrt{2}$ is irrational. We can find rational numbers a/b for which $(a/b) \cdot (a/b)$ is close to 2, but we cannot find such a number a/b for which $(a/b) \cdot (a/b)$ is *exactly* 2.

Unless a whole number is a perfect square its square root is irrational. For example, $\sqrt{9}$ and $\sqrt{25}$ are rational, but all of the following are irrational:

$$\sqrt{3}, \qquad -\sqrt{14}, \qquad \sqrt{45}.$$

There are also many irrational numbers that we do not obtain by taking square roots. The number π is an example.* Decimal notation for π does not end and does not repeat.

Examples   All of these are irrational.

18. $\pi = 3.1415926535\ldots$ Numeral does not repeat.

19. $1.101001000100001000001\ldots$ Numeral does not repeat.

20. $\sqrt{2} = 1.41421356\ldots$ Numeral does not repeat.

In Example 19 there is a pattern, but it is not a repeating pattern.

EXERCISE SET 6.3

Multiply.

1. $\sqrt{3}\sqrt{2}$ **2.** $\sqrt{5}\sqrt{7}$ **3.** $\sqrt[3]{2}\sqrt[3]{5}$ **4.** $\sqrt[3]{7}\sqrt[3]{2}$

5. $\sqrt[4]{8}\sqrt[4]{9}$ **6.** $\sqrt[4]{6}\sqrt[4]{3}$ **7.** $\sqrt{3a}\sqrt{10b}$ **8.** $\sqrt{2x}\sqrt{13y}$

9. $\sqrt[5]{9t^2}\sqrt[5]{2t}$ **10.** $\sqrt[5]{8y^3}\sqrt[5]{10y}$ **11.** $\sqrt{x-a}\sqrt{x+a}$ **12.** $\sqrt{y-b}\sqrt{y+b}$

---

* 3.14 and $\frac{22}{7}$ are only rational approximations to π.

Simplify by factoring. Assume that all expressions represent nonnegative numbers.

13. $\sqrt{8}$ **14.** $\sqrt{18}$ **15.** $\sqrt{24}$ **16.** $\sqrt{20}$

17. $\sqrt{40}$ **18.** $\sqrt{90}$ **19.** $\sqrt{180x^4}$ **20.** $\sqrt{175y^6}$

21. $\sqrt[3]{54x^8}$ **22.** $\sqrt[3]{40y^3}$

Multiply and simplify by factoring. Assume that all expressions represent nonnegative numbers.

23. $\sqrt{15}\sqrt{6}$ **24.** $\sqrt{2}\sqrt{32}$

25. $\sqrt[3]{3}\sqrt[3]{18}$ **26.** $\sqrt[3]{2}\sqrt[3]{20}$

27. $\sqrt{45}\sqrt{60}$ **28.** $\sqrt{24}\sqrt{75}$

29. $\sqrt{5b^3}\sqrt{10c^4}$ **30.** $\sqrt{2x^3y}\sqrt{12xy}$

31. $\sqrt[3]{y^4}\sqrt[3]{16y^5}$ **32.** $\sqrt[3]{5^2t^4}\sqrt[3]{5^4t^6}$

33. $\sqrt[3]{(b+3)^4}\sqrt[3]{(b+3)^2}$ **34.** $\sqrt[3]{(x+y)^3}\sqrt[3]{(x+y)^5}$

35. $\sqrt{12a^3b}\sqrt{8a^4b^2}$ **36.** $\sqrt{18x^2y^3}\sqrt{6xy^4}$

Approximate to the nearest tenth.

37. $\sqrt{180}$ **38.** $\sqrt{124}$ **39.** $\dfrac{8+\sqrt{480}}{4}$ **40.** $\dfrac{12-\sqrt{450}}{3}$

Which of the following are irrational numbers? Which are rational?

41. $-\dfrac{5}{6}$ **42.** 8.93 **43.** $8.23\overline{23}$ **44.** $\sqrt{5}$

45. $\sqrt{36}$ **46.** $-\sqrt{15}$ **47.** $2.101001\ldots$ (Numeral does not repeat.)   **48.** $3.1415926\ldots$ (Numeral does not repeat.)

☆ ─────────────────────────────────

49. *Speed of a skidding car.* After an accident, police can estimate the speed at which a car was traveling by measuring its skid marks. The formula

$$r = 2\sqrt{5L}$$

can be used, where r is the speed in mph, and L is the length of the skid marks in feet. Estimate the speed of a car that left skid marks (a) 20 ft long; (b) 70 ft long; (c) 90 ft long.

Multiply and simplify.

50. $\sqrt{1.6 \times 10^3}\sqrt{36 \times 10^{-8}}$ **51.** $\sqrt{968}\sqrt{1014}$ **52.** $\sqrt[3]{48}\sqrt[3]{63}\sqrt[3]{196}$

6.4 DIVIDING AND SIMPLIFYING WITH RADICALS

ROOTS OF QUOTIENTS

Notice that $\sqrt[3]{\dfrac{27}{8}} = \dfrac{3}{2}$ and that $\dfrac{\sqrt[3]{27}}{\sqrt[3]{8}} = \dfrac{3}{2}$. This example suggests that we can take the root of a quotient by taking the root of the numerator and denominator separately. This is true.

> **Rule A:** For any nonnegative number a and any positive number b, and any index k,
>
> $$\sqrt[k]{\dfrac{a}{b}} = \dfrac{\sqrt[k]{a}}{\sqrt[k]{b}}.$$ **(We can take the kth root of the numerator and denominator separately.)**

Examples Simplify by taking roots of numerator and denominator. Assume that all expressions under radicals represent positive numbers.

1. $\sqrt[3]{\dfrac{27}{125}} = \dfrac{\sqrt[3]{27}}{\sqrt[3]{125}} = \dfrac{3}{5}$

   We take the cube root of the numerator and denominator.

2. $\sqrt{\dfrac{25}{y^2}} = \dfrac{\sqrt{25}}{\sqrt{y^2}} = \dfrac{5}{y}$

   We take the square root of the numerator and denominator.

3. $\sqrt{\dfrac{16x^3}{y^4}} = \dfrac{\sqrt{16x^3}}{\sqrt{y^4}} = \dfrac{\sqrt{16x^2 \cdot x}}{\sqrt{y^4}} = \dfrac{4x\sqrt{x}}{y^2}$

4. $\sqrt[3]{\dfrac{27y^5}{343x^3}} = \dfrac{\sqrt[3]{27y^5}}{\sqrt[3]{343x^3}} = \dfrac{\sqrt[3]{27y^3 \cdot y^2}}{\sqrt[3]{343x^3}} = \dfrac{\sqrt[3]{27y^3} \cdot \sqrt[3]{y^2}}{\sqrt[3]{343x^3}} = \dfrac{3y\sqrt[3]{y^2}}{7x}$

We are assuming that no expression represents 0 or a negative number. Thus we need not be concerned about zero denominators.

DIVIDING RADICAL EXPRESSIONS

Reversing Rule A shows a way to divide radical expressions as follows.

> **Rule B:** $\dfrac{\sqrt[k]{a}}{\sqrt[k]{b}} = \sqrt[k]{\dfrac{a}{b}}$ **To divide, we divide the radicands. After doing this we can sometimes simplify by taking roots.**

Examples Divide. Then simplify by taking roots if possible. Assume that all expressions under radicals represent positive numbers.

5. $\dfrac{\sqrt{80}}{\sqrt{5}} = \sqrt{\dfrac{80}{5}} = \sqrt{16} = 4$ We divide the radicands.

6. $\dfrac{3\sqrt{2}}{5\sqrt{3}} = \dfrac{3}{5} \cdot \dfrac{\sqrt{2}}{\sqrt{3}} = \dfrac{3}{5} \cdot \sqrt{\dfrac{2}{3}}$

7. $\dfrac{5\sqrt[3]{32}}{\sqrt[3]{2}} = 5\sqrt[3]{\dfrac{32}{2}} = 5\sqrt[3]{16} = 5\sqrt[3]{8 \cdot 2} = 5\sqrt[3]{8}\sqrt[3]{2} = 5 \cdot 2\sqrt[3]{2} = 10\sqrt[3]{2}$

8. $\dfrac{\sqrt{72xy}}{2\sqrt{2}} = \dfrac{1}{2}\dfrac{\sqrt{72xy}}{\sqrt{2}} = \dfrac{1}{2}\sqrt{\dfrac{72xy}{2}} = \dfrac{1}{2}\sqrt{36xy} = \dfrac{1}{2}\sqrt{36}\sqrt{xy}$

$= \dfrac{1}{2} \cdot 6\sqrt{xy} = 3\sqrt{xy}$

9. $\dfrac{\sqrt[4]{33a^5b^3}}{\sqrt[4]{2b^{-1}}} = \sqrt[4]{\dfrac{33a^5b^3}{2b^{-1}}} = \sqrt[4]{\dfrac{33}{2}a^5b^4} = \sqrt[4]{a^4b^4}\sqrt[4]{\dfrac{33}{2}a} = ab\sqrt[4]{\dfrac{33}{2}a}$

POWERS AND ROOTS COMBINED

Consider the following.

$$\sqrt[3]{8^2} = \sqrt[3]{64} = 4$$
$$(\sqrt[3]{8})^2 = (2)^2 = 4$$

This suggests another important property of radical expressions.

> **Rule C:** For any nonnegative number a, any index k, and any natural number m,
>
> $\sqrt[k]{a^m} = (\sqrt[k]{a})^m.$ (We can raise to a power and then take a root, or we can take a root and then raise to a power.)

In some cases one way of calculating is easier than the other.

Examples Calculate as shown. Then use Rule C to calculate another way. Assume that all expressions under radicals are positive.

10. **a)** $\sqrt[3]{27^2} = \sqrt[3]{729} = 9$ Using Table 1

   b) $(\sqrt[3]{27})^2 = (3)^2 = 9$

11. a) $\sqrt[3]{2^6} = \sqrt[3]{64} = 4$

   b) $(\sqrt[3]{2})^6 = \sqrt[3]{2}\sqrt[3]{2}\sqrt[3]{2}\sqrt[3]{2}\sqrt[3]{2}\sqrt[3]{2} = 2 \cdot 2 = 4$

12. a) $(\sqrt{5x})^3 = \sqrt{5x}\sqrt{5x}\sqrt{5x} = \sqrt{5^3 x^3} = \sqrt{5^2 x^2}\sqrt{5x} = 5x\sqrt{5x}$

   b) $\sqrt{(5x)^3} = \sqrt{5^3 x^3} = \sqrt{5^2 x^2}\sqrt{5x} = 5x\sqrt{5x}$

EXERCISE SET 6.4

Simplify by taking roots of numerator and denominator. Assume that all expressions under radicals represent positive numbers.

1. $\sqrt{\dfrac{16}{25}}$ **2.** $\sqrt{\dfrac{100}{81}}$ **3.** $\sqrt[3]{\dfrac{64}{27}}$ **4.** $\sqrt[3]{\dfrac{343}{512}}$ **5.** $\sqrt{\dfrac{49}{y^2}}$

6. $\sqrt{\dfrac{121}{x^2}}$ **7.** $\sqrt{\dfrac{25y^3}{x^4}}$ **8.** $\sqrt{\dfrac{36a^5}{b^6}}$ **9.** $\sqrt[3]{\dfrac{8x^5}{27y^3}}$ **10.** $\sqrt[3]{\dfrac{64x^7}{216y^6}}$

Divide. Then simplify by taking roots if possible. Assume that all expressions under radicals represent positive numbers.

11. $\dfrac{\sqrt{21a}}{\sqrt{3a}}$ **12.** $\dfrac{\sqrt{28y}}{\sqrt{4y}}$ **13.** $\dfrac{\sqrt[3]{54}}{\sqrt[3]{2}}$ **14.** $\dfrac{\sqrt[3]{40}}{\sqrt[3]{5}}$

15. $\dfrac{\sqrt{40xy^3}}{\sqrt{8x}}$ **16.** $\dfrac{\sqrt{56ab^3}}{\sqrt{7a}}$ **17.** $\dfrac{\sqrt[3]{96a^4b^2}}{\sqrt[3]{12a^2b}}$ **18.** $\dfrac{\sqrt[3]{189x^5y^7}}{\sqrt[3]{7x^2y^2}}$

19. $\dfrac{\sqrt{72xy}}{2\sqrt{2}}$ **20.** $\dfrac{\sqrt{75ab}}{3\sqrt{3}}$ **21.** $\dfrac{\sqrt{x^3 - y^3}}{\sqrt{x - y}}$ **22.** $\dfrac{\sqrt{r^3 + s^3}}{\sqrt{r + s}}$

Hint: To divide $x^3 - y^3$ by $x - y$, factor $x^3 - y^3$. Then simplify the fractional expression by removing a factor of 1.

Calculate as shown. Then use Rule C to calculate another way. Assume that all expressions under radicals represent positive numbers.

23. $\sqrt{(6a)^3}$ **24.** $\sqrt{(7y)^3}$ **25.** $(\sqrt[3]{16b^2})^2$ **26.** $(\sqrt[3]{25r^2})^2$

27. $\sqrt{(18a^2b)^3}$

28. $\sqrt{(12x^2y)^3}$

29. $(\sqrt[3]{12c^2d})^2$

30. $(\sqrt[3]{9x^2y})^2$

☆

31. *Pendulums.* The *period* of a pendulum is the time it takes to complete one cycle, swinging to and fro. If a pendulum consists of a ball on a string, the period T is given by the formula

$$T = 2\pi\sqrt{\frac{L}{980}},$$

where T is in seconds and L is the length of the pendulum in centimeters. Find the period of a pendulum of length (a) 65 cm; (b) 98 cm; (c) 120 cm. Use 3.14 for π.

Divide and simplify.

32. $\dfrac{7\sqrt{a^2b}\sqrt{25xy}}{5\sqrt{a^{-4}b^{-1}}\sqrt{49x^{-1}y^{-3}}}$

33. $\dfrac{(\sqrt[3]{81mn^2})^2}{(\sqrt[3]{mn})^2}$

34. $\dfrac{\sqrt{44x^2y^9z}\sqrt{22y^9z^6}}{(\sqrt{11xy^8z^2})^2}$

6.5 ADDITION AND SUBTRACTION INVOLVING RADICALS

Any two real numbers can be added. For instance, the sum of 7 and $\sqrt{3}$ can be named

$$7 + \sqrt{3}.$$

We cannot simplify this name for the sum. However, when we have *like radicals* (radicals having the same index and radicand), we can use the distributive laws to simplify by collecting like radical terms.

Examples   Add or subtract. Simplify by collecting like radical terms if possible.

1. $6\sqrt{7} + 4\sqrt{7} = (6 + 4)\sqrt{7}$ Using the distributive law (factoring out $\sqrt{7}$)

$$= 10\sqrt{7}$$

2. $8\sqrt[3]{2} - 7x\sqrt[3]{2} + 5\sqrt[3]{2} = (8 - 7x + 5)\sqrt[3]{2}$ Factoring out $\sqrt[3]{2}$

$$= (13 - 7x)\sqrt[3]{2}$$

3. $6\sqrt[5]{4x} + 4\sqrt[5]{4x} - \sqrt[3]{4x} = (6 + 4)\sqrt[5]{4x} - \sqrt[3]{4x}$

$$= 10\sqrt[5]{4x} - \sqrt[3]{4x}$$

Sometimes we need to factor in order to have terms with like radicals.

Examples Add or subtract. Simplify by collecting like radical terms if possible.

4. $3\sqrt{8} - 5\sqrt{2} = 3\sqrt{4 \cdot 2} - 5\sqrt{2}$ Factoring 8

$\qquad\qquad\qquad = 3\sqrt{4} \cdot \sqrt{2} - 5\sqrt{2}$ Factoring $\sqrt{4 \cdot 2}$ into two radicals

$\qquad\qquad\qquad = 3 \cdot 2\sqrt{2} - 5\sqrt{2}$ Taking the square root of 4

$\qquad\qquad\qquad = 6\sqrt{2} - 5\sqrt{2}$

$\qquad\qquad\qquad = (6 - 5)\sqrt{2}$

$\qquad\qquad\qquad = \sqrt{2}$ Collecting like radical terms

5. $5\sqrt{2} - 4\sqrt{3}$ No simplification possible

6. $5\sqrt[3]{16y^4} + 7\sqrt[3]{2y} = 5\sqrt[3]{8y^3 \cdot 2y} + 7\sqrt[3]{2y}$ $\Bigg\}$ Factoring the first radical

$\qquad\qquad\qquad = 5\sqrt[3]{8y^3} \cdot \sqrt[3]{2y} + 7\sqrt[3]{2y}$ $\Bigg\}$

$\qquad\qquad\qquad = 5 \cdot 2y \cdot \sqrt[3]{2y} + 7\sqrt[3]{2y}$ Taking cube root

$\qquad\qquad\qquad = 10y\sqrt[3]{2y} + 7\sqrt[3]{2y}$

$\qquad\qquad\qquad = (10y + 7)\sqrt[3]{2y}$ Collecting like radical terms

EXERCISE SET 6.5

Add or subtract. Simplify by collecting like radical terms if possible, assuming that all expressions under radicals represent nonnegative numbers.

1. $6\sqrt{3} + 2\sqrt{3}$ **2.** $8\sqrt{5} + 9\sqrt{5}$

3. $9\sqrt[3]{5} - 6\sqrt[3]{5}$ **4.** $14\sqrt[5]{2} - 6\sqrt[5]{2}$

5. $4\sqrt[3]{y} + 9\sqrt[3]{y}$ **6.** $6\sqrt[4]{t} - 3\sqrt[4]{t}$

7. $8\sqrt{2} - 6\sqrt{2} + 5\sqrt{2}$ **8.** $2\sqrt{6} + 8\sqrt{6} - 3\sqrt{6}$

9. $4\sqrt[3]{3} - \sqrt{5} + 2\sqrt[3]{3} + \sqrt{5}$ **10.** $5\sqrt{7} - 8\sqrt[4]{11} + \sqrt{7} + 9\sqrt[4]{11}$

11. $8\sqrt{27} - 3\sqrt{3}$ **12.** $9\sqrt{50} - 4\sqrt{2}$

13. $8\sqrt{45} + 7\sqrt{20}$ **14.** $9\sqrt{12} + 16\sqrt{27}$

15. $18\sqrt{72} + 2\sqrt{98}$ **16.** $12\sqrt{45} - 8\sqrt{80}$

17. $3\sqrt[3]{16} + \sqrt[3]{54}$ **18.** $\sqrt[3]{27} - 5\sqrt[3]{8}$

19. $2\sqrt{128} - \sqrt{18} + 4\sqrt{32}$ **20.** $5\sqrt{50} - 2\sqrt{18} + 9\sqrt{32}$

21. $\sqrt{5a} + 2\sqrt{45a^3}$ **22.** $4\sqrt{3x^3} - \sqrt{12x}$

23. $\sqrt[3]{24x} - \sqrt[3]{3x^4}$

24. $\sqrt[3]{54x} - \sqrt[3]{2x^4}$

25. $\sqrt{8y - 8} + \sqrt{2y - 2}$

26. $\sqrt{12t + 12} + \sqrt{3t + 3}$

27. $\sqrt{x^3 - x^2} + \sqrt{9x - 9}$

28. $\sqrt{4x - 4} - \sqrt{x^3 - x^2}$

☆

29. Find as many pairs of numbers, a and b, as you can for which

$\sqrt{a} + \sqrt{b} = \sqrt{a + b}$.

Find several pairs of numbers for which the above equation is false.

Add or subtract by collecting like terms.

30. $\sqrt{432} - \sqrt{6125} + \sqrt{845} - \sqrt{4800}$

31. $\sqrt{1250x^3y} - \sqrt{1800xy^3} - \sqrt{162x^3y^3}$

32. $\dfrac{1}{2}\sqrt{36a^5bc^4} - \dfrac{1}{2}\sqrt[3]{64a^4bc^6} + \dfrac{1}{6}\sqrt{144a^3bc^2}$

33. $7x\sqrt{(x + y)^3} - 5xy\sqrt{x + y} - 2y\sqrt{(x + y)^3}$

6.6 MORE ABOUT MULTIPLICATION WITH RADICALS

To multiply expressions in which some factors contain more than one term, we use the procedures for multiplying polynomials.

Examples   Multiply.

1. $\boxed{\sqrt{3}}\,(x - \sqrt{5}) = \boxed{\sqrt{3}} \cdot x - \boxed{\sqrt{3}} \cdot \sqrt{5}$ Using the distributive law

$\qquad\qquad\quad = x\sqrt{3} - \sqrt{15}$ Multiplying radicals

2. $\boxed{\sqrt[3]{y}}\,(\sqrt[3]{y^2} + \sqrt[3]{2}) = \boxed{\sqrt[3]{y}} \cdot \sqrt[3]{y^2} + \boxed{\sqrt[3]{y}} \cdot \sqrt[3]{2}$ Using the distributive law

$\qquad\qquad\qquad = \sqrt[3]{y^3} + \sqrt[3]{2y}$ Multiplying radicals

$\qquad\qquad\qquad = y + \sqrt[3]{2y}$ Simplifying $\sqrt[3]{y^3}$

Example 3   Multiply.

$$\overset{\text{F}\qquad\qquad\text{O}\qquad\qquad\text{I}\qquad\quad\text{L}}{(4\sqrt{3} + \sqrt{2})(\sqrt{3} - 5\sqrt{2}) = 4(\sqrt{3})^2 - 20\sqrt{3} \cdot \sqrt{2} + \sqrt{2} \cdot \sqrt{3} - 5(\sqrt{2})^2}$$

$$= 4 \cdot 3 - 20\sqrt{6} + \sqrt{6} - 5 \cdot 2$$

$$= 12 - 20\sqrt{6} + \sqrt{6} - 10 = 2 - 19\sqrt{6}$$

Example 4 Multiply. Assume that all expressions under radicals represent positive numbers.

$$(\sqrt{a} + \sqrt{3})(\sqrt{b} + \sqrt{3}) = \sqrt{a}\sqrt{b} + \sqrt{a}\sqrt{3} + \sqrt{3}\sqrt{b} + \sqrt{3}\sqrt{3}$$
$$= \sqrt{ab} + \sqrt{3a} + \sqrt{3b} + 3$$

Example 5 Multiply.

$$(\sqrt{5} + \sqrt{7})(\sqrt{5} - \sqrt{7}) = (\sqrt{5})^2 - (\sqrt{7})^2 \quad \text{This is now a difference of two squares.}$$

$$= 5 - 7$$
$$= -2$$

Example 6 Multiply.

$$(\sqrt{3} + x)^2 = (\sqrt{3})^2 + 2x\sqrt{3} + x^2 \quad \text{Squaring a binomial}$$
$$= 3 + 2x\sqrt{3} + x^2$$

EXERCISE SET 6.6

Multiply.

1. $\sqrt{6}(2 - 3\sqrt{6})$

2. $\sqrt{3}(4 + \sqrt{3})$

3. $\sqrt{2}(\sqrt{3} - \sqrt{5})$

4. $\sqrt{5}(\sqrt{5} - \sqrt{2})$

5. $\sqrt{3}(2\sqrt{5} - 3\sqrt{4})$

6. $\sqrt{2}(3\sqrt{10} - 2\sqrt{2})$

7. $\sqrt[3]{2}(\sqrt[3]{4} - 2\sqrt[3]{32})$

8. $\sqrt[3]{3}(\sqrt[3]{9} - 4\sqrt[3]{21})$

9. $\sqrt[3]{a}(\sqrt[3]{2a^2} + \sqrt[3]{16a^2})$

10. $\sqrt[3]{x}(\sqrt[3]{3x^2} - \sqrt[3]{81x^2})$

11. $(\sqrt{3} - \sqrt{2})(\sqrt{3} + \sqrt{2})$

12. $(\sqrt{5} + \sqrt{6})(\sqrt{5} - \sqrt{6})$

13. $(\sqrt{8} + 2\sqrt{5})(\sqrt{8} - 2\sqrt{5})$

14. $(\sqrt{18} + 3\sqrt{7})(\sqrt{18} - 3\sqrt{7})$

For the following exercises assume that all expressions under radicals represent positive numbers.

15. $(\sqrt{a} + \sqrt{b})(\sqrt{a} - \sqrt{b})$

16. $(\sqrt{x} - \sqrt{y})(\sqrt{x} + \sqrt{y})$

17. $(3 - \sqrt{5})(2 + \sqrt{5})$

18. $(2 + \sqrt{6})(4 - \sqrt{6})$

19. $(\sqrt{3} + 1)(2\sqrt{3} + 1)$

20. $(4\sqrt{3} + 5)(\sqrt{3} - 2)$

21. $(2\sqrt{7} - 4\sqrt{2})(3\sqrt{7} + 6\sqrt{2})$

22. $(4\sqrt{5} + 3\sqrt{3})(3\sqrt{5} - 4\sqrt{3})$

23. $(\sqrt{a} + \sqrt{2})(\sqrt{a} + \sqrt{3})$

24. $(2 - \sqrt{x})(1 - \sqrt{x})$

25. $(2\sqrt[3]{3} + \sqrt[3]{2})(\sqrt[3]{3} - 2\sqrt[3]{2})$

26. $(3\sqrt[4]{7} + \sqrt[4]{6})(2\sqrt[3]{9} - 3\sqrt[4]{6})$

27. $(2 + \sqrt{3})^2$

28. $(\sqrt{5} + 1)^2$

29. a) Calculate $(3 + \sqrt{2})^2$ by first finding $\sqrt{2}$, then adding 3, and then squaring.

    b) Find $(3 + \sqrt{2})^2$ by expanding, to get $9 + 6\sqrt{2} + 2$, and then evaluating. Compare your answers.

Multiply and simplify.

30. $\sqrt{9 + 3\sqrt{5}}\sqrt{9 - 3\sqrt{5}}$ **31.** $(\sqrt{x + 2} - \sqrt{x - 2})^2$ **32.** $(\sqrt{3} + \sqrt{5} - \sqrt{6})^2$

33. $\sqrt[3]{y}(1 - \sqrt[3]{y})(1 + \sqrt[3]{y})$ **34.** $(\sqrt[3]{9} - 2)(\sqrt[3]{9} + 4)$ **35.** $[\sqrt{3} + \sqrt{2} + \sqrt{1}]^4$

36. $(\sqrt{y + 10} - \sqrt{10})(\sqrt{y + 10} + \sqrt{10})$ **37.** $(8\sqrt{a + 3} - a\sqrt{3})(2\sqrt{a + 3} + 5a\sqrt{3})$

38. $(8\sqrt{3} + 5\sqrt{2})(6\sqrt{7} - 2\sqrt{5})$ **39.** $(\sqrt{17} + \sqrt{2} - \sqrt{19})(\sqrt{17} + \sqrt{2} + \sqrt{19})$

40. Evaluate: $\sqrt{7 + 4\sqrt{3}} - \sqrt{7 - 4\sqrt{3}}$.

6.7 RATIONALIZING NUMERATORS OR DENOMINATORS

RATIONALIZING DENOMINATORS

Let us consider several ways of finding $\sqrt{\dfrac{1}{2}}$.

1. $\sqrt{\dfrac{1}{2}} = \dfrac{\sqrt{1}}{\sqrt{2}} = \dfrac{1}{\sqrt{2}}$

We find an approximation to $\sqrt{2}$, in a table or by computing, to be 1.414. Then

$$\dfrac{1}{\sqrt{2}} \approx \dfrac{1}{1.414}.$$

By long division we can then find decimal notation for the answer. The division is rather lengthy.

2. $\sqrt{\dfrac{1}{2}} = \sqrt{\dfrac{1}{2} \cdot \dfrac{2}{2}}$ Multiplying by $\dfrac{2}{2}$

$$= \sqrt{\dfrac{2}{4}} = \dfrac{\sqrt{2}}{\sqrt{4}} = \dfrac{\sqrt{2}}{2} \longleftarrow \boxed{\dfrac{\sqrt{2}}{2} \text{ is often written } \dfrac{1}{2}\sqrt{2}}$$

We find an approximation to $\sqrt{2}$ and then compute.

$$\dfrac{\sqrt{2}}{2} \approx \dfrac{1.414}{2} = 0.707$$

This time the division is easy.

3. $\sqrt{\dfrac{1}{2}} = \sqrt{0.5}$ Finding decimal notation for $\dfrac{1}{2}$

We can look up the square root of 0.5 in a table or compute it with a calculator. Before the days of calculators Method (3) was not very practical, so we had a choice between Methods (1) and (2). Method (2) is much easier, hence it was preferred. The procedure used in Method (2) involves finding a denominator that has no radicals in it. That procedure is called *rationalizing the denominator*. Today, rationalizing the denominator is not as important as it once was because we have calculators. However, the procedure is still important in some other ways. It is more or less standard to write radical expressions with the denominator rationalized.

Example 1   Rationalize the denominator: $\sqrt[3]{\dfrac{7}{9}}$.

$$\sqrt[3]{\frac{7}{9}} = \sqrt[3]{\frac{7}{3\cdot 3}\cdot\frac{3}{3}}$$ Multiplying by $\dfrac{3}{3}$ to make the denominator a perfect cube

$$= \sqrt[3]{\frac{21}{3\cdot 3\cdot 3}} = \frac{\sqrt[3]{21}}{\sqrt[3]{3^3}} = \frac{\sqrt[3]{21}}{3}$$

The idea in rationalizing a denominator is to multiply by 1 in order to make the denominator a perfect power. In Example 1 we multiplied by 1 under the radical. We can also multiply by 1 outside the radical, as in Example 2.

Example 2   Rationalize the denominator: $\sqrt{\dfrac{2a}{5b}}$. Assume that all expressions under radicals represent positive numbers.

$$\sqrt{\frac{2a}{5b}} = \frac{\sqrt{2a}}{\sqrt{5b}}$$ Converting to a quotient of radicals

$$= \frac{\sqrt{2a}}{\sqrt{5b}}\cdot\frac{\sqrt{5b}}{\sqrt{5b}}$$ Multiplying by 1

$$= \frac{\sqrt{10ab}}{\sqrt{25b^2}}$$ The radicand in the denominator is a perfect square.

$$= \frac{\sqrt{10ab}}{5b}$$ Taking the square root of $25b^2$

Example 3   Rationalize the denominator: $\dfrac{\sqrt[3]{a}}{\sqrt[3]{9x}}$.

$$\frac{\sqrt[3]{a}}{\sqrt[3]{9x}} = \frac{\sqrt[3]{a}}{\sqrt[3]{9x}} \cdot \frac{\sqrt[3]{3x^2}}{\sqrt[3]{3x^2}} \qquad \text{Multiplying by 1}$$

To choose the symbol for 1, we look at the radicand $9x$. This is $3 \cdot 3 \cdot x$. To make it a cube we need another 3 and two more x's. Thus we multiply by $\sqrt[3]{3x^2}/\sqrt[3]{3x^2}$. We have

$$\frac{\sqrt[3]{3ax^2}}{\sqrt[3]{27x^3}}, \quad \text{which simplifies to} \quad \frac{\sqrt[3]{3ax^2}}{3x}.$$

RATIONALIZING NUMERATORS

Sometimes in advanced mathematics it is necessary to rationalize a numerator. We use a similar procedure.

Example 4   Rationalize the numerator: $\dfrac{\sqrt{7}}{\sqrt{5}}$.

$$\frac{\sqrt{7}}{\sqrt{5}} = \frac{\sqrt{7}}{\sqrt{5}} \cdot \frac{\sqrt{7}}{\sqrt{7}} \qquad \text{Multiplying by 1}$$

$$= \frac{\sqrt{49}}{\sqrt{35}} \qquad \text{The radicand in the numerator is a perfect square.}$$

$$= \frac{7}{\sqrt{35}}$$

Example 5   Rationalize the numerator: $\dfrac{\sqrt[3]{5}}{\sqrt[3]{3}}$.

$$\frac{\sqrt[3]{5}}{\sqrt[3]{3}} = \frac{\sqrt[3]{5}}{\sqrt[3]{3}} \cdot \frac{\sqrt[3]{5^2}}{\sqrt[3]{5^2}} \qquad \text{Multiplying by 1}$$

$$= \frac{\sqrt[3]{5^3}}{\sqrt[3]{3 \cdot 5^2}}$$

$$= \frac{\sqrt[3]{5^3}}{\sqrt[3]{75}}$$

$$= \frac{5}{\sqrt[3]{75}}$$

RATIONALIZING WHEN THERE ARE TWO TERMS

When the denominator to be rationalized has two terms, choose a symbol for 1 as illustrated below.

Example 6   Rationalize the denominator: $\dfrac{4 + \sqrt{2}}{\sqrt{5} - \sqrt{2}}$.

$$\frac{4 + \sqrt{2}}{\sqrt{5} - \sqrt{2}} = \frac{4 + \sqrt{2}}{\sqrt{5} - \sqrt{2}} \cdot \frac{\sqrt{5} + \sqrt{2}}{\sqrt{5} + \sqrt{2}} \qquad \text{Multiplying by 1}$$

$$= \frac{(4 + \sqrt{2})(\sqrt{5} + \sqrt{2})}{(\sqrt{5} - \sqrt{2})(\sqrt{5} + \sqrt{2})} \qquad \begin{array}{l}\text{Multiplying numerators}\\\text{and denominators}\end{array}$$

$$= \frac{4\sqrt{5} + 4\sqrt{2} + \sqrt{2}\sqrt{5} + (\sqrt{2})^2}{(\sqrt{5})^2 - (\sqrt{2})^2} \qquad \text{Using FOIL}$$

$$= \frac{4\sqrt{5} + 4\sqrt{2} + \sqrt{10} + 2}{5 - 2} \qquad \begin{array}{l}\text{Squaring in the}\\\text{denominator}\end{array}$$

$$= \frac{4\sqrt{5} + 4\sqrt{2} + \sqrt{10} + 2}{3}$$

Note that the denominator in this example was $\sqrt{5} - \sqrt{2}$. We chose a symbol for 1 that had $\sqrt{5} + \sqrt{2}$ in the numerator and denominator. If the denominator had been $\sqrt{5} + \sqrt{2}$ we would have chosen $\dfrac{5 - \sqrt{2}}{5 - \sqrt{2}}$ for 1.

Examples   What symbol for 1 would you use to rationalize the denominator?

Expression	Symbol for 1		Expression	Symbol for 1
7. $\dfrac{3}{x + \sqrt{7}}$	$\dfrac{x - \sqrt{7}}{x - \sqrt{7}}$		8. $\dfrac{\sqrt{7} + 4}{3 - 2\sqrt{5}}$	$\dfrac{3 + 2\sqrt{5}}{3 + 2\sqrt{5}}$

Example 9   Rationalize the denominator: $\dfrac{4}{\sqrt{3} + x}$.

$$\frac{4}{\sqrt{3} + x} = \frac{4}{\sqrt{3} + x} \cdot \frac{\sqrt{3} - x}{\sqrt{3} - x}$$

$$= \frac{4(\sqrt{3} - x)}{(\sqrt{3} + x)(\sqrt{3} - x)} = \frac{4(\sqrt{3} - x)}{(\sqrt{3})^2 - x^2} = \frac{4(\sqrt{3} - x)}{3 - x^2}$$

We can also rationalize numerators. Of course we choose our symbol for 1 a little differently.

Examples What symbol for 1 would you use to rationalize the numerator?

	Expression	Symbol for 1

10. $\dfrac{\sqrt{5} + \sqrt{2}}{1 - \sqrt{3}}$ $\dfrac{\sqrt{5} - \sqrt{2}}{\sqrt{5} - \sqrt{2}}$

11. $\dfrac{x - \sqrt{7}}{3}$ $\dfrac{x + \sqrt{7}}{x + \sqrt{7}}$

12. $\dfrac{3 - 2\sqrt{5}}{\sqrt{7} + 4}$ $\dfrac{3 + 2\sqrt{5}}{3 + 2\sqrt{5}}$

Example 13 Rationalize the numerator: $\dfrac{4 + \sqrt{2}}{\sqrt{5} - \sqrt{2}}$.

$$\frac{4 + \sqrt{2}}{\sqrt{5} - \sqrt{2}} = \frac{4 + \sqrt{2}}{\sqrt{5} - \sqrt{2}} \cdot \frac{4 - \sqrt{2}}{4 - \sqrt{2}}$$

$$= \frac{16 - (\sqrt{2})^2}{4\sqrt{5} - \sqrt{5}\sqrt{2} - 4\sqrt{2} + (\sqrt{2})^2}$$

$$= \frac{14}{4\sqrt{5} - \sqrt{10} - 4\sqrt{2} + 2}$$

EXERCISE SET 6.7

Rationalize the denominator.

1. $\sqrt{\dfrac{6}{5}}$ **2.** $\sqrt{\dfrac{11}{6}}$ **3.** $\sqrt{\dfrac{10}{7}}$ **4.** $\sqrt{\dfrac{22}{3}}$

5. $\dfrac{6\sqrt{5}}{5\sqrt{3}}$ **6.** $\dfrac{2\sqrt{3}}{5\sqrt{2}}$ **7.** $\sqrt[3]{\dfrac{16}{9}}$ **8.** $\sqrt[3]{\dfrac{3}{9}}$

9. $\dfrac{\sqrt[3]{3a}}{\sqrt[3]{5c}}$ **10.** $\dfrac{\sqrt[3]{7x}}{\sqrt[3]{3y}}$ **11.** $\dfrac{\sqrt[3]{2y^4}}{\sqrt[3]{6x^4}}$ **12.** $\dfrac{\sqrt[3]{3a^4}}{\sqrt[3]{7b^2}}$

13. $\dfrac{1}{\sqrt[3]{xy}}$ **14.** $\dfrac{1}{\sqrt[3]{ab}}$

Rationalize the numerator.

15. $\dfrac{\sqrt{7}}{\sqrt{3x}}$ **16.** $\dfrac{\sqrt{6}}{\sqrt{5x}}$ **17.** $\sqrt{\dfrac{14}{21}}$ **18.** $\sqrt{\dfrac{12}{15}}$ **19.** $\dfrac{4\sqrt{13}}{3\sqrt{7}}$

20. $\dfrac{5\sqrt{21}}{2\sqrt{6}}$ **21.** $\dfrac{\sqrt[3]{7}}{\sqrt[3]{2}}$ **22.** $\dfrac{\sqrt[3]{5}}{\sqrt[3]{4}}$ **23.** $\sqrt{\dfrac{7x}{3y}}$ **24.** $\sqrt{\dfrac{6a}{2b}}$

25. $\dfrac{\sqrt[3]{5y^4}}{\sqrt[3]{6x^5}}$ **26.** $\dfrac{\sqrt[3]{3a^5}}{\sqrt[3]{7b^2}}$ **27.** $\dfrac{\sqrt{ab}}{3}$ **28.** $\dfrac{\sqrt{xy}}{5}$

Rationalize the denominator. Assume that all expressions under radicals represent positive numbers.

29. $\dfrac{5}{8 - \sqrt{6}}$ **30.** $\dfrac{7}{9 + \sqrt{10}}$ **31.** $\dfrac{-4\sqrt{7}}{\sqrt{5} - \sqrt{3}}$ **32.** $\dfrac{-3\sqrt{2}}{\sqrt{3} - \sqrt{5}}$

33. $\dfrac{\sqrt{5} - 2\sqrt{6}}{\sqrt{3} - 4\sqrt{5}}$ **34.** $\dfrac{\sqrt{6} - 3\sqrt{5}}{\sqrt{3} - 2\sqrt{7}}$ **35.** $\dfrac{\sqrt{x} - \sqrt{y}}{\sqrt{x} + \sqrt{y}}$ **36.** $\dfrac{\sqrt{a} + \sqrt{b}}{\sqrt{a} - \sqrt{b}}$

37. $\dfrac{5\sqrt{3} - 3\sqrt{2}}{3\sqrt{2} - 2\sqrt{3}}$ **38.** $\dfrac{7\sqrt{2} + 4\sqrt{3}}{4\sqrt{3} - 3\sqrt{2}}$

Rationalize the numerator. Assume that all expressions under radicals represent positive numbers.

39. $\dfrac{\sqrt{3} + 5}{8}$ **40.** $\dfrac{3 - \sqrt{2}}{5}$ **41.** $\dfrac{\sqrt{3} - 5}{\sqrt{2} + 5}$ **42.** $\dfrac{\sqrt{6} - 3}{\sqrt{3} + 7}$

43. $\dfrac{\sqrt{x} - \sqrt{y}}{\sqrt{x} + \sqrt{y}}$ **44.** $\dfrac{\sqrt{x} + \sqrt{y}}{\sqrt{x} - \sqrt{y}}$ **45.** $\dfrac{4\sqrt{6} - 5\sqrt{3}}{2\sqrt{3} + 7\sqrt{6}}$ **46.** $\dfrac{8\sqrt{2} + 5\sqrt{3}}{5\sqrt{3} - 7\sqrt{2}}$

☆ _____

Pythagorean formula in three dimensions. The length d of a diagonal of a rectangular box is given by $d = \sqrt{a^2 + b^2 + c^2}$, where a, b, and c are the lengths of the sides.

47. ▦ Find the length of a diagonal of a box whose sides have lengths 2, 4, and 5.

48. ▦ Will a baton 57 cm long fit inside an attache case that is 12.5 cm by 45 cm by 31.5 cm?

Rationalize the denominator.

49. $\dfrac{\sqrt{5} + \sqrt{10} - \sqrt{6}}{\sqrt{50}}$ **50.** $\dfrac{3\sqrt{y} + 4\sqrt{yz}}{5\sqrt{y} - 2\sqrt{z}}$

51. $\dfrac{b + \sqrt{b}}{1 + b + \sqrt{b}}$

52. $\dfrac{12}{3 - \sqrt{3 - y}}$

53. $\dfrac{36a^2 b}{\sqrt[3]{6a^2 b}}$

Rationalize the numerator.

54. $\dfrac{\sqrt{y + 18} - \sqrt{y}}{18}$

55. $\dfrac{\sqrt{x + 6} - 5}{\sqrt{x + 6} + 5}$

Simplify. (*Hint:* Rationalize the denominator.)

56. $\sqrt{a^2 - 3} - \dfrac{a^2}{\sqrt{a^2 - 3}}$

57. $5 \sqrt{\dfrac{x}{y}} + 4 \sqrt{\dfrac{y}{x}} - \dfrac{3}{\sqrt{xy}}$

58. $\dfrac{\dfrac{1}{\sqrt{w}} - \sqrt{w}}{\dfrac{\sqrt{w} + 1}{\sqrt{w}}}$

59. $\dfrac{1}{4 + \sqrt{3}} + \dfrac{1}{\sqrt{3}} + \dfrac{1}{\sqrt{3} - 4}$

6.8 RATIONAL NUMBERS AS EXPONENTS

FRACTIONAL EXPONENTS

Expressions like $a^{1/2}$, $5^{-1/4}$, and $(2y)^{4/5}$ have not yet been defined. We shall define such expressions in such a way that the usual properties of exponents hold.

Consider $a^{1/2} \cdot a^{1/2}$. If we still want to multiply by adding exponents it must follow that $a^{1/2} \cdot a^{1/2} = a^{1/2 + 1/2}$ or a^1. Thus we should define $a^{1/2}$ to be a square root of a. Similarly, $a^{1/3} \cdot a^{1/3} \cdot a^{1/3} = a^{1/3 + 1/3 + 1/3}$ or a^1, so $a^{1/3}$ should be defined to mean $\sqrt[3]{a}$.

> **For any nonnegative number a and any index n, $a^{1/n}$ means $\sqrt[n]{a}$ (the nonnegative nth root of a).**

Whenever we use fractional exponents we assume that the bases are nonnegative.

Examples   Rewrite without fractional exponents.

1. $x^{1/2} = \sqrt{x}$ **2.** $27^{1/3} = \sqrt[3]{27}$, or 3 **3.** $(abc)^{1/5} = \sqrt[5]{abc}$

Examples   Rewrite with fractional exponents.

4. $\sqrt[5]{7xy} = (7xy)^{1/5}$

5. $\sqrt[7]{\dfrac{x^3y}{9}} = \left(\dfrac{x^3y}{9}\right)^{1/7}$

How should we define $a^{2/3}$? If the usual properties of exponents are to hold, we have $a^{2/3} = (a^{1/3})^2$, or $(\sqrt[3]{a})^2$, or $\sqrt[3]{a^2}$. We make our definition accordingly.

> **For any natural numbers m and n ($n \ne 1$) and any nonnegative number a,**
>
> $$a^{m/n} \quad \text{means} \quad \sqrt[n]{a^m}, \text{ or } (\sqrt[n]{a})^m.$$

Examples   Rewrite without fractional exponents.

6. $(27)^{2/3} = \sqrt[3]{27^2}$, or $(\sqrt[3]{27})^2$
           $= 3^2$, or 9

7. $4^{3/2} = \sqrt[2]{4^3}$, or $(\sqrt[2]{4})^3$
           $= 2^3$, or 8.

Examples   Rewrite with fractional exponents.

8. $\sqrt[3]{9^4} = (\sqrt[3]{9})^4 = 9^{4/3}$

9. $(\sqrt[4]{7xy})^5 = (7xy)^{5/4}$

NEGATIVE FRACTIONAL EXPONENTS

Negative fractional exponents have a meaning similar to that of negative integer exponents. Changing the sign of an exponent amounts to taking a reciprocal, as follows.

> **For any rational number m/n and any positive real number a,**
>
> $$a^{-m/n} \quad \text{means} \quad \frac{1}{a^{m/n}},$$
>
> **that is, $a^{m/n}$ and $a^{-m/n}$ are reciprocals.**

Examples   Rewrite with positive exponents.

10. $4^{-1/2} = \dfrac{1}{4^{1/2}}$ $4^{-1/2}$ is the reciprocal of $4^{1/2}$

Since $4^{1/2} = \sqrt{4} = 2$, the answer simplifies to $\frac{1}{2}$.

11. $(5xy)^{-4/5} = \dfrac{1}{(5xy)^{4/5}}$ $(5xy)^{-4/5}$ is the reciprocal of $(5xy)^{4/5}$

> Don't make the mistake of thinking that a negative exponent means that the entire expression is negative.

LAWS OF EXPONENTS

The same laws hold for rational number exponents as for integer exponents. We list them for review.

> **For any real number a and any rational exponents m and n:**
>
> **1.** $a^m \cdot a^n = a^{m+n}$ **In multiplying, we can add exponents if the bases are the same.**
>
> **2.** $\dfrac{a^m}{a^n} = a^{m-n}$ **In dividing, we can subtract exponents if the bases are the same.**
>
> **3.** $(a^m)^n = a^{m \cdot n}$ **To raise a power to a power, we can multiply the exponents.**

Examples   Use the laws of exponents to simplify.

12. $3^{1/5} \cdot 3^{3/5} = 3^{1/5+3/5} = 3^{4/5}$ Adding exponents

13. $5^{2/3} \cdot 5^{1/4} = 5^{2/3+1/4} = 5^{8/12+3/12} = 5^{11/12}$ Adding exponents

14. $\dfrac{7^{1/4}}{7^{1/2}} = 7^{1/4-1/2} = 7^{1/4-2/4} = 7^{-1/4}$ Subtracting exponents

15. $(7.2^{2/3})^{3/4} = 7.2^{2/3 \cdot 3/4} = 7.2^{6/12}$ Multiplying exponents
$$= 7.2^{1/2}$$

SIMPLIFYING RADICAL EXPRESSIONS

Fractional exponents can be used to simplify some radical expressions. The procedure is as follows.

1. **Convert radical expressions to exponential expressions.**
2. **Use arithmetic and the laws of exponents to simplify.**
3. **Convert back to radical notation when appropriate.**

Important: **This works only when all expressions under radicals are nonnegative since fractional exponents are not defined otherwise. No absolute value signs will be needed.**

Example 16 Use fractional exponents to simplify.

$$\sqrt[6]{x^3} = x^{3/6} \qquad \text{Converting to an exponential expression}$$
$$= x^{1/2} \qquad \text{Using arithmetic to simplify the exponent}$$
$$= \sqrt{x} \qquad \text{Converting back to radical notation}$$

Example 17 Use fractional exponents to simplify.

$$\sqrt[8]{a^2b^4} = (a^2b^4)^{1/8} \qquad \text{Converting to exponential notation}$$
$$= a^{2/8} \cdot b^{4/8} \qquad \text{Using the third law of exponents}$$
$$= a^{1/4} \cdot b^{2/4} \qquad \text{Using arithmetic to simplify the exponents}$$
$$= (ab^2)^{1/4} \qquad \text{Using the third law of exponents (in reverse)}$$
$$= \sqrt[4]{ab^2} \qquad \text{Converting back to radical notation}$$

We can use properties of fractional exponents to write a single radical expression for a product or quotient.

Examples Use fractional exponents to write a single radical expression.

18. $\sqrt[3]{5} \cdot \sqrt{2} = 5^{1/3} \cdot 2^{1/2} \qquad \text{Converting to exponential notation}$
$$= 5^{2/6} \cdot 2^{3/6} \qquad \text{Rewriting so that exponents have a common denominator}$$
$$= (5^2 \cdot 2^3)^{1/6} \qquad \text{Using the third law of exponents}$$
$$= \sqrt[6]{5^2 \cdot 2^3} \qquad \text{Converting back to radical notation}$$
$$= \sqrt[6]{200} \qquad \text{Multiplying under the radical}$$

19. $\dfrac{\sqrt[4]{(x+y)^3}}{\sqrt{x+y}} = \dfrac{(x+y)^{3/4}}{(x+y)^{1/2}}$ Converting to exponential notation

$\qquad\qquad = (x+y)^{3/4-1/2}$ Using the second law of exponents

$\qquad\qquad = (x+y)^{1/4}$ Subtracting exponents

$\qquad\qquad = \sqrt[4]{x+y}$ Converting back to radical notation

Example 20 Write a single radical expression.

$a^{1/2}b^{-1/2}c^{5/6} = a^{3/6}b^{-3/6}c^{5/6}$ Rewriting exponents with a common denominator

$\qquad\qquad = (a^3b^{-3}c^5)^{1/6}$ Using the third law of exponents

$\qquad\qquad = \sqrt[6]{a^3b^{-3}c^5}$ Converting to radical notation

We have now seen several different methods of simplifying radical expressions. We list them.

Some Ways to Simplify Radical Expressions:

1. *Simplifying by factoring.* **We factor the radicand, looking for factors that are perfect powers.**

   Example: $\sqrt[3]{16} = \sqrt[3]{8}\,\sqrt[3]{2}$

   $\qquad\qquad\quad = 2\sqrt[3]{2}$

2. *Rationalizing denominators.* **Radical expressions are usually considered simpler if there are no radicals in the denominator.**

   Example: $\dfrac{1}{\sqrt{2}} = \dfrac{\sqrt{2}}{2}$

3. *Collecting like radical terms.*

   Example: $\sqrt{8} + 3\sqrt{2} = \sqrt{4}\cdot\sqrt{2} + 3\sqrt{2}$

   $\qquad\qquad\qquad\quad = 2\sqrt{2} + 3\sqrt{2}$

   $\qquad\qquad\qquad\quad = 5\sqrt{2}$

4. *Using fractional exponents to simplify.* **We convert to exponential notation and then use arithmetic and the laws of exponents to simplify the exponents. Then we convert back to radical notation.** *Caution:* **This works only when there are only nonnegative expressions under radical signs.**

EXERCISE SET 6.8

Rewrite without fractional exponents.

1. $x^{1/4}$ **2.** $y^{1/5}$ **3.** $(8)^{1/3}$ **4.** $(16)^{1/2}$

5. $(a^2b^2)^{1/5}$ **6.** $(x^3y^3)^{1/4}$ **7.** $16^{3/4}$ **8.** $4^{7/2}$

Rewrite with fractional exponents.

9. $\sqrt[3]{20}$ **10.** $\sqrt[3]{19}$ **11.** $\sqrt[5]{xy^2z}$ **12.** $\sqrt[7]{x^3y^2z^2}$

13. $(\sqrt{3mn})^3$ **14.** $(\sqrt[3]{7xy})^4$ **15.** $(\sqrt[7]{8x^2y})^5$ **16.** $(\sqrt[6]{2a^5b})^7$

Rewrite with positive exponents.

17. $x^{-1/3}$ **18.** $y^{-1/4}$ **19.** $\dfrac{1}{x^{-2/3}}$ **20.** $\dfrac{1}{x^{-5/6}}$

Use the laws of exponents to simplify.

21. $5^{3/4} \cdot 5^{1/8}$ **22.** $11^{2/3} \cdot 11^{1/2}$ **23.** $\dfrac{7^{5/8}}{7^{3/8}}$ **24.** $\dfrac{9^{9/11}}{9^{7/11}}$

25. $\dfrac{8.3^{3/4}}{8.3^{2/5}}$ **26.** $\dfrac{3.9^{3/5}}{3.9^{1/4}}$ **27.** $(10^{3/8})^{2/5}$ **28.** $(5^{5/9})^{3/7}$

Use fractional exponents to simplify.

29. $\sqrt[6]{a^4}$ **30.** $\sqrt[6]{y^2}$ **31.** $\sqrt[3]{8y^6}$ **32.** $\sqrt{x^4y^6}$

33. $\sqrt[4]{32}$ **34.** $\sqrt[8]{81}$ **35.** $\sqrt[6]{4x^2}$ **36.** $\sqrt[4]{16x^4y^2}$

37. $\sqrt[5]{32c^{10}d^{15}}$ **38.** $\sqrt[4]{16x^{12}y^{16}}$ **39.** $\sqrt[6]{\dfrac{m^{12}n^{24}}{64}}$ **40.** $\sqrt[5]{\dfrac{x^{15}y^{20}}{32}}$

41. $\sqrt[8]{r^4s^2}$ **42.** $\sqrt[12]{64t^6s^6}$ **43.** $\sqrt[3]{27a^3b^9}$ **44.** $\sqrt[4]{81x^8y^8}$

Use fractional exponents to write a single radical expression.

45. $\sqrt[3]{5} \cdot \sqrt{2}$ **46.** $\sqrt[3]{7} \cdot \sqrt[4]{5}$ **47.** $\sqrt{x}\sqrt[3]{2x}$ **48.** $\sqrt[3]{y}\sqrt[5]{3y}$

49. $\sqrt{x}\sqrt[3]{x-2}$ **50.** $\sqrt[4]{3x}\sqrt{y+4}$ **51.** $\dfrac{\sqrt[3]{(a+b)^2}}{\sqrt{(a+b)}}$ **52.** $\dfrac{\sqrt[3]{(x+y)^2}}{\sqrt[4]{(x+y)^3}}$

53. $a^{2/3} \cdot b^{3/4}$ **54.** $x^{1/3} \cdot y^{1/4} \cdot z^{1/6}$ **55.** $\dfrac{s^{7/12} \cdot t^{5/6}}{s^{1/3} \cdot t^{-1/6}}$ **56.** $\dfrac{x^{8/15} \cdot y^{4/5}}{x^{1/3} \cdot y^{-1/5}}$

☆

Use fractional exponents to write a single radical expression and simplify.

57. $\sqrt[5]{yx^2}\sqrt{xy}$

58. $\sqrt{x^5\sqrt[3]{x^4}}$

59. $\dfrac{\sqrt{(a+b)^3}\sqrt[3]{(a+b)^2}}{\sqrt[4]{a+b}}$

60. $\sqrt[4]{\sqrt[3]{8x^3y^6}}$

Simplify.

61. $(-\sqrt[4]{7}\sqrt[3]{w})^{12}$

62. $\sqrt[12]{p^2+2pq+q^2}$

63. $\dfrac{1}{\sqrt[3]{3}-\sqrt[3]{2}}$

64. $[\sqrt[10]{\sqrt[5]{x^{15}}}]^5[\sqrt[5]{\sqrt[10]{x^{15}}}]^5$

65. $\sqrt[p]{x^{5p}y^{7p+1}z^{p+3}}$

6.9 SOLVING RADICAL EQUATIONS

THE PRINCIPLE OF POWERS

These are radical equations:

$$\sqrt[3]{2x}+1=5, \qquad \sqrt{x}+\sqrt{4x-2}=7.$$

A radical equation has variables in one or more radicands. To solve such equations we need a new principle. Suppose an equation $a = b$ is true. If we square both sides, we get another true equation: $a^2 = b^2$. This can be generalized.

> *The Principle of Powers:* **If an equation $a = b$ is true, then $a^n = b^n$ is true for any natural number n.**

If an equation $a^n = b^n$ is true, it *may not* be true that $a = b$. For example, $3^2 = (-3)^2$ is true, but $3 = -3$ is not true.

Example 1 Solve: $\sqrt{x} - 3 = 4$.

$$\sqrt{x} = 7 \qquad \text{Adding to isolate the radical}$$

$$x = 7^2, \text{ or } 49 \qquad \text{Using the principle of powers}$$

Check: $\sqrt{x} - 3 = 4$

$$\begin{array}{c|c} \sqrt{49} - 3 & 4 \\ 7 - 3 & \\ 4 & \end{array}$$

The solution is 49.

The principle of powers does not always produce equivalent equations.

Example 2   Solve: $\sqrt{x} = -3$.

We might observe at the outset that this equation has no solution because the principal square root of a number is never negative. Let us continue as above, for comparison.

$x = (-3)^2$, or 9   Principle of powers (squaring)

Check: $\sqrt{x} = -3$

$$\begin{array}{c|c} \sqrt{9} & -3 \\ 3 & \end{array}$$

The number 9 does not check. Hence the equation has no solution.

> **In solving radical equations, possible solutions found using the principle of powers *must* be checked!**

To solve an equation with a radical term, we first isolate the radical term on one side of the equation. Then we use the principle of powers.

Example 3   Solve: $x = \sqrt{x + 7} + 5$.

$x - 5 = \sqrt{x + 7}$ Adding -5 to isolate the radical term

$(x - 5)^2 = (\sqrt{x + 7})^2$ Principle of powers: squaring both sides

$x^2 - 10x + 25 = x + 7$

$x^2 - 11x + 18 = 0$

$(x - 9)(x - 2) = 0$ Factoring

$x = 9$ or $x = 2$ Using the principle of zero products

The possible solutions are 9 and 2. Let us check.

For 9:
$$\begin{array}{c|c} & x = \sqrt{x+7} + 5 \\ \hline 9 & \sqrt{9+7} + 5 \\ & 9 \end{array}$$

For 2:
$$\begin{array}{c|c} & x = \sqrt{x+7} + 5 \\ \hline 2 & \sqrt{2+7} + 5 \\ & 8 \end{array}$$

Since 9 checks but 2 does not, the solution is 9.

Example 4   Solve: $\sqrt[3]{2x+1} + 5 = 0$.

$$\sqrt[3]{2x+1} = -5 \qquad \text{Adding } -5 \text{; this isolates the radical term}$$
$$2x + 1 = (-5)^3, \text{ or } -125 \qquad \text{Principle of powers; raising to the third power}$$
$$2x = -126 \qquad \text{Adding } -1$$
$$x = -63$$

Check:
$$\begin{array}{c|c} \sqrt[3]{2x+1} + 5 & = 0 \\ \hline \sqrt[3]{2 \cdot (-63) + 1} + 5 & 0 \\ \sqrt[3]{-125} + 5 & \\ -5 + 5 & \\ 0 & \end{array}$$

The solution is -63.

EQUATIONS WITH TWO RADICAL TERMS

A general strategy for solving equations with two radical terms is as follows.

> 1. **Isolate one of the radical terms.**
> 2. **Use the principle of powers.**
> 3. **If a radical remains, perform steps 1 and 2 again.**
> 4. **Check possible solutions.**

Example 5 Solve: $\sqrt{x-3} + \sqrt{x+5} = 4$.

$$\sqrt{x-3} = 4 - \sqrt{x+5} \qquad \text{Adding } -\sqrt{x+5}; \text{ this isolates}$$
$$\text{one of the radical terms}$$

$$(\sqrt{x-3})^2 = (4 - \sqrt{x+5})^2 \qquad \text{Principle of powers;}$$
$$\text{squaring both sides}$$

Here we are squaring a binomial. We square 4, then find twice the product of 4 and $\sqrt{x+5}$, and then the square of $\sqrt{x+5}$. (See the rule on p. 148.)

$$x - 3 = 16 - 8\sqrt{x+5} + (x+5)$$
$$-24 = -8\sqrt{x+5} \qquad \text{Isolating the remaining radical term}$$
$$3 = \sqrt{x+5}$$
$$3^2 = x + 5 \qquad \text{Squaring}$$
$$4 = x$$

The number 4 checks, so it is the solution.

A common error in solving equations like $\sqrt{x-3} + \sqrt{x+5} = 4$ is to square the left side, obtaining $(x-3) + (x+5)$. That is wrong because the square of a sum is *not* the sum of the squares.
Example: $\sqrt{9} + \sqrt{16} = 7$, but $9 + 16 = 25$, not 7^2.

Example 6 Solve: $\sqrt{2x-5} = 1 + \sqrt{x-3}$.

$$(\sqrt{2x-5})^2 = (1 + \sqrt{x-3})^2 \qquad \text{One radical is already}$$
$$\text{isolated; we square both sides}$$

$$2x - 5 = 1 + 2\sqrt{x-3} + (x-3)$$
$$x - 3 = 2\sqrt{x-3} \qquad \text{Isolating the remaining}$$
$$\text{radical term}$$

$$(x-3)^2 = (2\sqrt{x-3})^2 \qquad \text{Squaring both sides}$$
$$x^2 - 6x + 9 = 4(x-3)$$
$$x^2 - 6x + 9 = 4x - 12$$
$$x^2 - 10x + 21 = 0$$
$$(x-7)(x-3) = 0 \qquad \text{Factoring}$$
$$x = 7 \quad \text{or} \quad x = 3 \qquad \text{Using the principle}$$
$$\text{of zero products}$$

The numbers 7 and 3 check, so they are the solutions.

EXERCISE SET 6.9

Solve.

1. $\sqrt{y + 1} - 5 = 8$

2. $\sqrt{x - 2} - 7 = -4$

3. $\sqrt{3y + 1} = 9$

4. $\sqrt{2y + 1} = 13$

5. $\sqrt[3]{x} = -3$

6. $\sqrt[3]{y} = -4$

7. $\sqrt{x + 2} = -4$

8. $\sqrt{y - 3} = -2$

9. $\sqrt[3]{6x + 9} + 8 = 5$

10. $\sqrt[3]{3y + 6} + 2 = 3$

11. $\sqrt{3y + 1} = \sqrt{2y + 6}$

12. $\sqrt{5x - 3} = \sqrt{2x + 3}$

13. $\sqrt{y - 5} + \sqrt{y} = 5$

14. $\sqrt{x - 9} + \sqrt{x} = 1$

15. $3 + \sqrt{z - 6} = \sqrt{z + 9}$

16. $\sqrt{4x - 3} = 2 + \sqrt{2x - 5}$

17. $\sqrt{20 - x} + 8 = \sqrt{9 - x} + 11$

18. $4 + \sqrt{10 - x} = 6 + \sqrt{4 - x}$

19. $\sqrt{4y + 1} - \sqrt{y - 2} = 3$

20. $\sqrt{y + 15} - \sqrt{2y + 7} = 1$

☆

21. Solve: $\dfrac{x + \sqrt{x + 1}}{x - \sqrt{x + 1}} = \dfrac{5}{11}$.

22. Refer to Exercise 31 on p. 257. Calculate the length of a pendulum that will beat seconds (having a period of two seconds). Make a pendulum like this and check it with a stopwatch.

Solve.

23. $\sqrt[3]{\dfrac{z}{4}} - 10 = 2$

24. $\sqrt[4]{z^2 + 17} = 3$

25. $\sqrt{\sqrt{y} + 49} - \sqrt{y} = \sqrt{7}$

26. $\sqrt[3]{x^2 + x + 15} - 3 = 0$

27. $x^2 - 5x - \sqrt{x^2 - 5x - 2} = 4$

28. $\sqrt{8 - b} = b\sqrt{8 - b}$

29. $\sqrt{x - 2} - \sqrt{x + 2} + 2 = 0$

30. $6\sqrt{y} + 6y^{-1/2} = 37$

31. $\sqrt{a^2 + 30a} = a + \sqrt{5a}$

32. $\sqrt{\sqrt{x} + 4} = \sqrt{x} - 2$

33. $\dfrac{x - 1}{\sqrt{x^2 + 3x + 6}} = \dfrac{1}{4}$

34. $\sqrt{x + 1} - \dfrac{2}{\sqrt{x + 1}} = 1$

35. $\sqrt{11x + \sqrt{6 + x}} = 6$

36. $\sqrt{2x - 2} + \sqrt{7x + 4} = \sqrt{13x + 10}$

6.10 THE COMPLEX NUMBERS

IMAGINARY AND COMPLEX NUMBERS

Negative numbers do not have square roots in the system of real numbers. *Imaginary** numbers were invented so that negative numbers would have square roots. We begin by making up a number that is a square root of -1. We call this new number i. Thus $i^2 = -1$, or $i = \sqrt{-1}$. All other imaginary numbers can be expressed as a product of i and a real number.

Examples   Express in terms of i.

1. $\sqrt{-5} = \sqrt{-1 \cdot 5} = \sqrt{-1} \cdot \sqrt{5} = i\sqrt{5}$ or $\sqrt{5}i$

2. $-\sqrt{-49} = -\sqrt{-1 \cdot 49} = -\sqrt{-1} \cdot \sqrt{49} = -i\sqrt{49} = -7i$

> This procedure works, even though the general rule for factoring radicals does not hold for negative radicands.

> **An imaginary number is a number that can be named bi, where b is some real number.**

To form the system of *complex numbers* we take the imaginary numbers and the real numbers, and all possible sums of real and imaginary numbers. These are complex numbers:

$$3 - 4i, \qquad -\sqrt{46} + 2i, \qquad 14, \qquad i\sqrt{2}.$$

> **A complex number is any number that can be named $a + bi$, where a and b are any real numbers. (Note that a or b or both can be 0.)**

ADDITION AND SUBTRACTION

The complex numbers follow the commutative and associative laws of addition. Thus we can add and subtract them as we do binomials in real numbers.

---

* Don't let the name *imaginary* fool you. The imaginary numbers are very important in such fields as engineering and the physical sciences.

Examples   Add or subtract.

3. $(8 + 6i) + (3 + 2i) = (8 + 3) + (6 + 2)i$ Collecting the real
parts and the
imaginary parts

$$= 11 + 8i$$

4. $(3 + 2i) - (5 - 2i) = (3 - 5) + [2 - (-2)]i$
$$= -2 + 4i$$

MULTIPLYING

The complex numbers also obey the commutative and associative
laws of multiplication and the distributive laws. Thus we multiply
them in much the same way as we do binomials in real numbers. We
must remember, however, that $i^2 = -1$.

Examples   Multiply.

5. $(-4i)(3 - 5i) = (-4i) \cdot 3 + (-4i)(-5i)$ Using a
distributive law

$$= -12i + 20i^2$$
$$= -12i - 20 \qquad i^2 = -1$$
$$= -20 - 12i$$

6. $(1 + 2i)(1 + 3i) = 1 + 3i + 2i + 6i^2$ Multiplying each term
of one number by every
term of the other FOIL

$$= 1 + 3i + 2i - 6 \qquad i^2 = -1$$
$$= -5 + 5i \qquad \text{Collecting like terms}$$

Examples   Simplify.

7. $i^3 = (i^2)i = -1 \cdot i$
$$= -i$$

8. $i^4 = (i^2)(i^2) = (-1) \cdot (-1)$
$$= 1$$

9. $(3 - 2i)^2 = 3^2 - 2 \cdot (3)(2i) + (2i)^2$
$$= 9 - 12i + 4i^2$$
$$= 9 - 12i - 4$$
$$= 5 - 12i$$

Caution! The property $\sqrt{a}\sqrt{b} = \sqrt{ab}$ does not hold for imaginary numbers. All square roots of negative numbers must be expressed in terms of i before multiplying. For example,

$$\sqrt{-2} \cdot \sqrt{-5} = i\sqrt{2} \cdot i\sqrt{5} = i^2\sqrt{10} = -\sqrt{10} \text{ is correct.}$$

But

$$\sqrt{-2} \cdot \sqrt{-5} = \sqrt{(-2)(-5)} = \sqrt{10} \text{ is wrong!}$$

CONJUGATES AND DIVISION

Conjugates of complex numbers are defined as follows.

> The *conjugate* **of a complex number** $a + bi$ **is** $a - bi$, **and the** *conjugate* **of** $a - bi$ **is** $a + bi$.

Examples   Find the conjugate.

10. $5 + 7i$ The conjugate is $5 - 7i$.

11. $14 - 3i$ The conjugate is $14 + 3i$.

12. $-3 - 9i$ The conjugate is $-3 + 9i$.

13. $4i$ The conjugate is $-4i$.

When we multiply a number by its conjugate, we get a real number.

Examples   Multiply.

14. $(5 + 7i)(5 - 7i) = 5^2 - (7i)^2$
$$= 25 - 49i^2$$
$$= 25 - (-49)$$
$$= 74$$

15. $(2 - 3i)(2 + 3i) = 2^2 - (3i)^2$
$$= 4 - 9i^2$$
$$= 4 - (-9)$$
$$= 13$$

We use conjugates in dividing complex numbers.

Example 16 Divide: $\dfrac{-5 + 9i}{1 - 2i}$.

$$\frac{-5 + 9i}{1 - 2i} \cdot \boxed{\frac{1 + 2i}{1 + 2i}} = \frac{(-5 + 9i)(1 + 2i)}{(1 - 2i)(1 + 2i)} \qquad \text{Multiplying by 1}$$

$$= \frac{-23 - i}{5}$$

$$= -\frac{23}{5} - \frac{1}{5}i$$

Note the similarity between this example and rationalizing denominators. The symbol for the number 1 was chosen with reference to the divisor, $1 - 2i$. The symbol for 1 was formed using the conjugate of the divisor for both numerator and denominator.

Example 17 What symbol for 1 would you use to divide?

Division to be done	*Symbol for 1*
$\dfrac{3 + 5i}{4 + 3i}$	$\dfrac{4 - 3i}{4 - 3i}$

SOLUTIONS OF EQUATIONS

The equation $x^2 + 1 = 0$ has no real-number solution, but it has two nonreal solutions.

Example 18 Determine whether i is a solution of the equation $x^2 + 1 = 0$.

We substitute i for x in the equation.

$$\begin{array}{c|c}
\multicolumn{2}{c}{x^2 + 1 = 0} \\ \hline
i^2 + 1 & 0 \\
-1 + 1 & \\
0 &
\end{array}$$

The number i is a solution.

Any equation consisting of a polynomial in one variable on one side and 0 on the other has complex-number solutions (some may be real). It is not always easy to find the solutions, but they always exist.

Example 19 Determine whether $1 + i$ is a solution of $x^2 - 2x + 2 = 0$.

We substitute $1 + i$ for x in the equation.

$$
\begin{array}{c|c}
x^2 - 2x + 2 = 0 & \\
\hline
(1 + i)^2 - 2(1 + i) + 2 & 0 \\
1 + 2i + i^2 - 2 - 2i + 2 & \\
1 + 2i - 1 - 2 - 2i + 2 & \\
(1 - 1 - 2 + 2) + (2 - 2)i & \\
0 + 0i & \\
0 &
\end{array}
$$

$1 + i$ is a solution.

Example 20 Determine whether $2i$ is a solution of $x^2 + 3x - 4 = 0$.

$$
\begin{array}{c|c}
x^2 + 3x - 4 = 0 & \\
\hline
(2i)^2 + 3(2i) - 4 & 0 \\
4i^2 + 6i - 4 & \\
-4 + 6i - 4 & \\
-8 + 6i &
\end{array}
$$

$2i$ is not a solution.

EXERCISE SET 6.10

Express in terms of i.

1. $\sqrt{-15}$ 2. $\sqrt{-17}$ 3. $\sqrt{-16}$

4. $\sqrt{-25}$ 5. $-\sqrt{-12}$ 6. $-\sqrt{-20}$

Add or subtract and simplify.

7. $(3 + 2i) + (5 - i)$ 8. $(-2 + 3i) + (7 + 8i)$ 9. $(4 - 3i) + (5 - 2i)$

10. $(-2 - 5i) + (1 - 3i)$ 11. $(9 - i) + (-2 + 5i)$ 12. $(6 + 4i) + (2 - 3i)$

13. $(3 - i) - (5 + 2i)$ 14. $(-2 + 8i) - (7 + 3i)$ 15. $(4 - 2i) - (5 - 3i)$

16. $(-2 - 3i) - (1 - 5i)$ 17. $(9 + 5i) - (-2 - i)$ 18. $(6 - 3i) - (2 + 4i)$

Simplify.

19. $(3 + 2i)(1 + i)$ **20.** $(4 + 3i)(2 + 5i)$ **21.** $(2 + 3i)(6 - 2i)$

22. $(5 + 6i)(2 - i)$ **23.** $(6 - 5i)(3 + 4i)$ **24.** $(5 - 6i)(2 + 5i)$

25. $(7 - 2i)(2 - 6i)$ **26.** $(-4 + 5i)(3 - 4i)$ **27.** i^6

28. $(-i)^4$ **29.** $(5i)^3$ **30.** $(-3i)^5$

31. $(3 - 2i)^2$ **32.** $(5 - 2i)^2$ **33.** $(2 + 3i)^2$

34. $(4 + 2i)^2$ **35.** $(-2 + 3i)^2$ **36.** $(-5 - 2i)^2$

Divide.

37. $\dfrac{3 + 2i}{2 + i}$ **38.** $\dfrac{4 + 5i}{5 - i}$ **39.** $\dfrac{5 - 2i}{2 + 5i}$

40. $\dfrac{3 - 2i}{4 + 3i}$ **41.** $\dfrac{8 - 3i}{7i}$ **42.** $\dfrac{3 + 8i}{5i}$

Determine whether the complex number is a solution of the equation.

43. $1 + 2i$;
$x^2 - 2x + 5 = 0$

44. $1 - 2i$;
$x^2 - 2x + 5 = 0$

45. $1 - i$;
$x^2 + 2x + 2 = 0$

46. $2 + i$;
$x^2 - 4x - 5 = 0$

☆ ──────────────────────────────

47. A function g is given by $g(x) = \dfrac{x^4 - x^2}{x - 1}$.   **48.** Evaluate $\dfrac{1}{w - w^2}$ when $w = \dfrac{1 - i}{10}$.

Find $g(2i)$.

Express in terms of i.

49. $\dfrac{1}{8}(-24 - \sqrt{-1024})$ **50.** $12\sqrt{-\dfrac{1}{32}}$ **51.** $7\sqrt{-64} - 9\sqrt{-256}$

Simplify.

52. $\dfrac{i^5 + i^6 + i^7 + i^8}{(1 - i)^4}$ **53.** $(1 - i)^3(1 + i)^3$ **54.** $\dfrac{5 - \sqrt{5}i}{\sqrt{5}i}$

55. $\dfrac{6}{1 + \dfrac{3}{i}}$ **56.** $\left(\dfrac{1}{2} - \dfrac{1}{3}i\right)^2 - \left(\dfrac{1}{2} + \dfrac{1}{3}i\right)^2$ **57.** $\dfrac{i - i^{38}}{1 + i}$

58. $\dfrac{2}{1 - 3i} - \dfrac{3}{4 + 2i}$ **59.** $\dfrac{2\sqrt{6} + 4\sqrt{5}i}{2\sqrt{6} - 4\sqrt{5}i}$ **60.** $\dfrac{1 - 4i}{4i(1 + 4i)^{-1}}$

TEST OR REVIEW—CHAPTER 6

1. Simplify: $\dfrac{12^{-3}x^{-4}y}{4^{-2}x^{-3}y^2}$.

2. Multiply and write scientific notation for your answer:
   $(3.2 \times 10^7)(4.2 \times 10^{-12})$.

In Questions 3–7 assume that letters can represent any real number.

Simplify.

3. $\sqrt{36y^2}$ 4. $\sqrt{x^2 + 10x + 25}$

5. Simplify: $\sqrt[3]{-8}$.   6. Simplify: $\sqrt[10]{(-4)^{10}}$.

7. Multiply and simplify by factoring: $\sqrt[3]{x^4}\sqrt[3]{8x^5}$.

8. Simplify and then approximate to the nearest tenth: $\sqrt{148}$.

In Questions 9 and 10 assume that all expressions under radical signs represent positive numbers.

9. Divide. Then simplify by taking roots if possible: $\sqrt{\dfrac{20x^3y}{4y}}$.

10. Simplify: $(\sqrt[3]{16a^2b})^2$.

11. Add. Then simplify by collecting like radical terms: $3\sqrt{128} + 2\sqrt{18} + 2\sqrt{32}$.

12. Multiply and simplify:
    $(\sqrt{20} + 2\sqrt{5})(\sqrt{20} - 3\sqrt{5})$.

13. Rationalize the denominator. Assume that a and b represent positive numbers.
    $\dfrac{\sqrt{a} + \sqrt{b}}{\sqrt{a} - \sqrt{b}}$.

14. Rewrite with fractional exponents:
    $(\sqrt{5xy^2})^5$.

15. Use fractional exponents to write a single radical expression.
    $\sqrt[4]{2y}\sqrt{x - 3}$

16. Solve: $\sqrt{y - 6} = \sqrt{y + 9} - 3$.

17. Express in terms of i and simplify: $\sqrt{-18}$.

18. Simplify: $(5 + 8i) - (-2 + 3i)$.

19. Determine whether $1 + 2i$ is a solution of $x^2 + 2x + 5 = 0$.

20. Divide: $\dfrac{-7 + 14i}{6 - 8i}$.

7

QUADRATIC EQUATIONS AND FUNCTIONS

7.1 QUADRATIC EQUATIONS

EQUATIONS OF THE TYPE $ax^2 = k$

Equations of the second degree are called *quadratic*. The standard form of a quadratic equation is

$$ax^2 + bx + c = 0.$$

Here a, b, and c are constants, and $a \neq 0$ (otherwise the equation would not be of second degree). The constants b and c may be 0.

We first consider equations in which $b = 0$, that is, equations $ax^2 + c = 0$ or $ax^2 = -c$, equations in which ax^2 is a constant.

Example 1   Solve: $3x^2 = 6$.

$$3x^2 = 6$$
$$x^2 = 2 \qquad \text{Multiplying by } \tfrac{1}{3}$$

Next, recall that every positive number has two square roots and that they are additive inverses of each other. Then proceed as follows:

$$x^2 = 2$$
$$x = \sqrt{2} \quad \text{or} \quad x = -\sqrt{2} \qquad \text{Taking square roots}$$

We often use the symbol $\pm\sqrt{2}$ to represent both of the numbers, $\sqrt{2}$ and $-\sqrt{2}$.

In this equation we can check both numbers at once.

$$
\begin{array}{c|c}
\multicolumn{2}{c}{3x^2 = 6} \\
\hline
3(\pm\sqrt{2})^2 & 6 \\
3 \cdot 2 & \\
6 &
\end{array}
$$

The solutions are $\sqrt{2}$ and $-\sqrt{2}$, or $\pm\sqrt{2}$.

Sometimes we rationalize denominators to simplify answers.

Example 2   Solve: $-5x^2 + 2 = 0$.

$$-5x^2 + 2 = 0$$
$$x^2 = \tfrac{2}{5} \qquad \text{Adding } -2 \text{ and multiplying by } -\tfrac{1}{5}$$
$$x = \sqrt{\tfrac{2}{5}} \quad \text{or} \quad x = -\sqrt{\tfrac{2}{5}} \qquad \text{Taking square roots}$$
$$x = \sqrt{\tfrac{2}{5} \cdot \tfrac{5}{5}} \quad \text{or} \quad x = -\sqrt{\tfrac{2}{5} \cdot \tfrac{5}{5}} \qquad \text{Rationalizing the denominators}$$
$$x = \sqrt{\tfrac{10}{5}} \quad \text{or} \quad x = -\sqrt{\tfrac{10}{5}}$$

Check:

$$-5x^2 + 2 = 0$$

$$-5\left(\pm\frac{\sqrt{10}}{5}\right)^2 + 2 \ \bigg|\ 0$$

$$-5\left(\frac{10}{25}\right) + 2$$

$$-2 + 2$$

$$0$$

The solutions are $\dfrac{\sqrt{10}}{5}$ and $-\dfrac{\sqrt{10}}{5}$, or $\pm\dfrac{\sqrt{10}}{5}$.

Sometimes we get solutions that are nonreal numbers.

Example 3   Solve: $4x^2 + 9 = 0$.

$$4x^2 + 9 = 0$$

$$x^2 = -\tfrac{9}{4} \qquad \text{Adding } -9 \text{ and multiplying by } \tfrac{1}{4}$$

$$x = \sqrt{\tfrac{9}{-4}} \quad \text{or} \quad x = -\sqrt{-\tfrac{9}{4}} \qquad \text{Taking square roots}$$

$$x = \tfrac{3}{2}i \quad \text{or} \quad x = -\tfrac{3}{2}i \qquad \text{Simplifying}$$

The numbers $\tfrac{3}{2}i$ and $-\tfrac{3}{2}i$ check, so they are solutions.

EQUATIONS OF THE TYPE $ax^2 + bx = 0$

When c is 0 but b is not, we can factor and use the principle of zero products.

Example 4   Solve: $3x^2 + 5x = 0$. ←

$$3x^2 + 5x = 0$$

$$x(3x + 5) = 0 \qquad\qquad \text{Factoring}$$

$$x = 0 \quad \text{or} \quad 3x + 5 = 0 \qquad \text{Principle of zero products}$$

$$x = 0 \quad \text{or} \qquad\ \ 3x = -5$$

$$x = 0 \quad \text{or} \qquad\quad\ x = -\frac{5}{3}$$

A common error is to multiply on both sides by $\frac{1}{3x}$. That procedure is wrong because one of the solutions is lost. You must factor and use the principle of zero products.

This time the checks cannot be done together.

Check: For 0:   For $-\frac{5}{3}$:

$$\begin{array}{c|c} 3x^2 + 5x = 0 \\ \hline 3 \cdot (0)^2 + 5(0) & 0 \\ 0 + 0 \\ 0 \end{array} \qquad \begin{array}{c|c} 3x^2 + 5x = 0 \\ \hline 3(-\frac{5}{3})^2 + 5(-\frac{5}{3}) & 0 \\ 3(\frac{25}{9}) + (-\frac{25}{3}) \\ \frac{25}{3} + (-\frac{25}{3}) \\ 0 \end{array}$$

The solutions are 0 and $-\frac{5}{3}$.

A quadratic equation of the above type will always have 0 as one solution and a nonzero number as the other solution.

EQUATIONS OF THE TYPE $ax^2 + bx + c = 0$

When none of the constants a, b, or c is 0, we can try factoring.

Example 5   Solve: $3x^2 + x - 2 = 0$.

$$3x^2 + x - 2 = 0$$

$$(3x - 2)(x + 1) = 0 \qquad \text{Factoring}$$

$$3x - 2 = 0 \quad \text{or} \quad x + 1 = 0$$

$$3x = 2 \quad \text{or} \qquad x = -1$$

$$x = \frac{2}{3} \quad \text{or} \qquad x = -1$$

Check: For -1:   For $\frac{2}{3}$:

$$\begin{array}{c|c} 3x^2 + x - 2 = 0 \\ \hline 3(-1)^2 + (-1) - 2 & 0 \\ 3 - 1 - 2 \\ 0 \end{array} \qquad \begin{array}{c|c} 3x^2 + x - 2 = 0 \\ \hline 3(\frac{2}{3})^2 + (\frac{2}{3}) - 2 & 0 \\ 3 \cdot \frac{4}{9} + \frac{2}{3} - 2 \\ \frac{12}{9} + \frac{6}{9} - 2 \\ \frac{18}{9} - 2 \\ 0 \end{array}$$

The solutions are -1 and $\frac{2}{3}$.

Sometimes it helps to find the standard form before factoring.

Example 6 Solve: $(x - 1)(x + 1) = 5(x - 1)$.

$$(x - 1)(x + 1) = 5(x - 1)$$

$$x^2 - 1 = 5x - 5 \qquad \text{Multiplying}$$

$$x^2 - 5x + 4 = 0 \qquad \text{Finding standard form}$$

$$(x - 4)(x - 1) = 0 \qquad \text{Factoring}$$

$$x = 4 \quad \text{or} \quad x = 1 \qquad \text{Principle of zero products}$$

The solutions are 4 and 1 (the check has been left for the reader).

APPLIED PROBLEMS

Sometimes when we translate a problem to mathematical language, we get a quadratic equation. If so, we proceed as usual to solve the problem. That is, we solve the equation and then check possible answers in the original problem.

Example 7 A rectangular garden is 60 ft by 80 ft. Part of the garden is torn up to install a sidewalk of uniform width around it. The new area of the garden is $\frac{1}{6}$ of the old area. How wide is the sidewalk?

First make a drawing and label it with the known information We don't know how wide the sidewalk is, so we have called the width x.

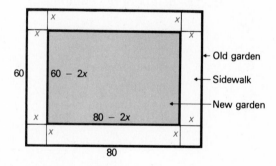

Remember, the area of a rectangle is $l \times w$ (length times width) Then:

Area of old garden $= 60 \cdot 80$

Area of new garden $= (60 - 2x)(80 - 2x)$

Since the new garden is $\frac{1}{6}$ of the old, we have

$$(60 - 2x)(80 - 2x) = \frac{1}{6} \cdot 60 \cdot 80.$$

We now solve this equation.

$$4800 - 120x - 160x + 4x^2 = 800 \qquad \text{Multiplying binomials on the left}$$

$$4x^2 - 280x + 4000 = 0 \qquad \text{Collecting like terms}$$

$$x^2 - 70x + 1000 = 0 \qquad \text{Multiplying on both sides by } \tfrac{1}{4}$$

$$(x - 20)(x - 50) = 0 \qquad \text{Factoring}$$

$$x = 20 \quad \text{or} \quad x = 50 \qquad \text{Principle of zero products}$$

We now check in the original problem. We see that 50 is not a solution because when $x = 50$, $60 - 2x = -40$, and the width of the garden cannot be negative.

Let's see whether 20 checks. If the sidewalk is 20 ft wide, the garden itself will have length $80 - 2 \cdot 20$, or 40 ft. The width will be $60 - 2 \cdot 20$, or 20 ft. The new area is thus $20 \cdot 40$, or 800 ft^2. The old area was $60 \cdot 80$, or 4800 ft^2. The new area of 800 ft^2 is $\tfrac{1}{6}$ of this, so the number 20 checks. The answer is that the sidewalk is 20 ft wide.

Example 8 A wire is stretched from the ground to the top of a building, as shown. The wire is 20 ft long. The height of the building is 4 ft greater than the distance d from the building. Find the distance d and the height of the building.

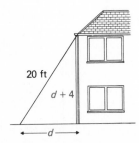

We first make a drawing and label it. We do not know d or the height of the building, but the height must be $d + 4$.

As we look at the drawing, we see that a right triangle is formed. We can use the Pythagorean theorem, which contains a right triangle formula.* It says that in any right triangle, the square of the longest

---

* For more detail on this formula, see p. 340.

side (the hypotenuse) is the sum of the squares of the other two sides (the legs).

$$c^2 = a^2 + b^2$$

In this problem then, we have

$$20^2 = d^2 + (d + 4)^2$$

We now solve this equation.

$$400 = d^2 + d^2 + 8d + 16 \qquad \text{Squaring}$$
$$2d^2 + 8d - 384 = 0 \qquad \text{Finding standard form}$$
$$d^2 + 4d - 192 = 0 \qquad \text{Multiplying by } \tfrac{1}{2}$$
$$(d + 16)(d - 12) = 0 \qquad \text{Factoring}$$
$$d + 16 = 0 \quad \text{or} \quad d - 12 = 0 \qquad \text{Principle of zero products}$$
$$d = -16 \quad \text{or} \qquad d = 12$$

We know that -16 is not an answer because distances are not negative. The number 12 checks (we leave the check to you), so the answer is that $d = 12$ ft, and the height of the building is 16 ft.

EXERCISE SET 7.1

Solve.

1. $4x^2 = 20$ 　　　　**2.** $3x^2 = 21$ 　　　　**3.** $25x^2 + 4 = 0$

4. $9x^2 + 16 = 0$ 　　**5.** $2x^2 - 3 = 0$ 　　**6.** $3x^2 - 7 = 0$

7. $14x^2 + 9x = 0$ 　　　　**8.** $19x^2 + 8x = 0$

9. $5x^2 + 10x = 0$ 　　　　**10.** $3x^2 + 12x = 0$

11. $x^2 - 6x + 5 = 0$ 　　**12.** $x^2 - 7x + 6 = 0$ 　　**13.** $6x^2 - x - 2 = 0$

14. $2x^2 + 13x + 15 = 0$ 　**15.** $x^2 + 8x + 15 = 0$ 　**16.** $x^2 + 9x + 14 = 0$

17. $4x(x - 2) - 5x(x - 1) = 2$ 　　**18.** $3x(x + 1) - 7x(x + 2) = 6$

19. $14(x - 4) - (x + 2) = (x + 2)(x - 4)$ 　　**20.** $11(x - 2) + (x - 5) = (x + 2)(x - 6)$

21. A picture frame measures 14 in. by 20 in.; 160 in² of picture shows. Find the width of the frame.

22. A picture frame measures 12 cm by 20 cm; 84 cm² of picture shows. Find the width of the frame.

23. The width of a rectangle is 4 ft less than the length. The area is 12 ft². Find the length and the width.

24. The width of a rectangle is 5 m less than the length. The area is 24 m². Find the length and the width.

25. The length of a rectangle is twice the width. The area is 288 yd^2. Find the length and the width.

26. The length of a rectangle is twice the width. The area is 338 cm^2. Find the length and the width.

27. The hypotenuse of a right triangle is 26 ft long. The length of one leg is 14 ft more than the other. Find the lengths of the legs.

28. The hypotenuse of a right triangle is 25 m long. The length of one leg is 17 m less than the other. Find the lengths of the legs.

29. The hypotenuse of a right triangle is 5 ft long. One leg is 1 ft shorter than the other. Find the lengths of the legs.

30. The hypotenuse of a right triangle is 13 m long. One leg is 7 m longer than the other. Find the lengths of the legs.

☆ ─────────────────────────────────

Solve.

31. ▦ $25.55x^2 - 1635.2 = 0$

32. $(3x^2 - 7x - 20)(2x - 5) = 0$

33. $(x)(2x^2 + 9x - 56)(3x + 10) = 0$

34. $\left(x - \dfrac{1}{3}\right)\left(x - \dfrac{1}{3}\right) + \left(x - \dfrac{1}{3}\right)\left(x + \dfrac{2}{9}\right) = 0$

35. Boats A and B leave the same point at the same time at right angles. B travels 7 km/h slower than A. After 4 hr they are 68 km apart. Find the speed of each boat.

36. Find three consecutive integers such that the square of the first plus the product of the other two is 67.

7.2 THE QUADRATIC FORMULA

SOLVING EQUATIONS WITH THE FORMULA

Some quadratic equations are difficult to solve by factoring. The following formula can be used to find the solutions.

The Quadratic Formula: **The solutions of a quadratic equation $ax^2 + bx + c = 0$ are given by**

$$x = \frac{-b \pm \sqrt{b^2 - 4ac}}{2a}.$$

To obtain this formula, we solve the equation $ax^2 + bx + c = 0$.

$$4a^2x^2 + 4abx + 4ac = 0$$

Multiplying on both sides by $4a$, to make the first term a square

$$4a^2x^2 + 4abx + (b^2 - b^2) + 4ac = 0$$

Adding $b^2 - b^2$ on the left, to complete the square

$$4a^2x^2 + 4abx + b^2 - (b^2 - 4ac) = 0$$

Rearranging

$$4a^2x^2 + 4abx + b^2 = b^2 - 4ac$$

Adding $b^2 - 4ac$ on both sides

$$(2ax + b)^2 = b^2 - 4ac$$

Factoring the trinomial square on the left

We now take square roots:

$$2ax + b = \sqrt{b^2 - 4ac} \quad \text{or} \quad 2ax + b = -\sqrt{b^2 - 4ac}.$$

Solving for x, we get

$$x = \frac{-b + \sqrt{b^2 - 4ac}}{2a} \quad \text{or} \quad x = \frac{-b - \sqrt{b^2 - 4ac}}{2a}.$$

These are the solutions. We abbreviate as follows:

$$x = \frac{-b \pm \sqrt{b^2 - 4ac}}{2a}.$$

Example 1   Solve: $5x^2 - 8x = 3$.

First find standard form and determine a, b, and c.

$$5x^2 - 8x - 3 = 0; \quad a = 5, b = -8, c = -3$$

Then substitute into the quadratic formula.

$$x = \frac{-(-8) \pm \sqrt{(-8)^2 - 4 \cdot 5 \cdot (-3)}}{2 \cdot 5}$$

$$x = \frac{8 \pm \sqrt{64 + 60}}{10} = \frac{8 \pm \sqrt{124}}{10}$$

$$x = \frac{8 \pm \sqrt{4 \cdot 31}}{10} = \frac{8 \pm 2\sqrt{31}}{10}$$

$$x = \frac{2(4 + \sqrt{31})}{2 \cdot 5} = \frac{4 \pm \sqrt{31}}{5}$$

Caution! To avoid a common error in simplifying, remember to *factor numerator and denominator* and then remove a factor of 1.

The checking of possible solutions such as these is cumbersome. When using the quadratic formula, no numbers are ever obtained that are not solutions of the original equation unless a mistake has been made. Therefore, it is usually better to check your work, rather than to check by substituting into the original equation.

The solutions are $\dfrac{4 + \sqrt{31}}{5}$ and $\dfrac{4 - \sqrt{31}}{5}$.

Some quadratic equations have solutions that are not real numbers.

Example 2   Solve: $x^2 + x + 1 = 0$.

$a = 1, b = 1, c = 1$

$$x = \frac{-b \pm \sqrt{b^2 - 4ac}}{2a}$$

$$x = \frac{-1 \pm \sqrt{1^2 - 4 \cdot 1 \cdot 1}}{2 \cdot 1} = \frac{-1 \pm \sqrt{1 - 4}}{2} = \frac{-1 \pm \sqrt{-3}}{2}$$

$$x = \frac{-1 \pm i\sqrt{3}}{2}$$

The solutions are $\dfrac{-1 + i\sqrt{3}}{2}$ and $\dfrac{-1 - i\sqrt{3}}{2}$.

Example 3   Solve: $2 + \dfrac{7}{x} = \dfrac{4}{x^2}$.

First find standard form.

$$2x^2 + 7x = 4 \qquad \text{Multiplying by } x^2, \text{ the LCM of the denominators}$$

$$2x^2 + 7x - 4 = 0 \qquad \text{Adding } -4$$

$a = 2, b = 7, c = -4$

$$x = \frac{-7 \pm \sqrt{7^2 - 4 \cdot 2 \cdot (-4)}}{2 \cdot 2}$$

$$x = \frac{-7 \pm \sqrt{49 + 32}}{4}$$

$$x = \frac{-7 \pm \sqrt{81}}{4} = \frac{-7 \pm 9}{4}$$

$$x = \frac{-7 + 9}{4} = \frac{1}{2} \quad \text{or} \quad x = \frac{-7 - 9}{4} = -4$$

In this case, since we started with a fractional equation, we *do* need to check.* At least we need to show that neither of the numbers makes a denominator 0. Since neither of them does, the solutions are $\frac{1}{2}$ and -4.

The solutions of a quadratic equation can *always* be found using the quadratic formula. They are not always easy to find by factoring. A general strategy for solving quadratic equations is as follows.

To solve a quadratic equation:

a) Try factoring.

b) If factoring seems difficult, use the quadratic formula; it *always works!*

APPROXIMATING SOLUTIONS

A square root table or a calculator can be used to approximate solutions.

Example 4 Approximate to the nearest tenth the solutions of the equation in Example 1.

From Table 1 in the back of the book, $\sqrt{31} \approx 5.568$.

$$\frac{4 + \sqrt{31}}{5} \approx \frac{4 + 5.568}{5}$$

$$\approx \frac{9.568}{5} \quad \text{Adding}$$

$$\approx 1.913 \quad \text{Dividing}$$

$$\approx 1.9 \quad \text{Rounding to the nearest tenth}$$

$$\frac{4 - \sqrt{31}}{5} \approx \frac{4 - 5.568}{5}$$

$$\approx \frac{-1.568}{5} \quad \text{Subtracting}$$

$$\approx -0.313 \quad \text{Dividing}$$

$$\approx -0.3 \quad \text{Rounding to the nearest tenth}$$

---

* The quadratic formula always gives correct results when we start with the standard form. In such cases we need check only to detect errors. In this example, however, we cleared of fractions before obtaining standard form, and this step could possibly introduce numbers that do not check in the original equation.

EXERCISE SET 7.2

Solve.

1. $x^2 + 6x + 4 = 0$

2. $x^2 - 6x - 4 = 0$

3. $3p^2 = -8p - 5$

4. $3u^2 = 18u - 6$

5. $x^2 - x + 1 = 0$

6. $x^2 + x + 2 = 0$

7. $x^2 + 13 = 4x$

8. $x^2 + 13 = 6x$

9. $r^2 + 3r = 8$

10. $h^2 + 4 = 6h$

11. $1 + \dfrac{2}{x} + \dfrac{5}{x^2} = 0$

12. $1 + \dfrac{5}{x^2} = \dfrac{2}{x}$

13. $3x + x(x - 2) = 0$

14. $4x + x(x - 3) = 0$

15. $(2t - 3)^2 + 17t = 15$

16. $2y^2 - (y + 2)(y - 3) = 12$

17. $(x - 2)^2 + (x + 1)^2 = 0$

18. $(x + 3)^2 + (x - 1)^2 = 0$

Use the square root table, Table 1, in the back of the book to approximate solutions to the nearest tenth.

19. $x^2 + 4x - 7 = 0$

20. $x^2 + 6x + 4 = 0$

21. $2x^2 - 3x - 7 = 0$

22. $3x^2 - 3x - 2 = 0$

☆ ————————————————————————————

Solve.

23. ▦ $2.2x^2 + 0.5x - 1 = 0$

24. ▦ $5.33x^2 - 8.23x - 3.24 = 0$

25. Solve for x in terms of y: $3x^2 + xy + 4y^2 - 9 = 0$.

Solve.

26. $\dfrac{5}{x} + \dfrac{x}{4} = \dfrac{11}{7}$

27. $2x^2 - x - \sqrt{5} = 0$

28. $\sqrt{3}x^2 + 6x + \sqrt{3} = 0$

29. $ix^2 - x - 1 = 0$

30. $(1 + \sqrt{3})x^2 - (3 + 2\sqrt{3})x + 3 = 0$

31. Solve $3x^2 + xy + 4y^2 - 9 = 0$ for x in terms of y.

32. One solution of $kx^2 + 3x - k = 0$ is -2. Find the other.

7.3 FRACTIONAL EQUATIONS AND APPLIED PROBLEMS

FRACTIONAL EQUATIONS THAT ARE QUADRATIC

In a fractional equation, when we clear of fractions, we sometimes get a quadratic equation.

Example 1 Solve: $\dfrac{14}{x + 2} - \dfrac{1}{x - 4} = 1.$

We multiply by the LCM of the denominators, which is $(x + 2)(x - 4)$.

$$(x + 2)(x - 4) \cdot \left[\frac{14}{x + 2} - \frac{1}{x - 4} \right] = (x + 2)(x - 4) \cdot 1$$

$$(x + 2)(x - 4) \cdot \frac{14}{x + 2} - (x + 2)(x - 4) \cdot \frac{1}{x - 4}$$

$$= (x + 2)(x - 4) \cdot 1 \qquad \text{Using a distributive law}$$

$14(x - 4) - (x + 2) = (x + 2)(x - 4)$ Simplifying by removing factors of 1

$14x - 56 - x - 2 = x^2 - 2x - 8$ Multiplying

$13x - 58 = x^2 - 2x - 8$ Collecting like terms

$0 = x^2 - 15x + 50$ Writing standard form

$0 = (x - 5)(x - 10)$ Factoring

$x = 5 \quad \text{or} \quad x = 10$ Using the principle of zero products

Since we started with a fractional equation, we might have introduced some numbers that are not solutions of the original equation when we cleared of fractions. This time, we must check.

Check: For 5:   For 10:

For 5:

$$\frac{14}{x + 2} - \frac{1}{x - 4} = 1$$

$$\frac{14}{5 + 2} - \frac{1}{5 - 4} \;\bigg|\; 1$$

$$\frac{14}{7} - 1$$

$$2 - 1$$

$$1$$

For 10:

$$\frac{14}{x + 2} - \frac{1}{x - 4} = 1$$

$$\frac{14}{10 + 2} - \frac{1}{10 - 4} \;\bigg|\; 1$$

$$\frac{14}{12} - \frac{1}{6}$$

$$\frac{14}{12} - \frac{2}{12}$$

$$1$$

The solutions are 5 and 10.

If for a fractional equation, clearing of fractions produces a quadratic equation, we solve that resulting equation as we would solve any quadratic equation. However, it is necessary to check possible solutions by substituting into the original equation.

APPLIED PROBLEMS

Some problems translate to equations like that in Example 1.

Example 2   Fred Fast and Sam Slow work together and mow a lawn in 4 hours. It would take Sam 6 hours more than Fred to do the job alone. How long would each need to do the job working alone?

Let x represent the time it takes Fred to work alone. Then $x + 6$ is the amount of time it takes Sam to do the work alone. Thus Fred can do $1/x$ of the job in 1 hour and Sam can do $1/(x + 6)$ of the job in 1 hour. They can do $(1/x) + [1/(x + 6)]$ of the job in 1 hour. They can also do $\frac{1}{4}$ of it in 1 hour. Thus we get the equation

$$\frac{1}{x} + \frac{1}{x + 6} = \frac{1}{4}.$$

Now we solve.

$$4x(x + 6) \cdot \left[\frac{1}{x} + \frac{1}{x + 6} \right] = 4x(x + 6) \cdot \frac{1}{4} \qquad \text{Multiplying by the LCM}$$

$$4x(x + 6) \cdot \frac{1}{x} + 4x(x + 6) \cdot \frac{1}{x + 6} = 4x(x + 6) \cdot \frac{1}{4}$$

$$4(x + 6) + 4x = x(x + 6)$$
$$4x + 24 + 4x = x^2 + 6x$$
$$8x + 24 = x^2 + 6x$$
$$0 = x^2 - 2x - 24 \qquad \text{Standard form}$$
$$0 = (x - 6)(x + 4) \qquad \text{Factoring}$$
$$x = 6 \quad \text{or} \quad x = -4 \qquad \text{Principle of zero products}$$

Since negative time has no meaning in this problem, -4 is not a solution. Let's see if 6 checks. This is the time for Fred to do the job alone. Then Sam would take 12 hours alone. Thus Fred would do $\frac{4}{6}$ of the job in 4 hours and Sam would do $\frac{4}{12}$ of it. Thus in 4 hours they would do $\frac{4}{6} + \frac{4}{12}$, or $\frac{8}{12} + \frac{4}{12}$ of the job. This is all of it, so the numbers check. The answer is that it would take Fred 6 hours alone and Sam 12 hours alone.

Example 3   A man leaves Homeville and travels the 600 miles to Fartown at a certain speed. He makes the return trip 10 mph slower.

His total time for the round trip was 22 hours. How fast did he travel on each part of the trip?

We first make a drawing and label it with the known information.

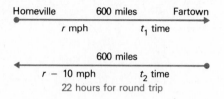

We let r represent the speed on the first part of the trip. Then $r - 10$ is the speed on the return trip. We can organize these facts in a chart.

	Distance	Speed	Time
Trip out	600	r	t_1
Return trip	600	$r - 10$	t_2

Since we do not know either time, but only the total, we have called the times t_1 and t_2. Since the total was 22 hours, we can write one equation at once:

$$t_1 + t_2 = 22.$$

Looking for a way to substitute into this equation, we recall the definition of speed, $r = d/t$, which we solve for t to get $t = d/r$. Now we go back to the chart. For each line we write $t = d/r$:

$$t_1 = \frac{600}{r}$$

and

$$t_2 = \frac{600}{r - 10}.$$

Substituting into our first equation, we get

$$\frac{600}{r} + \frac{600}{r - 10} = 22.$$

Next, we solve. The LCM is $r(r - 10)$.

$$r(r - 10) \cdot \left[\frac{600}{r} + \frac{600}{r - 10}\right] = r(r - 10) \cdot 22 \quad \text{Multiplying by the LCM}$$

$$r(r - 10) \cdot \frac{600}{r} + r(r - 10) \cdot \frac{600}{r - 10} = r(r - 10) \cdot 22$$

$$600(r - 10) + 600r = 22r^2 - 220r$$

$$600r - 6000 + 600r = 22r^2 - 220r$$

$$300r - 3000 + 300r = 11r^2 - 110r \quad \text{Multiplying by } \tfrac{1}{2}$$

$$0 = 11r^2 - 710r + 3000 \quad \text{Standard form}$$

$$0 = (11r - 50)(r - 60) \quad \text{Factoring}$$

$$r = 60 \quad \text{or} \quad r = \frac{50}{11}, \text{ or } 4\frac{6}{11} \quad \begin{array}{l}\text{Principle of}\\\text{zero products}\end{array}$$

$4\frac{6}{11}$ is not a solution because it would make the return speed negative, but 60 checks. The answer is that the speed out was 60 mph, and the speed back was 50 mph.

Example 4   A car travels 300 miles at a certain speed. If the speed had been 10 mph faster the trip would have been made in 1 hour less time. Find the speed.

We make a drawing, labeling it with the known information. We can also organize the information in a table.

	300 miles	
Time t		Speed r

	300 miles	
Time $t - 1$		Speed $r + 10$

Distance	Speed	Time
300	r	t
300	$r + 10$	$t - 1$

From the first line of the table, we obtain

$$r = \frac{300}{t}.$$

From the second line we get

$$r + 10 = \frac{300}{t - 1}.$$

We substitute for r from the first equation into the second, and solve.

$$\frac{300}{t} + 10 = \frac{300}{t - 1}$$

$$t(t - 1)\left[\frac{300}{t} + 10\right] = t(t - 1) \cdot \frac{300}{t - 1} \quad \begin{array}{l}\text{Multiplying} \\ \text{by the LCM}\end{array}$$

$$t(t - 1) \cdot \frac{300}{t} + t(t - 1) \cdot 10 = t(t - 1) \cdot \frac{300}{t - 1}$$

$$300(t - 1) + 10(t^2 - t) = 300t$$

$$10t^2 - 10t - 300 = 0 \qquad \text{Standard form}$$

$$t^2 - t - 30 = 0 \qquad \text{Multiplying by } \tfrac{1}{10}$$

$$(t - 6)(t + 5) = 0 \qquad \text{Factoring}$$

$$t = 6 \quad \text{or} \quad t = -5 \qquad \begin{array}{l}\text{Principle of zero} \\ \text{products}\end{array}$$

Since negative time has no meaning in this problem, we try 6 hours. Remembering that $r = d/t$, we get $r = 300/6 = 50$ mph.

To check, we take the speed 10 mph faster, which is 60 mph, and see how long the trip would have taken at that speed.

$$t = \frac{d}{r} = \frac{300}{60} = 5 \text{ hours}$$

This is one hour less than the trip actually took, so we have an answer to the problem. The answer is that the speed was 50 mph.

EXERCISE SET 7.3

Solve.

1. $\dfrac{1}{x} = \dfrac{x - 2}{24}$

2. $\dfrac{x + 3}{14} = \dfrac{2}{x}$

3. $\dfrac{1}{2x - 1} - \dfrac{1}{2x + 1} = \dfrac{1}{4}$

4. $\dfrac{1}{4 - x} - \dfrac{1}{2 + x} = \dfrac{1}{4}$

5. $\dfrac{50}{x} - \dfrac{50}{x - 5} = -\dfrac{1}{2}$

6. $3x + \dfrac{10}{x - 1} = \dfrac{16}{x - 1}$

7. $\dfrac{x + 2}{x} = \dfrac{x - 1}{2}$

8. $\dfrac{x}{3} - \dfrac{6}{x} = 1$

9. $x - 6 = \dfrac{1}{x + 6}$

10. $x + 7 = \dfrac{1}{x - 7}$

11. During the first part of a trip a canoeist travels 80 miles at a certain speed. The canoeist travels 35 miles on the second part of the trip at a speed 5 mph slower. The total time for the trip is 3 hr. What was the speed on each part of the trip?

12. During the first part of a trip a car travels 120 miles at a certain speed. It travels 100 miles on the second part of the trip at a speed 10 mph slower. The total time for the trip is 4 hr. What was the speed on each part of the trip?

13. A car travels 280 miles at a certain speed. If the speed has been 5 mph faster the trip would have been made in 1 hour less time. Find the speed.

14. A car travels 200 miles at a certain speed. If the speed had been 10 mph faster the trip would have been made in 1 hour less time. Find the speed.

15. Airplane A travels 2800 km at a certain speed. Plane B travels 2000 km at a speed 50 km/h faster than plane A in 3 hr less time. Find the speed of each plane.

16. Airplane A travels 600 miles at a certain speed. Plane B travels 1000 miles at a speed 50 mph faster, but takes 1 hour longer. Find the speed of each plane.

17. Two pipes are connected to the same tank. When working together they can fill the tank in 2 hours. The larger pipe, working alone, can fill the tank in 3 hours less time than the smaller one. How long would the smaller one take, working alone, to fill the tank?

18. Two hoses are connected to a swimming pool. When working together they can fill the pool in 4 hours. The larger hose, working alone, can fill the pool in 6 hours less time than the smaller one. How long would the smaller one take, working alone, to fill the pool?

19. A boat travels 1 km upstream and 1 km back. The time for the roundtrip is 1 hour. The speed of the stream is 2 km/hr. What is the speed of the boat in still water?

20. A boat travels 10 miles upstream and 10 miles back. The time required for the roundtrip is 3 hours. The speed of the stream is 5 mph. What is the speed of the boat in still water?

☆ ──────────────────────────────

21. 🖩 At a factory, smokestack A pollutes the air 2.13 times as fast as smokestack B. Together they yield a certain amount of pollution in 16.3 hr. How long would it take each to pollute this much alone?

22. 🖩 At Pizza Perfect, Ron Italio can make 100 large pizza crusts in 1.2 hours less time than Chester Mosserelli. Together they can do the job in 1.8 hours. How long does it take each to do the job alone?

Solve.

23. $\dfrac{12}{x^2 - 9} = 1 + \dfrac{3}{x - 3}$

24. $\dfrac{x^2}{x - 2} - \dfrac{x + 4}{2} + \dfrac{2 - 4x}{x - 2} + 1 = 0$

25. $\dfrac{4}{2x + i} - \dfrac{1}{x - i} = \dfrac{2}{x + i}$

26. Find a when the reciprocal of $a - 1$ is $a + 1$.

27. A discount store bought a quantity of beach towels for $250 and sold all but 15 at a profit of $3.50 per towel. With the total amount received the manager could buy 4 more than twice as many as were bought before. Find the cost per towel.

7.4 SOLUTIONS OF QUADRATIC EQUATIONS

THE DISCRIMINANT

We now consider quadratic equations with real-number coefficients.

From the quadratic formula we know that the solutions x_1 and x_2 of a quadratic equation are given by

$$x_1 = \frac{-b + \sqrt{b^2 - 4ac}}{2a} \quad \text{and} \quad x_2 = \frac{-b - \sqrt{b^2 - 4ac}}{2a}.$$

The expression $b^2 - 4ac$ shows the nature of the solutions. This expression is called the *discriminant*. If it is 0, then it doesn't matter whether we choose the plus or minus sign in the formula; hence there is just one solution. Since there is no negative radicand, the solutions must be real. If the discriminant is positive, there will be two real solutions. If it is negative, we will be taking the square root of a negative number; hence there will be two nonreal solutions, and they will be complex conjugates. We summarize.

Discriminant $b^2 - 4ac$	Nature of solutions
0	Only one solution; it is a real number.
Positive	Two different real-number solutions.
Negative	Two different nonreal number solutions (complex conjugates).

Example 1   Determine the nature of the solutions of $9x^2 - 12x + 4 = 0$.

$$a = 9, \quad b = -12, \quad \text{and} \quad c = 4$$

We compute the discriminant.

$$b^2 - 4ac = (-12)^2 - 4 \cdot 9 \cdot 4$$
$$= 144 - 144 = 0$$

There is just one solution and it is a real number.

Example 2   Determine the nature of the solutions of $x^2 + 5x + 8 = 0$.

$$a = 1, \quad b = 5, \quad \text{and} \quad c = 8$$

We compute the discriminant.

$$b^2 - 4ac = 5^2 - 4 \cdot 1 \cdot 8$$
$$= 25 - 32 = -7$$

Since the discriminant is negative, there are two nonreal solutions.

Example 3   Determine the nature of the solutions of $x^2 + 5x + 6 = 0$.

$$a = 1, \quad b = 5, \quad \text{and} \quad c = 6$$
$$b^2 - 4ac = 5^2 - 4 \cdot 1 \cdot 6 = 1$$

Since the discriminant is positive, there are two solutions and they are real numbers.

WRITING EQUATIONS FROM SOLUTIONS

We know by the principle of zero products that $(x - 2)(x + 3) = 0$ has solutions 2 and -3. If we know the solutions of an equation, we can write the equation, using this principle in reverse.

Example 4   Write a quadratic equation whose solutions are $2i$ and $-2i$.

$$x = 2i \quad \text{or} \quad x = -2i$$

$x - 2i = 0 \quad$ or $\quad x + 2i = 0$ Getting 0's on one side

$$(x - 2i)(x + 2i) = 0 \quad \text{Principle of zero products (multiplying)}$$

$$x^2 + 2ix - 2ix - (2i)^2 = x^2 - (2i)^2 = 0 \quad \text{Using FOIL}$$

$$x^2 + 4 = 0$$

Example 5   Find a quadratic equation whose solutions are 3 and $-\frac{2}{5}$.

$$x = 3 \quad \text{or} \qquad x = -\frac{2}{5}$$

$$x - 3 = 0 \quad \text{or} \quad \left(x + \frac{2}{5}\right) = 0 \qquad \text{Getting 0's on one side}$$

$$(x - 3)\left(x + \frac{2}{5}\right) = 0 \qquad \begin{array}{l}\text{Principle of zero products} \\ \text{(multiplying)}\end{array}$$

$$x^2 + \frac{2}{5}x - 3x - 3 \cdot \frac{2}{5} = 0 \qquad \text{Using FOIL}$$

$$x^2 - \frac{13}{5}x - \frac{6}{5} = 0, \quad \text{or}$$

$$5x^2 - 13x - 6 = 0 \qquad \begin{array}{l}\text{Collecting like terms and} \\ \text{multiplying by 5}\end{array}$$

Example 6   Write a quadratic equation whose solutions are $\sqrt{3}$ and $-2\sqrt{3}$.

$$x = \sqrt{3} \quad \text{or} \qquad x = -2\sqrt{3}$$

$$x - \sqrt{3} = 0 \quad \text{or} \quad x + 2\sqrt{3} = 0 \qquad \text{Getting 0's on one side}$$

$$(x - \sqrt{3})(x + 2\sqrt{3}) = 0 \qquad \text{Principle of zero products}$$

$$x^2 + 2\sqrt{3}x - \sqrt{3}x - 2(\sqrt{3})^2 = 0 \qquad \text{Using FOIL}$$

$$x^2 + \sqrt{3}x - 6 = 0 \qquad \text{Collecting like terms}$$

EXERCISE SET 7.4

Determine the nature of the solutions of each equation.

1. $x^2 - 6x + 9 = 0$

2. $x^2 + 10x + 25 = 0$

3. $x^2 + 7 = 0$

4. $x^2 + 2 = 0$

5. $x^2 - 2 = 0$

6. $x^2 - 5 = 0$

7. $4x^2 - 12x + 9 = 0$

8. $4x^2 + 8x - 5 = 0$

9. $x^2 - 2x + 4 = 0$

10. $x^2 + 3x + 4 = 0$

11. $9t^2 - 3t = 0$

12. $4m^2 + 7m = 0$

13. $y^2 = \frac{1}{2}y + \frac{3}{5}$

14. $y^2 + \frac{9}{4} = 4y$

15. $4x^2 - 4\sqrt{3}x + 3 = 0$

16. $6y^2 - 2\sqrt{3}y - 1 = 0$

Write a quadratic equation having the given numbers as solutions.

17. $-11, \quad 9$

18. $-4, \quad 4$

19. 7, only solution
    [*Hint:* It must be a double solution,
    that is, $(x - 7)(x - 7) = 0$.]

20. -5, only solution

21. $-\dfrac{2}{5}, \quad \dfrac{6}{5}$

22. $-\dfrac{1}{4}, \quad -\dfrac{1}{2}$

23. $\dfrac{c}{2}, \quad \dfrac{d}{2}$

24. $\dfrac{k}{3}, \quad \dfrac{m}{4}$

25. $\sqrt{2}, \quad 3\sqrt{2}$

26. $-\sqrt{3}, \quad 2\sqrt{3}$

☆
___

For each equation under the given condition, (a) find k, and (b) find the other solution.

27. $kx^2 - 17x + 33 = 0$;
    one solution is 3

28. $kx^2 - 2x + k = 0$;
    one solution is -3

29. Prove:

  a) The sum of the solutions of $ax^2 + bx + c = 0$ is $-\dfrac{b}{a}$.

  b) The product of the solutions of $ax^2 + bx + c = 0$ is $\dfrac{c}{a}$.

For each equation under the given condition, a) find k, and b) find the other solution.

30. $kx^2 - 17x + 33 = 0$; one solution is 3.

31. $kx^2 - 2x + k = 0$; one solution is -3.

32. Find k if $kx^2 - 4x + (2k - 1) = 0$ and the product of the solutions is 3.

33. Find a quadratic equation for which the sum of the solutions is $\sqrt{3}$ and the product is 8.

34. Write a quadratic equation having the given numbers as solutions:

  a) $\dfrac{2 + \sqrt{3}}{2}, \dfrac{2 - \sqrt{3}}{2}$

  b) $\dfrac{g}{h}, \quad -\dfrac{h}{g}$

  c) $2 - 5i, 2 + 5i$

35. Find h and k, where $3x^2 - hx + 4k = 0$ and the sum of the solutions is -12 and the product of the solutions is 20.

36. One solution of the equation $p(q - r)y^2 + q(r - p)y + r(p - q) = 0$ is 2. Find the other.

37. The sum of the squares of the solutions of $x^2 + 2kx - 5 = 0$ is 26. Find the absolute value of k.

38. The solutions of $x^2 + px + q = 0$ are p and q, where $p \neq 0$ and $q \neq 0$. Find the sum of the solutions.

39. When the solutions of each of the equations $x^2 + kx + 8 = 0$ and $x^2 - kx + 8 = 0$ are listed in increasing order, each solution of the second equation is 6 more than the corresponding solution of the first equation. Find k.
___

7.5 FORMULAS

To solve a formula for a certain letter, we use the principles for solving equations to get that letter alone on one side. When square roots appear, we can usually get rid of the radical signs by squaring both sides.

Example 1 (*A pendulum formula*) Solve $T = 2\pi\sqrt{\dfrac{l}{g}}$ for l.

$$T = 2\pi\sqrt{\frac{l}{g}}$$

$$T^2 = \left(2\pi\sqrt{\frac{l}{g}}\right)^2 \qquad \text{Principle of powers (squaring)}$$

$$T^2 = 2^2\pi^2\frac{l}{g}$$

$$gT^2 = 4\pi^2 l \qquad \text{Clearing of fractions}$$

$$\frac{gT^2}{4\pi^2} = l \qquad \text{Multiplying by } \frac{1}{4\pi^2}$$

We now have l alone on one side, so the formula is solved for l.

In most formulas the letters represent nonnegative numbers, so we do not need to use absolute value signs when taking square roots.

Example 2 (*A right triangle formula*) Solve $c^2 = a^2 + b^2$ for a.

$$c^2 - b^2 = a^2 \qquad \text{Adding } -b^2 \text{ to get } a^2 \text{ alone}$$

$$\sqrt{c^2 - b^2} = a \qquad \text{Taking square root}$$

Example 3 (*A motion formula*) Solve $s = gt + 16t^2$ for t.

This time we use the quadratic formula to get t alone on one side of the equation.

$$16t^2 + gt - s = 0 \qquad \text{Writing the equation in standard form}$$

$$a = 16, b = g, \text{ and } c = -s$$

$$t = \frac{-g \pm \sqrt{g^2 - 4 \cdot 16 \cdot (-s)}}{2 \cdot 16} \qquad \text{Using the quadratic formula}$$

Since taking the negative square root would result in a negative answer, we take the positive one.

$$t = \frac{-g + \sqrt{g^2 + 64s}}{32}$$

The following list of steps should help you when solving formulas for a given letter. Try to remember that when solving a formula you do the same things you would do to solve any equation. The same principles hold.

> **To Solve a Formula for a Letter, say b:**
>
> 1. **Clear of fractions and use the principle of powers, as needed, to eliminate radicals and until b does not appear in any denominator. (In some cases you may clear of fractions first, and in some cases you may use the principle of powers first. But do these until radicals are gone and b is not in any denominator.)**
> 2. **Collect all terms with b^2 in them. Also collect all terms with b in them.**
> 3. **If b^2 does not appear, you can finish by using just the addition and multiplication principles. Get all terms containing b on one side of the equation, then factor out b. Dividing on both sides will then get b alone.**
> 4. **If b^2 appears but b does not appear to be the first power, solve the equation for b^2. Then take the square root on both sides.**
> 5. **If there are terms containing both b and b^2, put the equation in standard form and use the quadratic formula.**

Example 4   Solve the following formula for a: $q = \dfrac{a}{\sqrt{a^2 + b^2}}$.

In this case, we could either clear of fractions first or use the principle of powers first. Let us clear of fractions. We then have

$$q\sqrt{a^2 + b^2} = a.$$

Now we square both sides and then continue.

$$\begin{aligned}
(q\sqrt{a^2 + b^2})^2 &= a^2 &&\text{Squaring}\\
q^2(a^2 + b^2) &= a^2 \\
q^2a^2 + q^2b^2 &= a^2 \\
q^2b^2 &= a^2 - q^2a^2 &&\text{Getting all } a^2 \text{ terms together}\\
q^2b^2 &= a^2(1 - q^2) &&\text{Factoring out } a^2\\
\frac{q^2b^2}{1 - q^2} &= a^2 &&\text{Multiplying by } \frac{1}{1 + q^2}\\
\sqrt{\frac{q^2b^2}{1 - q^2}} &= a &&\text{Taking the square root}
\end{aligned}$$

> *Important!* *a* appears alone on one side of the equation and *does not* appear on the other side as well.

Now that we have *a* alone on one side, we have the formula solved for *a*.

EXERCISE SET 7.5

Solve each formula for the given letter. Assume that all variables represent nonnegative numbers.

1. $A = 6s^2$, for s
   (Area of a cube)

2. $A = 4\pi r^2$, for r
   (Area of a sphere)

3. $F = \dfrac{Gm_1m_2}{r^2}$, for r

   (Law of gravity)

4. $N = \dfrac{kQ_1Q_2}{s^2}$, for s

   (Number of phone calls between two cities)

5. $E = mc^2$, for c
   (Energy-mass relationship)

6. $A = \pi r^2$, for r
   (Area of a circle)

7. $a^2 + b^2 = c^2$, for b
   (Pythagorean formula in two dimensions)

8. $a^2 + b^2 + c^2 = d^2$, for c
   (Pythagorean formula in three dimensions)

9. $N = \dfrac{k^2 - 3k}{2}$, for k

   (Number of diagonals of a polygon)

10. $s = v_0t + \dfrac{gt^2}{2}$, for t

    (A motion formula)

11. $A = 2\pi r^2 + 2\pi rh$, for r
    (Area of a cylinder)

12. $A = \pi r^2 + \pi rs$, for r
    (Area of a cone)

13. $T = 2\pi\sqrt{\dfrac{l}{g}}$, for g

    (A pendulum formula)

14. $W = \sqrt{\dfrac{1}{LC}}$, for L

    (An electricity formula)

15. $P_1 - P_2 = \dfrac{32LV}{gD^2}$, for D

16. $N + p = \dfrac{6.2A^2}{pR^2}$, for R

17. $m = \dfrac{m_0}{\sqrt{1 - \dfrac{v^2}{c^2}}}$, for v

    (A relativity formula)

18. Solve the formula given in Exercise 17 for c.

☆

19. *Stopping distance.* The formula $s = -4t^2 + 40t + 2$, giving the stopping distance s of a car in terms of the time after the brakes are applied, can be solved for t. Then it will give the time required to stop in a certain distance. Solve the formula for t. Then use the formula to determine how long it will take to stop in a distance of 50 ft.

7.6 EQUATIONS REDUCIBLE TO QUADRATIC

Certain equations that are not really quadratic can be thought of in such a way that they can be solved as quadratic. For example, consider this fourth-degree equation.

$$x^4 \quad - \quad 9x^2 \quad + 8 = 0$$

$$(x^2)^{\,2} - 9\,(x^2) \quad + 8 = 0 \qquad \text{Thinking of } x^4 \text{ as } (x^2)^{\,2}$$

$$u^2 \quad - \quad 9u \quad + 8 = 0 \qquad \text{To make this clearer, write } u \text{ instead of } x^2.$$

The equation $u^2 - 9u + 8 = 0$ can be solved by factoring or the quadratic formula. After that, we can find x by remembering that $x^2 = u$. Equations that can be solved like this are said to be *reducible to quadratic*.

Example 1   Solve: $x^4 - 9x^2 + 8 = 0$.

Let $u = x^2$. Then we solve the equation found by substituting u for x^2.

$$u^2 - 9u + 8 = 0$$
$$(u - 8)(u - 1) = 0 \qquad \text{Factoring}$$
$$u - 8 = 0 \quad \text{or} \quad u - 1 = 0 \qquad \text{Principle of zero products}$$
$$u = 8 \quad \text{or} \qquad u = 1$$

Now we substitute x^2 for u and solve these equations.

$$x^2 = 8 \qquad \text{or} \quad x^2 = 1$$
$$x = \pm\sqrt{8} \quad \text{or} \quad x = \pm 1$$
$$x = \pm 2\sqrt{2} \quad \text{or} \quad x = \pm 1$$

To check first note that when $x = 2\sqrt{2}$, $x^2 = 8$ and $x^4 = 64$. Also, when $x = -2\sqrt{2}$, $x^2 = 8$ and $x^4 = 64$. Similarly, when $x = 1$, $x^2 = 1$, and $x^4 = 1$, and when $x = -1$, $x^2 = 1$, and $x^4 = 1$. Thus instead of making four checks we need make only two.*

Check:

For $\pm 2\sqrt{2}$:

$$
\begin{array}{c|c}
x^4 - 9x^2 + 8 = 0 \\
\hline
(\pm 2\sqrt{2})^4 - 9(\pm 2\sqrt{2})^2 + 8 & 0 \\
64 - 9 \cdot 8 + 8 \\
0
\end{array}
$$

For ± 1:

$$
\begin{array}{c|c}
x^4 - 9x^2 + 8 = 0 \\
\hline
(\pm 1)^4 - 9(\pm 1)^2 + 8 & 0 \\
1 - 9 + 8 \\
0
\end{array}
$$

A common error is to solve for u and then forget to solve for x. Remember that you must find values for the *original* variable!

The solutions are 1, -1, $2\sqrt{2}$, and $-2\sqrt{2}$.

Example 2 Solve: $x - 3\sqrt{x} - 4 = 0$.

Let $u = \sqrt{x}$. Then we solve the equation found by substituting u for $\sqrt{x}$ (and, of course, u^2 for x).

$$u^2 - 3u - 4 = 0$$
$$(u - 4)(u + 1) = 0$$
$$u = 4 \quad \text{or} \quad u = -1$$

Now we substitute $\sqrt{x}$ for u and solve these equations.

$$\sqrt{x} = 4 \quad \text{or} \quad \sqrt{x} = -1$$

Squaring the first equation we get $x = 16$. The second equation has no real solution since principal square roots are never negative.

The number 16 checks, so it is the solution.

Example 3 Solve: $(x^2 - 1)^2 - (x^2 - 1) - 2 = 0$.

Let $u = x^2 - 1$. Then we solve the equation found by substituting u for $x^2 - 1$.

$$u^2 - u - 2 = 0$$
$$(u - 2)(u + 1) = 0$$
$$u = 2 \quad \text{or} \quad u = -1$$

---

* A check by substituting is necessary when solving equations reducible to quadratic. Reducing the equation to quadratic can sometimes introduce numbers that are not solutions of the original equation.

Now we substitute $x^2 - 1$ for u and solve these equations.

$$x^2 - 1 = 2 \quad \text{or} \quad x^2 - 1 = -1$$
$$x^2 = 3 \quad \text{or} \quad x^2 = 0$$
$$x = \pm\sqrt{3} \quad \text{or} \quad x = 0$$

The numbers $\sqrt{3}$, $-\sqrt{3}$, and 0 check. They are the solutions.

Example 4   Solve: $y^{-2} - y^{-1} - 2 = 0$.

Let $u = y^{-1}$. Then we solve the equation found by substituting u for y^{-1} and u^2 for y^{-2}.

$$u^2 - u - 2 = 0$$
$$(u - 2)(u + 1) = 0$$
$$u = 2 \quad \text{or} \quad u = -1$$

Now we substitute y^{-1} or $\dfrac{1}{y}$ for u and solve these equations:

$$\frac{1}{y} = 2 \quad \text{or} \quad \frac{1}{y} = -1.$$

Solving, we get

$$y = \frac{1}{2} \quad \text{or} \quad y = \frac{1}{(-1)} = -1.$$

The numbers $\dfrac{1}{2}$ and -1 both check. They are the solutions.

EXERCISE SET 7.6

Solve.

1. $x - 10\sqrt{x} + 9 = 0$

2. $2x - 9\sqrt{x} + 4 = 0$

3. $x^4 - 10x^2 + 25 = 0$

4. $x^4 - 3x^2 + 2 = 0$

5. $(x^2 - 6x)^2 - 2(x^2 - 6x) - 35 = 0$

6. $(1 + \sqrt{x})^2 + (1 + \sqrt{x}) - 6 = 0$

7. $(y^2 - 5y)^2 - 2(y^2 - 5y) - 24 = 0$

8. $(2t^2 + t)^2 - 4(2t^2 + t) + 3 = 0$

9. $w^4 - 4w^2 - 2 = 0$

10. $t^4 - 5t^2 + 5 = 0$

11. $x^{-2} - x^{-1} - 6 = 0$

12. $4x^{-2} - x^{-1} - 5 = 0$

13. $2x^{-2} + x^{-1} - 1 = 0$

14. $m^{-2} + 9m^{-1} - 10 = 0$

☆ ───

Solve. Check possible solutions by substituting into the original equation.

15. ▦ $6.75x - 35\sqrt{x} - 5.36 = 0$

16. ▦ $\pi x^4 - \pi^2 x^2 - \sqrt{99.3} = 0$

Solve.

17. $\left(\dfrac{y^2 - 1}{y}\right)^2 - 4\left(\dfrac{y^2 - 1}{y}\right) - 12 = 0$

18. $\left(\sqrt{\dfrac{x}{x - 3}}\right)^2 - 24 = 10\sqrt{\dfrac{x}{x - 3}}$

19. $\sqrt{x - 3} - \sqrt[4]{x - 3} = 12$

20. $a^3 - 26a^{3/2} - 27 = 0$

───

7.7 QUADRATIC VARIATION

FINDING EQUATIONS OF VARIATION

We have already studied direct and inverse variation in Chapter 5. We now consider variation when the equations involved are of second degree and, in some cases, higher. Consider the equation for the area of a circle, in which A and r are variables and π is a constant:

$$A = \pi r^2.$$

We say that the area varies directly as the square of the radius. When (r_1, A_1) and (r_2, A_2) are solutions of the equation we have $A_1 = \pi r_1^2$ and $A_2 = \pi r_2^2$. Then dividing we get

$$\frac{A_1}{A_2} = \frac{\pi r_1^2}{\pi r_2^2} = \frac{r_1^2}{r_2^2}.$$

A proportion like $\dfrac{A_1}{A_2} = \dfrac{r_1^2}{r_2^2}$ can be helpful in solving problems.

> y **varies directly as the square of** x **if there is some positive constant** k **such that** $y = kx^2$.

Example 1   Find an equation of variation where y varies directly as the square of x, and $y = 12$ when $x = 2$.

We write an equation of variation and find k.

$$y = kx^2, \quad \text{so } 12 = k \cdot 2^2 \quad \text{and} \quad 3 = k.$$

Thus $y = 3x^2$.

From the law of gravity, we know that the weight W of an object varies inversely as the square of its distance d from the center of the earth:

$$W = \frac{k}{d^2}.$$

When (d_1, W_1) and (d_2, W_2) are solutions of the equation we have $W_1 = k/d_1^2$ and $W_2 = k/d_2^2$. Then

$$\frac{W_1}{W_2} = \frac{\dfrac{k}{d_1^2}}{\dfrac{k}{d_2^2}} = \frac{k}{d_1^2} \cdot \frac{d_2^2}{k} = \frac{d_2^2}{d_1^2}.$$

A proportion like $\dfrac{W_1}{W_2} = \dfrac{d_2^2}{d_1^2}$ can be helpful in solving problems.

> **y varies inversely as the square of x if there is some positive constant k such that $y = k/x^2$.**

Example 2   Find an equation of variation where W varies inversely as the square of d and $W = 3$ when $d = 5$.

$$W = \frac{k}{d^2}, \quad \text{so } 3 = \frac{k}{5^2} \quad \text{and} \quad k = 75$$

Thus $W = \dfrac{75}{d^2}$.

Consider the equation for the area A of a triangle with height h and base b:

$$A = \frac{1}{2}bh.$$

We say that the area varies *jointly* as the height and the base. From two solutions of the equation we get the proportion

$$\frac{A_1}{A_2} = \frac{\frac{1}{2}b_1 h_1}{\frac{1}{2}b_2 h_2} = \frac{b_1 h_1}{b_2 h_2}.$$

> **y varies jointly as x and z if there is some positive constant k such that $y = kxz$.**

Example 3 Find an equation of variation where y varies jointly as x and z, and $y = 42$ when $x = 2$ and $z = 3$.

$y = kxz$, so $42 = k \cdot 2 \cdot 3$ and $k = 7$

Thus $y = 7xz$.

The equation

$$y = k \cdot \frac{xz^2}{w}$$

asserts that y varies jointly as x and the square of z, and inversely as w.

Example 4 Find an equation of variation where y varies jointly as x and z and inversely as the square of w, and $y = 105$ when $x = 3$, $z = 20$, and $w = 2$.

$$y = k \cdot \frac{xz}{w^2}, \quad \text{so } 105 = k \cdot \frac{3 \cdot 20}{2^2} \quad \text{and} \quad k = 7.$$

Thus $y = 7 \cdot \dfrac{xz}{w^2}$.

SOLVING PROBLEMS

Many problem situations can be described with equations of variation.

Example 5 The volume of wood V in a tree varies jointly as the height h and the square of the girth g (girth is distance around). If the volume of a redwood tree is 216 m^3 when the height is 30 m and the girth is 1.5 m, what is the height of a tree whose volume is 960 m^3 and girth is 2 m?

Method 1. First find k using the first set of data. Then solve for h using the second set of data.

$$V = khg^2$$
$$216 = k \cdot 30 \cdot 1.5^2$$
$$3.2 = k$$

Then

$$960 = 3.2 \cdot h \cdot 2^2$$
$$75 = h.$$

Method 2. Let h_1 represent the height of the first tree and h_2 the height of the second tree. Use a proportion to solve for h_2 without first finding the variation constant.

$$\frac{V_1}{V_2} = \frac{h_1 \cdot g_1^2}{h_2 \cdot g_2^2}$$

$$\frac{216}{960} = \frac{30 \cdot 1.5^2}{h_2 \cdot 2^2}$$

$$h_2 = \frac{960 \cdot 67.5}{4 \cdot 216}$$

$$h_2 = 75 \text{ m}$$

Example 6 The intensity I of a TV signal varies inversely as the square of the distance d from the transmitter. If the intensity is 23 watts per square meter (W/m^2) at a distance of 2 km, what is the intensity at a distance of 6 km?

We use the proportion.

$$\frac{I_1}{I_2} = \frac{d_2^2}{d_1^2}$$

$$\frac{I_2}{I_1} = \frac{d_1^2}{d_2^2}$$

$$\frac{I_2}{23} = \frac{2^2}{6^2}$$

$$I_2 = \frac{4 \cdot 23}{36}$$

$$I_2 = 2.56 \text{ W/m}^2 \qquad \text{Rounded to the nearest hundredth}$$

EXERCISE SET 7.7

Find an equation of variation in which:

1. y varies directly as the square of x and $y = 0.15$ when $x = 0.1$.

2. y varies directly as the square of x and $y = 6$ when $x = 3$.

3. y varies inversely as the square of x and $y = 0.15$ when $x = 0.1$.

4. y varies inversely as the square of x and $y = 6$ when $x = 3$.

5. y varies jointly as x and z and $y = 56$ when $x = 7$ and $z = 8$.

6. y varies directly as x and inversely as z and $y = 4$ when $x = 12$ and $z = 15$.

7. y varies jointly as x and the square of z and $y = 105$ when $x = 14$ and $z = 5$.

8. y varies jointly as x and z and inversely as w and $y = \frac{3}{2}$ when $x = 2$, $z = 3$, and $w = 4$.

9. y varies jointly as x and z and inversely as the product of w and p, and $y = \frac{3}{28}$ when $x = 3$, $z = 10$, $w = 7$, and $p = 8$.

10. y varies jointly as x and z and inversely as the square of w, and $y = \frac{12}{5}$ when $x = 16$, $z = 3$, and $w = 5$.

Solve.

11. *Stopping distance of a car.* The stopping distance d of a car after the brakes are applied varies directly as the square of the speed r. If a car traveling 60 mph can stop in 200 ft, how many feet will it take the same car to stop when it is traveling 80 mph?

12. *Area of a cube.* The area of a cube varies directly as the square of the length of a side. If a cube has an area 168.54 in² when the length of a side is 5.3 in., what will the area be when the length of a side is 10.2 in.?

13. *Weight of an astronaut.* The weight W of an object varies inversely as the square of the distance d from the center of the earth. At sea level (6400 km from the center of the earth) an astronaut weighs 100 kg. Find his weight when he is 200 km above the surface of the earth and the spacecraft is not in motion.

14. *Intensity of light.* The intensity I of light from a light bulb varies inversely as the square of the distance d from the bulb. Suppose I is 90 W/m² when the distance is 5 m. Find the intensity at a distance of 10 m.

15. *Earned run average.* A pitcher's earned run average A varies directly as the number R of earned runs allowed and inversely as the number I of innings pitched. In a recent year Tom Seaver had an earned run average of 2.92. He gave up 85 earned runs in 262 innings. How many earned runs would he have given up had he pitched 300 innings with the same average? Round to the nearest whole number.

16. *Volume of a gas.* The volume V of a given mass of a gas varies directly as the temperature T and inversely as the pressure P. If $V = 231$ cm³ when $T = 42°$ and $P = 20$ kg/cm², what is the volume when $T = 30°$ and $P = 15$ kg/cm²?

☆

17. Suppose y varies directly as x and x is tripled. What is the effect on y?

18. Suppose y varies inversely as the cube of x and x is multiplied by 0.5. What is the effect on y?

19. *The Gravity Model in Sociology.* It has been determined that the average number of telephone calls in a day N, between two cities, is directly proportional to the populations P_1 and P_2 of the cities and inversely proportional to the square of the distance between the cities. That is, $N = kP_1P_2/d^2$.

    a) The population of Indianapolis is 744,624 and the population of Cincinnati is 452,524, and the distance between the cities is 174 km. The average number of daily phone calls between the two cities is 11,153. Find the value k and write the equation of variation.

    b) The average number of daily phone calls between Indianapolis and New York is 4270 and the population of New York is 7,895,563. Find the distance between Indianapolis and New York.

7.8 QUADRATIC FUNCTIONS AND THEIR GRAPHS

GRAPHS OF $f(x) = ax^2$

In Chapter 4 the notion of a *function* was introduced. We graphed equations such as $f(x) = 3x + 2$, or $y = 3x + 2$. Recall that the letter y acts very much like the symbol $f(x)$. In Chapter 4, we considered graphs of linear equations, or linear *functions*, that is, graphs of equations like $y = mx + b$. We now consider equations (or functions) in which the right-hand side is a quadratic polynomial,

$$y = ax^2 + bx + c.$$

Such functions, or equations, are called *quadratic*. To graph such a function, we choose values for x and compute y.*

Example 1　Graph $y = x^2$.

We choose some numbers for x, some positive and some negative, and for each number we compute y.

x	0	1	2	3	-1	-2	-3
y, or x^2	0	1	4	9	1	4	9

---

* Or $f(x)$. In this section we will sometimes use y and sometimes $f(x)$. For most purposes they are interchangeable.

We plot these ordered pairs and connect them with a smooth curve.

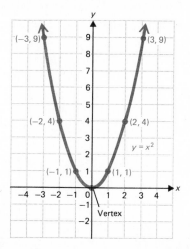

All quadratic functions have graphs similar to the one above.

Such curves are called *parabolas*. They are smooth, cup-shaped curves, with a line of symmetry. In the graph above, the vertical axis is the *line of symmetry*. If the paper were folded on this line, the two halves of the curve would match. The point $(0, 0)$ is known as the *vertex* of the parabola.

Some common errors in drawing graphs of quadratic functions are shown in the figure below.

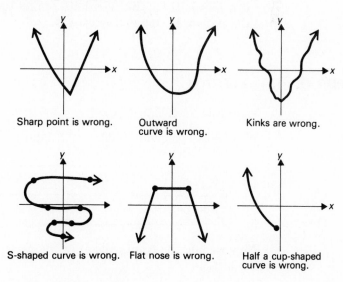

Sharp point is wrong.   Outward
                         curve is wrong.   Kinks are wrong.

S-shaped curve is wrong. Flat nose is wrong.   Half a cup-shaped
                                                  curve is wrong.

The graph of $y = \frac{1}{2}x^2$ is a flatter parabola than the graph of $y = x^2$. The graph of $y = 2x^2$ is narrower, as shown here, but the vertex and line of symmetry have not changed.

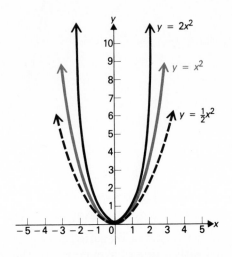

The graph of $y = ax^2$ **is a parabola with the vertical axis as its line of symmetry and its vertex at the origin.**

If $|a|$ is greater than 1, the parabola is narrower than $y = x^2$.

If $|a|$ is between 0 and 1, the parabola is flatter than $y = x^2$.

If a is positive, the parabola opens upward; if a is negative, the parabola opens downward.

GRAPHS OF $f(x) = a(x - h)^2$

Example 2   Graph $f(x) = 2(x - 3)^2$.

We choose some values of x and compute $f(x)$.

x	0	1	2	3	4	5	6
$f(x)$, or $2(x - 3)^2$	18	8	2	0	2	8	18

Now we plot the points and draw the curve.

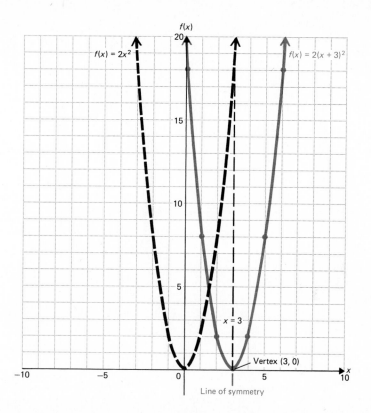

The curve in Example 2 looks just like the graph of $f(x) = 2x^2$, except that it is moved three units to the right. The line of symmetry is now the line $x = 3$ and the vertex is the point $(3, 0)$.

> **The graph of $y = a(x - h)^2$ looks just like the graph of $y = ax^2$, except that it is moved to the right or left. If h is positive, it is moved to the right. If h is negative, it is moved to the left. The vertex is now $(h, 0)$ and the line of symmetry is $x = h$.**

Example 3   Graph $f(x) = 2(x + 3)^2$.

We know that the graph looks like that of $f(x) = 2x^2$, but moved right or left. Think of $x + 3$ as $x - (-3)$. The number we are *subtracting* is negative, so we move *left*. We can draw $f(x) = 2x^2$ lightly, then move it three units to the left. See the graph at the top of the following page. We should compute a few values as a check.

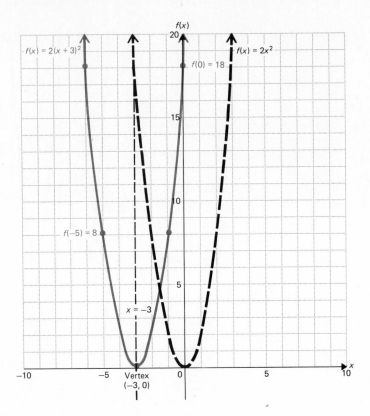

Example 4   Graph $y = -2(x + 2)^2$.

We know that the graph looks like that of $y = 2x^2$, but moved to the left two units, and it will also open downward, because of the negative coefficient, -2. See the graph below.

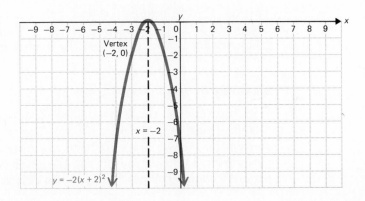

GRAPHS OF $f(x) = a(x - h)^2 + k$

Given a graph of $f(x) = a(x - h)^2$, what happens to it if we add a constant k? Suppose we add 2. This increases each function value $f(x)$ by 2, so the curve is moved up. If we add a negative number the curve is moved down. The vertex of the parabola will be at the point (h, k) and the line of symmetry will be $x = h$.

Note that if a parabola opens upward $(a > 0)$ the function value, or y-value, at the vertex is a least, or *minimum* value. That is, it is less than the y-value at any other point. If the parabola opens downward $(a < 0)$ the function value at the vertex will be a greatest, or *maximum* value.

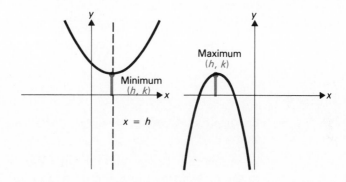

The graph of $f(x) = a(x - h)^2 + k$ looks just like the graph of $f(x) = a(x - h)^2$, except that it is moved up or down. If k is positive the curve is moved up k units. If k is negative the curve is moved down. The vertex is at (h, k) and the line of symmetry is $x = h$. If $a > 0$ then k is the minimum function value. If $a < 0$ then k is the maximum function value.

Example 5   Graph $f(x) = 2(x + 3)^2 - 5$ and find the minimum function value.

We know that the graph looks like that of $f(x) = 2(x + 3)^2$ (this is shown in Example 3) but moved down five units. The vertex is now $(-3, -5)$, and the minimum function value is -5.

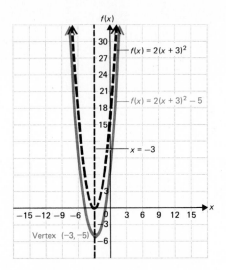

Example 6   Graph $f(x) = \frac{1}{2}(x - 3)^2 + 5$ and find the minimum function value.

We know that the graph looks just like that of $f(x) = \frac{1}{2}x^2$, but moved to the right three units and up five units. The vertex is $(3, 5)$ and the line of symmetry is the line $x = 3$. We draw $f(x) = \frac{1}{2}x^2$ and then move the curve over and up. We plot a few points as a check. The minimum function value is 5. (See Figure 1 on page 327.)

Example 7   Graph $y = -2(x + 3)^2 + 5$ and find the maximum function value.

We know that the graph looks like that of $f(x) = 2x^2$, but moved to the left three units and up five units, except that it now opens downward. The vertex is $(-3, 5)$. This time there is a *greatest* (maximum) function value, 5. (See Figure 2 on page 327.)

x	$f(x)$
0	$9\frac{1}{2}$
1	7
2	$5\frac{1}{2}$
6	$9\frac{1}{2}$

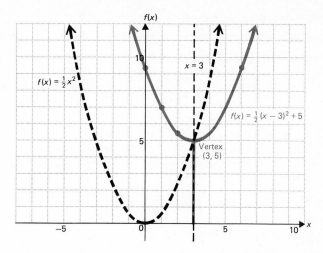

Figure 1

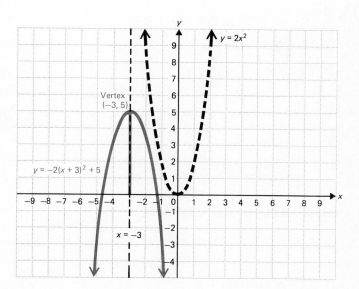

Figure 2

EXERCISE SET 7.8

Graph these functions.

1. $f(x) = 5x^2$ **2.** $f(x) = 4x^2$ **3.** $f(x) = \frac{1}{4}x^2$ **4.** $f(x) = \frac{1}{3}x^2$

5. $f(x) = -\frac{1}{2}x^2$ **6.** $f(x) = -\frac{1}{4}x^2$ **7.** $f(x) = -4x^2$ **8.** $f(x) = -3x^2$

For each of the following, graph the function, label the vertex, and draw the line of symmetry.

9. $f(x) = (x - 3)^2$ **10.** $f(x) = (x - 5)^2$

Graph each of the following.

11. $y = 2(x - 4)^2$ **12.** $y = -4(x - 6)^2$ **13.** $f(x) = -2(x + 6)^2$

14. $f(x) = 2(x + 4)^2$ **15.** $f(x) = 3(x - 1)^2$ **16.** $f(x) = -4(x - 2)^2$

For each of the following, graph the function, label the vertex, and draw the line of symmetry.

17. $f(x) = (x - 3)^2 + 1$ **18.** $f(x) = (x + 2)^2 - 3$

For each of the following, graph the function and find the maximum or minimum function value.

19. $y = \frac{1}{2}(x - 1)^2 - 3$ **20.** $y = \frac{1}{2}(x + 1)^2 + 4$

21. $f(x) = -3(x + 4)^2 + 1$ **22.** $f(x) = -2(x - 5)^2 - 3$

23. $y = -2(x + 2)^2 - 3$ **24.** $y = 3(x - 4)^2 + 2$

25. $f(x) = -(x + 1)^2 - 2$ **26.** $f(x) = -(x + 1)^2 + 1$

7.9 MORE ABOUT GRAPHING QUADRATIC FUNCTIONS

COMPLETING THE SQUARE

The procedures discussed in Section 7.8 enable us to graph any quadratic function $f(x) = ax^2 + bx + c$. By *completing the square*, we can always rewrite the polynomial $ax^2 + bx + c$ in the form

$$a(x - h)^2 + k.$$

Example 1   Graph: $f(x) = x^2 - 6x + 4$.

$$f(x) = x^2 - 6x + 4$$
$$= (x^2 - 6x) + 4$$

We complete the square inside the parentheses. We take half the x-coefficient:

$$\frac{1}{2} \cdot (-6) = -3.$$

We square it:

$$(-3)^2 = 9.$$

Then we add $9 - 9$ inside the parentheses.

$$f(x) = (x^2 - 6x + 9 - 9) + 4$$
$$= (x^2 - 6x + 9) + (-9 + 4) \qquad \text{Rearranging terms}$$
$$= 1 \cdot (x - 3)^2 - 5$$

The vertex is $(3, -5)$.

The line of symmetry is $x = 3$.

We plot a few points as a check, and draw the curve.

x	$f(x)$
0	4
1	−1
5	−1
6	4

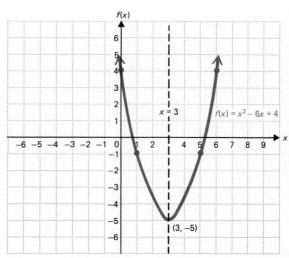

Example 2   Graph: $f(x) = -2x^2 + 10x - 7$.

We first factor the expression $-2x^2 + 10x$. Then we proceed as before:

$$f(x) = -2x^2 + 10x - 7.$$

We "remove" -2 from the first two terms by factoring. This makes the coefficient of x^2 inside the parentheses 1.

$$f(x) = -2(x^2 - 5x) - 7$$

We take half the x-coefficient and square it, to get $\frac{25}{4}$. Then we add $\frac{25}{4} - \frac{25}{4}$ inside the parentheses.

$$f(x) = -2\left(x^2 - 5x + \frac{25}{4} - \frac{25}{4}\right) - 7$$

$$= -2\left(x^2 - 5x + \frac{25}{4}\right) + 2\left(\frac{25}{4}\right) - 7 \qquad \text{Multiplying by } -2, \text{ using the distributive law, and rearranging terms}$$

$$= -2\left(x - \frac{5}{2}\right)^2 + \frac{11}{2}$$

The vertex is $(\frac{5}{2}, \frac{11}{2})$.

The line of symmetry is $x = \frac{5}{2}$.

The coefficient -2 is negative, so the graph opens downward.

We plot a few points as a check and draw the curve.

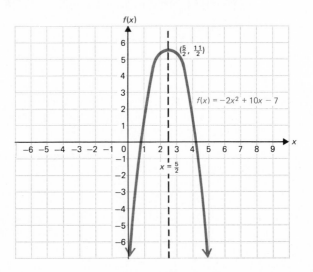

INTERCEPTS (Optional)

The points at which a graph crosses the x-axis are called its *x-intercepts*. These are of course the points at which $y = 0$.

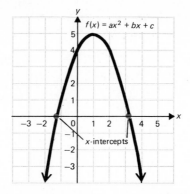

To find the x-intercepts of a quadratic function $f(x) = ax^2 + bx + c$ we solve the equation

$$0 = ax^2 + bx + c.$$

Example 3 Find the x-intercepts of the graph of $f(x) = x^2 - 2x - 2$.

We solve the equation

$$0 = x^2 - 2x - 2.$$

The equation is difficult to factor, so we use the quadratic formula and get $x = 1 \pm \sqrt{3}$. Thus the x-intercepts are $(1 - \sqrt{3}, 0)$ and $(1 + \sqrt{3}, 0)$. For plotting, we approximate, to get $(-0.7, 0)$ and $(2.7, 0)$. We sometimes refer to the x-coordinates as intercepts.

It is useful to have these points when graphing a function.

The discriminant, $b^2 - 4ac$, tells us how many real-number solutions the equation $0 = ax^2 + bx + c$ has, so it also indicates how many intercepts there are. Compare.

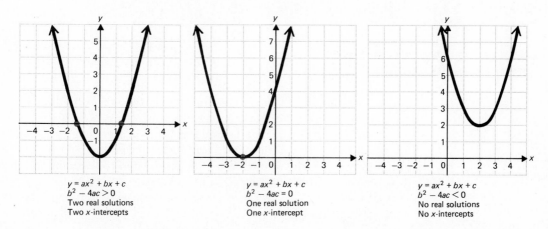

EXERCISE SET 7.9

Complete the square. Then graph.

1. $f(x) = x^2 - 2x - 3$

2. $f(x) = x^2 + 2x - 5$

3. $f(x) = -x^2 + 4x + 1$

4. $f(x) = -x^2 - 4x + 3$

5. $f(x) = 3x^2 - 24x + 50$

6. $f(x) = 4x^2 + 8x - 3$

7. $f(x) = -2x^2 + 2x + 1$

8. $f(x) = -2x^2 - 2x + 3$

Find the x-intercepts.

9. $f(x) = x^2 - 4x + 1$

10. $f(x) = x^2 + 6x + 10$

11. $f(x) = -x^2 + 2x + 3$

12. $f(x) = -x^2 + 3x + 4$

13. $f(x) = 2x^2 - 4x + 6$

14. $f(x) = 2x^2 + 4x - 1$

15. $f(x) = 4x^2 + 12x + 9$

16. $f(x) = 3x^2 - 6x + 1$

☆
___

Find the x-intercepts.

17. ▦ $f(x) = -0.978x^2 + 2.101x + 3.121$ **18.** ▦ $f(x) = 2.899x^2 - 5.901x + 0.969$

19. Let $y = (x - p)^2 + (x - q)^2$, where p and q are constants. For what value of x is y a minimum?

___

7.10 APPLICATIONS OF QUADRATIC FUNCTIONS

MAXIMUM AND MINIMUM PROBLEMS (Optional)

For a quadratic function, the value $f(x)$ at the vertex will either be greater than any other $f(x)$ or less than all other $f(x)$. If the graph opens upward, $f(x)$ will be a minimum. If the graph opens downward, $f(x)$ will be a maximum. In certain problems we want to find a maximum or a minimum. If the situation can be translated to a quadratic function, we can solve by finding $f(x)$ at the vertex.

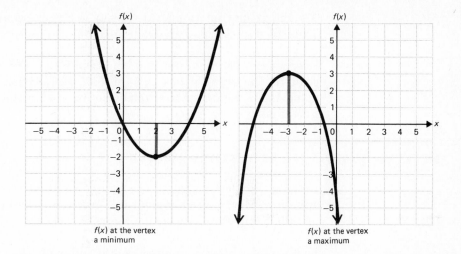

f(x) at the vertex
a minimum

f(x) at the vertex
a maximum

Example 1   What are the dimensions of the largest rectangular pen that can be enclosed with 64 meters of fence?

We make a drawing and label it. The perimeter must be 64 m, so we have

$$2w + 2l = 64. \qquad (1)$$

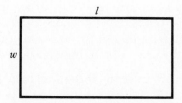

We wish to find the maximum area, so we try to find a quadratic function for the area that is an expression in one variable. We know that

$$A = lw. \qquad (2)$$

Solving (1) for l, we get $l = 32 - w$. Substituting in (2), we get a quadratic function:

$$A = (32 - w)w$$
$$A = -w^2 + 32w.$$

Completing the square, we get

$$A = -(w - 16)^2 + 256.$$

The maximum function value (area) is 256. It occurs when $w = 16$. Thus the dimensions are 16 m by 16 m.

Every quadratic function is of the type $f(x) = ax^2 + bx + c$. If a is positive, the graph opens upward and the function has a minimum value. If a is negative, the graph opens downward and the function has a maximum value.

FITTING QUADRATIC FUNCTIONS TO DATA

In many problems, a quadratic function can be used to describe the situation. We can find a quadratic function if we know three inputs and their outputs. Each such ordered pair is called a *data point*.

Example 2   Pizza Unlimited has the following prices for pizzas.

Diameter	Price
8″	$3.00
12″	$4.25
16″	$5.75

Is price a quadratic function of diameter? It probably should be, because the price should be proportional to the area, and the area is a quadratic function of the diameter. (The area of a circular region is given by $A = \pi r^2$ or $(\pi/4) \cdot d^2$.) **(a)** Fit a quadratic function to the data points $(8, 3)$, $(12, 4.25)$, and $(16, 5.75)$. **(b)** Use the function to find the price of a 14-inch pizza.

a) We use the three data points to find a, b, and c in $f(x) = ax^2 + bx + c$.

$$3.00 = a(8)^2 + b(8) + c$$
$$4.25 = a(12)^2 + b(12) + c$$
$$5.75 = a(16)^2 + b(16) + c$$

Simplifying, we get the system of equations

$$3.00 = 64a + 8b + c$$
$$4.25 = 144a + 12b + c$$
$$5.75 = 256a + 16b + c.$$

We solve this system, obtaining $(0.0078125, 0.15625, 1.25)$. Thus the function we are looking for is

$$f(x) = 0.0078125x^2 + 0.15625x + 1.25.$$

b) To find the price of a 14-inch pizza, we find $f(14)$.

$$f(14) = 0.0078125(14)^2 + 0.15625(14) + 1.25$$
$$= 4.97, \quad \text{to the nearest cent.}$$

Note that this price function always gives results greater than \$1.25, since $f(0) = 1.25$, and diameters are never negative.

EXERCISE SET 7.10

Solve.

1. A rancher is fencing off a rectangular field with a fixed perimeter of 76 ft. What dimensions will yield the maximum area? What is the maximum area?

2. A carpenter is building a rectangular room with a fixed perimeter of 68 ft. What dimensions will yield the maximum area? What is the maximum area?

3. What is the maximum product of two numbers whose sum is 22? What numbers yield this product?

4. What is the maximum product of two numbers whose sum is 45? What numbers yield this product?

5. What is the minimum product of two numbers whose difference is 4? What are the numbers?

6. What is the minimum product of two numbers whose difference is 6? What are the numbers?

7. What is the minimum product of two numbers whose difference is 5? What are the numbers?

8. What is the minimum product of two numbers whose difference is 7? What are the numbers?

Find the quadratic function that fits each set of data points.

9. $(1, 4), (-1, -2), (2, 13)$

10. $(1, 4), (-1, 6), (-2, 16)$

11. $(1, 5), (2, 9), (3, 7)$

12. $(1, -4), (2, -6), (3, -6)$

13. *Predicting earnings.* A business earns \$38 in the first week, \$66 in the second week, and \$86 in the third week. The manager graphs the points $(1, 38)$, $(2, 66)$, and $(3, 86)$ and uses a quadratic function to describe the situation.

a) Find a quadratic function that fits the data.

b) Using the function, predict the earnings for the fourth week.

14. *Predicting earnings.* A business earns \$1000 in its first month, \$2000 in the second month, and \$8000 in the third month. The manager plots the points $(1, 1000)$, $(2, 2000)$, and $(3, 8000)$ and uses a quadratic function to describe the situation.

a) Find a quadratic function that fits the data.

b) Using the function, predict the earnings for the fourth month.

15. a) Find a quadratic function that fits the following data.

Travel speed in km/h	Number of daytime accidents (for every 200 million km)
60	200
80	130
100	100

b) Use the function to calculate the number of daytime accidents that occur at 50 km/h.

16. a) Find a quadratic function that fits the following data.

Travel speed in km/h	Number of nighttime accidents (for every 200 million km)
60	400
80	250
100	250

b) Use the function to calculate the number of nighttime accidents that occur at 50 km/h.

☆

 Find the maximum or minimum value for each function.

17. $f(x) = 2.31x^2 - 3.105x - 5.98$ **18.** $f(x) = -17.7x^2 + 6.29x + 6.08$

19. The sum of the base and the height of a triangle is 38 cm. Find the dimensions for which the area is a maximum, and find the maximum area.

20. The perimeter of a rectangle is 44 ft. Find the least possible length of a diagonal.

21. A horticulturist has 180 ft of fencing with which to form a rectangular garden. A greenhouse will provide one side of the garden, and the fencing will be used for the other three sides. What is the area of the largest region that can be enclosed?

TEST OR REVIEW—CHAPTER 7

1. Solve: $4x(x - 2) - 3x(x + 1) = -18$.

2. A rectangle is 4.5 inches longer than it is wide. Its area is 34 in². Find its length and width.

3. Solve: $x^2 + x + 1 = 0$.

4. Solve. Approximate the solutions to the nearest tenth.

$x^2 + 4x - 2 = 0$

5. Solve: $\dfrac{1}{4 - x} + \dfrac{1}{2 + x} = \dfrac{3}{4}$.

6. Two pipes can fill a tank together in $1\frac{1}{2}$ hours. One pipe requires 4 hours longer alone than the other. How long would it take for the faster pipe, working alone?

7. Determine the nature of the solutions of $x^2 + 5x + 17 = 0$.

8. Write a quadratic equation having solutions $\sqrt{3}$ and $3\sqrt{3}$.

9. Solve for r: $V = \dfrac{1}{3}\pi(R^2 + r^2)$.

10. Solve: $x^4 - 5x^2 + 5 = 0$.

11. The area of a balloon varies directly as the square of its radius. If the area is 3.4 cm² when the radius is 5 cm, what is the area when the radius is 7 cm?

12. a) Graph: $f(x) = 4(x - 3)^2 + 5$.
  b) Label the vertex.
  c) Draw the line of symmetry.
  d) Find the maximum or minimum function value.

13. Complete the square: $f(x) = 2x^2 + 4x - 6$.

14. Find the x-intercepts: $f(x) = x^2 - x - 6$.

15. What is the minimum product of two numbers having a difference of 8? What are the numbers?

16. Find the quadratic function fitting these data points.

$(0, 0)$, $(3, 0)$, $(5, 2)$

8

EQUATIONS OF SECOND DEGREE AND THEIR GRAPHS

8.1 THE DISTANCE FORMULA

THE PYTHAGOREAN THEOREM

A famous theorem about right triangles relates the lengths of the sides.

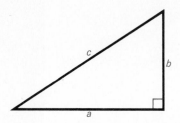

The Pythagorean Theorem: **In a right triangle, if c is the length of the hypotenuse and a and b are the lengths of the other two sides (the legs), then**

$$a^2 + b^2 = c^2.$$

If we know two sides of a right triangle, we can find the other.

Example 1   Find the length of the hypotenuse of this right triangle.

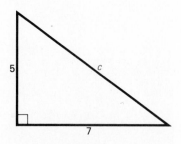

$$a^2 + b^2 = c^2 \qquad \text{Pythagorean theorem}$$
$$7^2 + 5^2 = c^2 \qquad \text{Substituting}$$
$$49 + 25 = c^2$$
$$74 = c^2$$
$$\sqrt{74} = c \qquad \text{Taking principal square root}$$

The answer is $\sqrt{74}$. We do not consider the negative square root, because lengths are not negative. From Table 1, we can approximate $\sqrt{74}$. We find $c \approx 8.602$.

Example 2   Find the length of the leg of this right triangle.

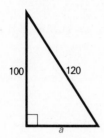

$$a^2 + b^2 = c^2 \qquad \text{Pythagorean theorem}$$
$$a^2 + 100^2 = 120^2 \qquad \text{Substituting}$$
$$a^2 + 10{,}000 = 14{,}400$$
$$a^2 = 4400$$
$$a = \sqrt{4400}, \quad \text{or} \quad 10\sqrt{44}$$

Approximating, using Table 1, we get

$$a \approx 10 \times 6.633, \quad \text{or} \quad 66.33.$$

THE DISTANCE FORMULA

We now develop a formula for finding the distance between any two points on a graph when we know their coordinates. First, we consider points on a vertical or horizontal line.

If points are on a vertical line, they have the same first coordinate.

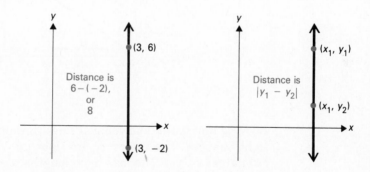

On the left above, we can find the distance between the points by subtracting their second coordinates.

$$6 - (-2) = 8 \qquad \text{The distance is 8.}$$

If we subtract the opposite way, we get

$$-2 - 6 = -8 \qquad \text{The distance is } |-8|, \text{ or } 8.$$

If we use absolute value, as on the right above, we need not worry about the order. We can subtract either way.

If points are on a horizontal line, we subtract their first coordinates and take the absolute value.

Examples   Find the distance between these points.

3. $(-5, 13)$ and $(-5, 2)$ **4.** $(7, -3)$ and $(-5, -3)$

We subtract second coordinates, since the first coordinates are the same. The distance is

Since the second coordinates are the same, we subtract first coordinates.

$$|13 - 2|, \quad \text{or} \quad 11.$$

$$|-5 - 7| = |-12|, \quad \text{or} \quad 12$$

Next we consider two points that are not on either a vertical or horizontal line, such as (x_1, y_1) and (x_2, y_2) in the figure below. By

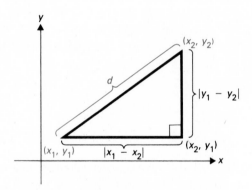

drawing horizontal and vertical lines through these points, we form a right triangle. The vertex of the right angle of this triangle has coordinates (x_2, y_1). The legs have lengths $|x_1 - x_2|$ and $|y_1 - y_2|$. The distance we want is the length of the hypotenuse. By the Pythagorean theorem,

$$d^2 = |x_1 - x_2|^2 + |y_1 - y_2|^2.$$

Since squares of numbers are never negative, we don't really need the absolute value signs. Thus we have

$$d^2 = (x_1 - x_2)^2 + (y_1 - y_2)^2.$$

Taking the principal square root we get the distance formula.

> *The Distance Formula:*   **The distance between any two points (x_1, y_1) and (x_2, y_2) is given by**
>
> $$d = \sqrt{(x_1 - x_2)^2 + (y_1 - y_2)^2}.$$

This formula holds even when the two points are on a vertical or horizontal line.

Example 5   Find the distance between $(4, -3)$ and $(-5, 4)$.

We substitute into the distance formula.

$$d = \sqrt{[4 - (-5)]^2 + (-3 - 4)^2} \qquad \text{Substituting}$$
$$d = \sqrt{9^2 + (-7)^2}$$
$$d = \sqrt{130}$$
$$d = \sqrt{10} \cdot \sqrt{13}$$
$$\approx 3.162 \times 3.606 \approx 11.402$$

> Note that the distance formula has a radical sign in it. Do not make the mistake of writing $d = (x_1 - x_2)^2 + (y_1 - y_2)^2$.

EXERCISE SET 8.1

In each exercise, the lengths of two sides of a right triangle are given. Find the length of the other side (c is always the length of the hypotenuse). Where appropriate, use Table 1 to find an approximation.

1. $a = 3, \quad b = 4$ **2.** $b = 8, \quad c = 10$ **3.** $a = 16, \quad c = 20$

4. $a = 5, \quad b = 12$ **5.** $a = 6, \quad b = 8$ **6.** $b = 20, \quad c = 25$

7. $a = 5, \quad b = 7$ **8.** $b = 11, \quad c = 12$ **9.** $a = 1, \quad b = 1$

10. $a = 4, \quad b = 7$ **11.** $a = 2, \quad b = 2\sqrt{2}$ **12.** $a = 3, \quad b = 3\sqrt{6}$

Find the distance between each pair of points. Where appropriate, use Table 1 to find an approximation.

13. $(9, 5)$ and $(6, 1)$ **14.** $(1, 10)$ and $(7, 2)$ **15.** $(0, -7)$ and $(3, -4)$

16. $(6, 2)$ and $(6, -8)$ **17.** $(2, 2)$ and $(-2, -2)$ **18.** $(5, 21)$ and $(-3, 1)$

☆
_____

The converse of the Pythagorean theorem is true. That is, if the sides of a triangle have lengths a, b, and c, and $a^2 + b^2 = c^2$, then the triangle is a right triangle. Determine whether the points are vertices of a right triangle.

19. $(9, 6)$, $(-1, 2)$, and $(1, -3)$ **20.** $(-8, -5)$, $(6, 1)$, and $(-4, 5)$

21. Explain why the distance formula holds even if two points are on a vertical or horizontal line.

22. Find the distance between these points:

a) $(-1, 3k)$ and $(6, 2k)$ b) $(0, \sqrt{7})$ and $(\sqrt{6}, 0)$

c) $(6m, -7n)$ and $(-2m, n)$ d) $(\sqrt{d}, -\sqrt{3c})$ and $(\sqrt{d}, \sqrt{3c})$

e) $\left(\dfrac{5}{7}, \dfrac{1}{14}\right)$ and $\left(\dfrac{1}{7}, \dfrac{11}{14}\right)$ f) $(-3\sqrt{3}, 1 - \sqrt{6})$ and $(\sqrt{3}, 1 + \sqrt{6})$

g) ▦$(5.989, 2.001)$ and $(3.712, -7.784)$

23. Prove: If the endpoints of a segment are (x_1, y_1) and (x_2, y_2), then the coordinates of the midpoint are $\left(\dfrac{x_1 + x_2}{2}, \dfrac{y_1 + y_2}{2}\right)$.

24. Find the midpoints of the segments having the following endpoints.

a) $(-12, -7)$ and $(-3, 13)$ b) $(-a, b)$ and (a, b)

c) $(2 - \sqrt{3}, 5\sqrt{2})$ and $(2 + \sqrt{3}, 3\sqrt{2})$ d) $\left(\dfrac{1}{3}, -\dfrac{7}{11}\right)$ and $\left(-\dfrac{2}{3}, \dfrac{3}{4}\right)$

25. Determine whether the points $(6, 3)$, $(-4, -1)$, and $(-2, -6)$ are vertices of a right triangle.

26. Find the point on the y-axis that is equidistant from $(2, 10)$ and $(6, 2)$.

_____

8.2 CONIC SECTIONS: CIRCLES AND ELLIPSES

CIRCLES

Circle

A circle is defined to be the set of all points in a plane that are at a fixed distance from a point in that plane. Circles are called *conic sections* because they can be formed by cutting a cone by a plane, as shown.

Let us find an equation for a circle. We call the center (a, b) and we suppose the radius has length r. Suppose (x, y) is any point on the circle. By the distance formula, we have

$$\sqrt{(x - a)^2 + (y - b)^2} = r$$ Squaring both sides gives us an equation of the circle in standard form.

> *Equation of a circle:* **A circle with center (a, b) and radius r has an equation**
>
> $$(x - a)^2 + (y - b)^2 = r^2.$$ **(Standard form)**

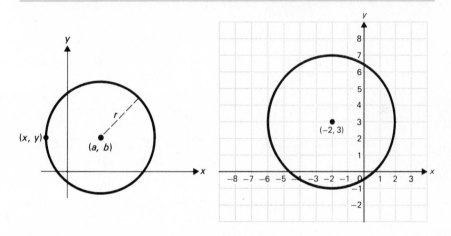

Figure 1   **Figure 2**

Example 1   Find the center and radius and graph this circle.

$$(x + 2)^2 + (y - 3)^2 = 16$$

We rewrite to get the standard form given above.

$$[x - (-2)]^2 + (y - 3)^2 = 4^2$$

Thus the center is $(-2, 3)$ and the radius is 4. The graph is as shown. It can be drawn by locating the center and then using a compass.

WRITING EQUATIONS

Example 2   Write an equation of a circle with center $(9, -5)$ and radius $\sqrt{2}$.

We use the standard form $(x - a)^2 + (y - b)^2 = r^2$.

$$(x - 9)^2 + [y - (-5)]^2 = (\sqrt{2})^2$$ Substituting
$$(x - 9)^2 + (y + 5)^2 = 2$$ Simplifying

If a circle is centered at the origin, then $(a, b) = (0, 0)$.

> **A circle centered at the origin with radius r has an equation**
>
> $$x^2 + y^2 = r^2.$$

EQUATIONS NOT IN STANDARD FORM

Example 3   Show that this is an equation of a circle. Find the center and radius.

$$x^2 + y^2 + 6x - 4y + 9 = 0$$

We first regroup the terms and then complete the square twice, once with $x^2 + 6x$ and once with $y^2 - 4y$.

$$(x^2 + 6x) + (y^2 - 4y) + 9 = 0$$
$$(x^2 + 6x + 9 - 9) + (y^2 - 4y + 4 - 4) + 9 = 0$$
$$(x^2 + 6x + 9) + (y^2 - 4y + 4) - 9 - 4 + 9 = 0$$
$$(x + 3)^2 + (y - 2)^2 = 4$$
$$[x - (-3)]^2 + (y - 2)^2 = 2^2$$

This shows that the graph of the equation is a circle. The center is $(-3, 2)$ and the radius is 2.

> Remember, from p. 345, that the standard form is
> $$(x - a)^2 + (y - b)^2 = r^2.$$

ELLIPSES

When a cone is cut at an angle, as shown, the conic section formed is an *ellipse*.

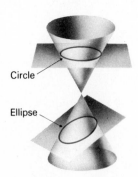

Circle

Ellipse

You can draw an ellipse by sticking two tacks in a piece of cardboard. Then tie a string to the tacks, place a pencil as shown, and draw.

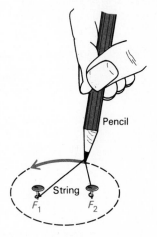

Pencil

String

F_1 F_2

The formal mathematical definition is related to this method of drawing the ellipse. The ellipse is defined to be the set of all points in a plane such that the *sum* of the distances from two fixed points F_1 and F_2 (called *foci*) is constant.

Ellipses have equations as follows.

Equation of an ellipse: **An ellipse with its center at the origin has an equation**

$$\frac{x^2}{a^2} + \frac{y^2}{b^2} = 1, \quad a, b > 0. \qquad \textbf{(Standard form)}$$

We can think of a circle as a special kind of ellipse. A circle is formed when the angle at which the cutting plane cuts the cone is 90°. It is also formed when the foci, F_1 and F_2, are the same point. An ellipse with its foci close together is very nearly a circle.

In graphing ellipses, it helps to first find the intercepts. If we replace x by 0, we can find the y-intercepts.

$$\frac{0^2}{a^2} + \frac{y^2}{b^2} = 1$$

$$\frac{y^2}{b^2} = 1$$

$$y^2 = b^2$$

$$y = \pm b$$

$(0, b)$

$(-a, 0)$

$(a, 0)$

$(0, -b)$

Thus the y-intercepts are $(0, b)$ and $(0, -b)$. Similarly, the x-intercepts are $(a, 0)$ and $(-a, 0)$. Plotting these points and filling in an oval-shaped curve we get a graph of the ellipse. If a more precise graph is desired, we can plot more points.

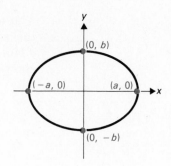

Example 4   Graph the ellipse

$$\frac{x^2}{4} + \frac{y^2}{9} = 1.$$

Note that

$$\frac{x^2}{4} + \frac{y^2}{9} = \frac{x^2}{2^2} + \frac{y^2}{3^2}.$$

Thus the x-intercepts are $(2, 0)$ and $(-2, 0)$, and the y-intercepts are $(0, 3)$ and $(0, -3)$.

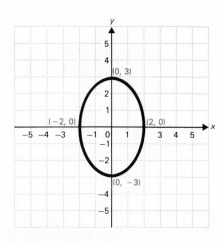

EXERCISE SET 8.2

Find the center and radius of each circle. Graph each circle.

1. $(x + 1)^2 + (y + 3)^2 = 4$

2. $(x - 2)^2 + (y + 3)^2 = 1$

3. $(x - 3)^2 + y^2 = 2$

4. $x^2 + (y - 1)^2 = 3$

5. $x^2 + y^2 = 25$

6. $x^2 + y^2 = 9$

Find an equation of the circle having center and radius as given.

7. Center $(0, 0)$, radius 7

8. Center $(0, 0)$, radius 4

9. Center $(-2, 7)$, radius $\sqrt{5}$

10. Center $(5, 6)$, radius $2\sqrt{3}$

Find the center and radius of each circle.

11. $x^2 + y^2 + 8x - 6y - 15 = 0$

12. $x^2 + y^2 + 6x - 4y - 15 = 0$

13. $x^2 + y^2 - 8x + 2y + 13 = 0$

14. $x^2 + y^2 + 6x + 4y + 12 = 0$

15. $x^2 + y^2 - 4x = 0$

16. $x^2 + y^2 + 10y - 75 = 0$

Graph each ellipse.

17. $\dfrac{x^2}{4} + \dfrac{y^2}{1} = 1$

18. $\dfrac{x^2}{1} + \dfrac{y^2}{4} = 1$

19. $\dfrac{x^2}{9} + \dfrac{y^2}{16} = 1$

20. $\dfrac{x^2}{25} + \dfrac{y^2}{16} = 1$

☆

21. Find an equation of a circle satisfying the given conditions:

a) Center $(0, 0)$, passing through $\left(\dfrac{1}{4}, \dfrac{\sqrt{31}}{4}\right)$.

b) Center $(-4, 1)$, passing through $(2, -5)$.

c) Center $(-3, -2)$ and tangent to the y-axis.

d) The endpoints of a diameter are $(7, 3)$ and $(-1, -3)$.

The equation (standard form) of an ellipse centered at the origin is

$$\frac{x^2}{a^2} + \frac{y^2}{b^2} = 1.$$

The vertices are $(a, 0)$, $(-a, 0)$, $(0, b)$, and $(0, -b)$. The equation (standard form) of an ellipse with center (h, k) is

$$\frac{(x - h)^2}{a^2} + \frac{(y - k)^2}{b^2} = 1.$$

The vertices are $(a + h, k)$, $(-a + h, k)$, $(h, b + k)$, and $(h, -b + k)$.

22. For each ellipse write the equation in standard form, and find the center and vertices. Then graph the ellipse.

a) $4x^2 + 9y^2 = 36$

b) $4(x - 1)^2 + 9(y + 2)^2 = 36$

c) $16x^2 + y^2 + 96x - 8y + 144 = 0$

d) $4x^2 + 25y^2 - 8x + 50y - 71 = 0$

23. Find the equation of the ellipse with the following vertices.

a) $(-8, 0)$, $(8, 0)$, $(0, -2)$, $(0, 2)$

b) $(0, 2)$, $(5, -1)$, $(10, 2)$, $(5, 5)$

24. Find an equation of the ellipse that contains the point $(1, 2\sqrt{3})$ and two of its vertices are $(2, 0)$ and $(-2, 0)$.

8.3 CONIC SECTIONS: PARABOLAS AND HYPERBOLAS

PARABOLAS

When a cone is cut as shown on the left below, the conic section formed is a *parabola*. If both parts of the cone are cut by a plane, as shown on the right, the conic section is called a *hyperbola*. We do not show it here, but graphs of quadratic functions (see Chapter 7) are actually parabolas. General equations of parabolas are quadratic.

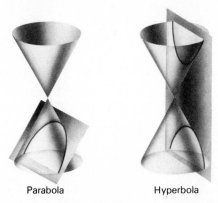

Parabola   Hyperbola

Parabolas: **Parabolas have equations as follows.**

$y = ax^2 + bx + c$ **(Line of symmetry parallel to y-axis)**
$x = ay^2 + by + c$ **(Line of symmetry parallel to x-axis)**

Example 1   Graph $y = x^2 - 4x + 8$. Then graph $x = y^2 - 4y + 8$.

a) We complete the square:

$$y = (x^2 - 4x + 4) + 4 = (x - 2)^2 + 4.$$

The vertex is at $(2, 4)$ and the curve opens upward, as shown in Figure 1.

b) To graph $x = y^2 - 4y + 8$ we complete the square on y:

$$x = (y - 2)^2 + 4.$$

Since x and y are merely interchanged, this graph looks just like the first one, but is turned horizontally. The vertex is $(4, 2)$. It helps to plot a few points. (See Figure 2.)

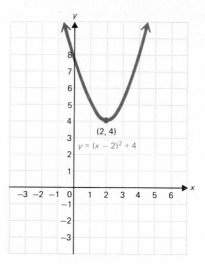

Figure 1

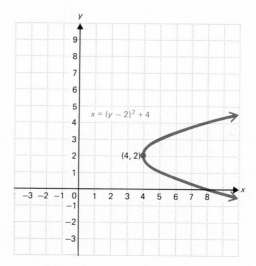

Figure 2

HYPERBOLAS (STANDARD FORM)

A hyperbola looks somewhat like a pair of parabolas, but the actual shapes are different. A hyperbola has two vertexes, and the line through the vertexes is known as an *axis*. The point halfway between the vertexes is called the *center*. We consider only hyperbolas centered at the origin.

Hyperbolas: **Hyperbolas with their centers at the origin have equations as follows.**

$$\frac{x^2}{a^2} - \frac{y^2}{b^2} = 1 \qquad \text{(Axis horizontal)}$$

$$\frac{y^2}{b^2} - \frac{x^2}{a^2} = 1 \qquad \text{(Axis vertical)}$$

Note carefully that these equations have a 1 on the right.

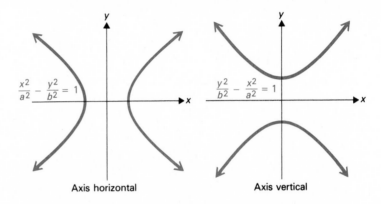

Axis horizontal   Axis vertical

To graph a hyperbola, it helps to begin by graphing the lines called *asymptotes*.

Asymptotes of a hyperbola: **For hyperbolas with equations as given above, the asymptotes are the lines**

$$y = \frac{b}{a}x \quad \text{and} \quad y = -\frac{b}{a}x.$$

As a hyperbola gets farther away from the origin, it gets closer and closer to its asymptotes.

Parabolas do not have asymptotes.

The next thing to do is to find the vertexes. Then it is easy to sketch the curve.

Example 2 Graph: $\dfrac{x^2}{4} - \dfrac{y^2}{9} = 1$.

a) Note that

$$\frac{x^2}{4} - \frac{y^2}{9} = \frac{x^2}{2^2} - \frac{y^2}{3^2},$$

so $a = 2$ and $b = 3$. The asymptotes are thus

$$y = \frac{3}{2}x \quad \text{and} \quad y = -\frac{3}{2}x.$$

We sketch them.

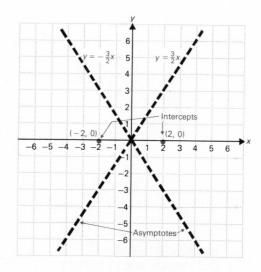

b) Next we find the intercepts. Let $y = 0$ (simply cover up the term containing y), and we see that $x^2/2^2 = 1$, so $x = \pm 2$. The intercepts are $(2, 0)$ and $(-2, 0)$.

There are intercepts only on one axis. If we cover up the term containing x, we see that $y^2/9 = -1$, and this equation has no real-number solutions.

c) We plot the intercepts and sketch the graph. Then through each intercept we draw a smooth curve that approaches the asymptotes closely.

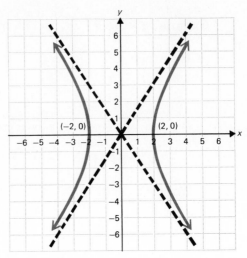

Example 3   Graph: $\dfrac{y^2}{36} - \dfrac{x^2}{4} = 1$.

a) Note that

$$\frac{y^2}{36} - \frac{x^2}{4} = \frac{y^2}{6^2} - \frac{x^2}{2^2} = 1.$$

> The intercept distance is this number, in the term without the minus sign. There is y in this term, so the intercepts are on the y-axis.

The slopes of the asymptotes are $\frac{6}{2}$ and $-\frac{6}{2}$.

Using the numbers 6 and 2 we can quickly sketch a rectangle to use as a guide. To find the corners, we go 6 units (up or down) in the y-direction and 2 units (right or left) in the x-direction. The asymptotes go through the corners.

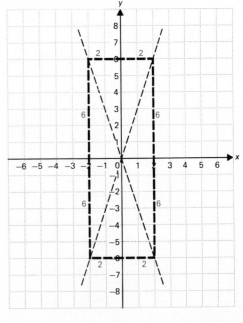

c) Now we draw curves through the intercepts toward the asymptotes as shown.

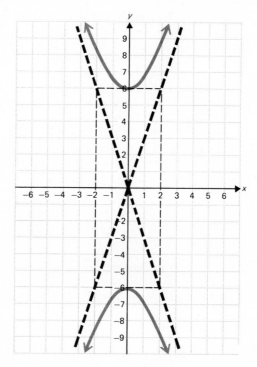

HYPERBOLAS (NONSTANDARD FORM)

The equations we have already seen for hyperbolas are the standard ones, but there are other hyperbolas. We consider some of them.

> **Hyperbolas having the x- and y-axes as asymptotes have equations as follows:**
>
> $$xy = c, \quad \text{where } c \text{ is a nonzero constant.}$$

Example 4   Graph: $xy = -8$.

First solve for y:

$$y = -\frac{8}{x}.$$

Next find some solutions, keeping the results in a table. (Note that we cannot use 0 for x.)

x	2	-2	4	-4	1	-1
y	-4	4	-2	2	-8	8

Now plot the points. Connect those points in the second quadrant with a smooth curve. Similarly, connect those in the fourth quadrant. Remember that the x-axis and the y-axis are asymptotes.

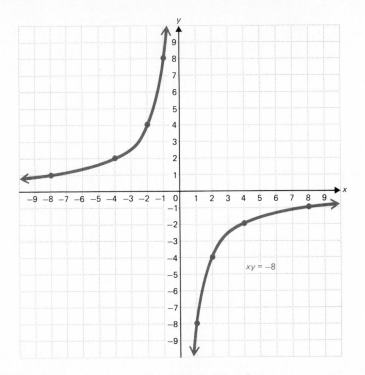

In Example 4, the two parts, or *branches*, of the hyperbola are in the second and fourth quadrants. If the constant c is positive the two branches will be in the first and third quadrants, as in the graph shown at the top of the following page.

Hyperbolas having equations $xy = c$ have the same shape as hyperbolas aligned with the coordinate axis, having equations

$$\frac{x^2}{a^2} - \frac{y^2}{a^2} = 1.$$

The graph is merely rotated 45° from the position of the latter.

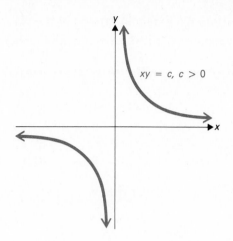

EXERCISE SET 8.3

Graph each of the following.

1. $y = x^2$

2. $x = y^2$

3. $x = y^2 + 4y + 1$

4. $y = x^2 - 2x + 3$

5. $y = -x^2 + 4x - 5$

6. $x = 4 - 3y - y^2$

7. $x = -3y^2 - 6y - 1$

8. $y = -5 - 8x - 2x^2$

9. $\dfrac{x^2}{16} - \dfrac{y^2}{16} = 1$

10. $\dfrac{y^2}{9} - \dfrac{x^2}{9} = 1$

11. $\dfrac{y^2}{16} - \dfrac{x^2}{9} = 1$

12. $\dfrac{x^2}{9} - \dfrac{y^2}{4} = 1$

13. $\dfrac{x^2}{25} - \dfrac{y^2}{36} = 1$

14. $\dfrac{y^2}{9} - \dfrac{x^2}{25} = 1$

15. $xy = 6$

16. $xy = -4$

17. $xy = -8$

18. $xy = 3$

☆

19. *Horsepower of an engine.* The horsepower of a certain kind of engine is given by the formula

$$H = \frac{D^2 N}{2.5},$$

where N is the number of cylinders and D is the diameter of each piston. Graph this equation, assuming that $N = 6$ (a 6-cylinder engine). Let D run from 2.5 to 8.

20. Find an equation of a parabola satisfying the given conditions:

    a) Line of symmetry parallel to y-axis and passing through the points $(0, 3)$, $(-1, 6)$, $(2, 9)$.

    b) Line of symmetry parallel to x-axis, vertex $(2, 1)$ and passing through the point $(18, -1)$.

21. Classify the graph of each of the following equations as a circle, an ellipse, a parabola, or a hyperbola.

    a) $x^2 + y^2 - 10x + 8y - 40 = 0$ b) $y + 1 = 2x^2$

    c) $9x^2 - 4y^2 - 36x + 24y - 36 = 0$ d) $1 - 3y = 2y^2 - x$

    e) $4x^2 + 25y^2 - 8x - 100y + 4 = 0$ f) $\dfrac{x^2}{7} + \dfrac{y^2}{7} = 1$

8.4 SYSTEMS OF EQUATIONS

ALGEBRAIC SOLUTIONS

We consider systems of one first-degree and one second-degree equation. For example, the graphs may be a circle and a line. If so, there are three possibilities. (See the figure below.)

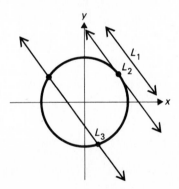

For L_1 there is no point of intersection, hence no solution in real numbers. For L_2 there is one point of intersection, hence one real-number solution. For L_3 there are two points of intersection, hence two real solutions.

These systems can be solved graphically by finding the points of intersection. In solving algebraically we use the substitution method.

Example 1   Solve.

$$x^2 + y^2 = 25 \qquad (1)$$
$$3x - 4y = 0 \qquad (2)$$

First solve the linear equation (2) for x:

$$x = \frac{4}{3}y. \qquad (3)$$

Then substitute $\frac{4}{3}y$ for x in equation (1) and solve for y:

$$\left(\frac{4}{3}y\right)^2 + y^2 = 25$$

$$\frac{16}{9}y^2 + y^2 = 25$$

$$\frac{25}{9}y^2 = 25$$

$$y^2 = 9$$

$$y = \pm 3.$$

Now substitute these numbers into equation (3) and solve for x:

$$x = \tfrac{4}{3}(3), \quad \text{or } 4; \qquad x = \tfrac{4}{3}(-3), \quad \text{or } -4.$$

The pairs $(4, 3)$ and $(-4, -3)$ check, so they are solutions.

Example 2   Solve the system.

$$y + 3 = 2x \qquad (1)$$
$$x^2 + 2xy = -1 \qquad (2)$$

First solve the linear equation (1) for y:

$$y = 2x - 3. \qquad (3)$$

Then substitute $2x - 3$ for y in equation (2) and solve for x:

$$x^2 + 2x(2x - 3) = -1$$
$$5x^2 - 6x + 1 = 0$$
$$(5x - 1)(x - 1) = 0 \qquad \text{Factoring}$$
$$5x - 1 = 0 \qquad x - 1 = 0 \qquad \text{Using the principle of}$$
$$\qquad\qquad\qquad\qquad\qquad\qquad \text{zero products}$$

$$x = \frac{1}{5} \quad \text{or} \qquad x = 1.$$

Now substitute these numbers into equation (3) and solve for y:

$$y = 2\left(\frac{1}{5}\right) - 3, \quad \text{or} \quad -\frac{13}{5},$$

$$y = 2(1) - 3, \quad \text{or} \quad -1.$$

The pairs $(\frac{1}{5}, -\frac{13}{5})$ and $(1, -1)$ check, so they are solutions.

SOLVING PROBLEMS

Example 3   The sum of two numbers is 14 and the sum of their squares is 106. What are the numbers?

We translate to equations.

The sum of two numbers is 14.

$$x + y \qquad = 14$$

The sum of the squares is 106.

$$x^2 + y^2 \qquad = 106$$

Now we solve the system

$$x + y = 14$$
$$x^2 + y^2 = 106.$$

$(14 - y)^2 + y^2 = 106$	Solving the first equation for x and substituting into the second
$14^2 - 28y + y^2 + y^2 = 106$	Squaring
$2y^2 - 28y + 90 = 0$	Standard form
$y^2 - 14y + 45 = 0$	Multiplying by $\frac{1}{2}$
$(y - 5)(y - 9) = 0$	Factoring
$y = 5 \quad \text{or} \quad y = 9$	Principle of zero products

If $y = 5$, then $x = 9$; if $y = 9$, then $x = 5$.

We get two solutions: $(5, 9)$ and $(9, 5)$. These check, and they give the same answer to the problem. The numbers are 5 and 9.

Example 4   The perimeter of a rectangular field is 204 m and the area is 2565 m². Find the dimensions of the field.

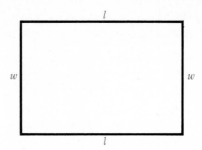

We first translate the conditions of the problem to equations, using w for the width and l for the length.

Perimeter: $2w + 2l = 204$

Area: $lw = 2565$

Now we solve the system

$$2w + 2l = 204$$
$$lw = 2565.$$

$$2w + 2\left(\frac{2565}{w}\right) = 204 \qquad \text{Solving the second equation for } l \text{ and substituting into the first}$$

$$2w^2 + 2 \times 2565 = 204w \qquad \text{Multiplying by } w$$

$$2w^2 - 204w + 2 \times 2565 = 0 \qquad \text{Standard form}$$

$$w^2 - 102w + 2565 = 0 \qquad \text{Multiplying by } \frac{1}{2}$$

$$w = \frac{102 \pm \sqrt{102^2 - 4 \cdot (1) \cdot 2565}}{2} \qquad \text{Quadratic formula}$$

$$w = \frac{102 \pm 12}{2}$$

$$w = 57 \quad \text{or} \quad w = 45$$

If $w = 57$, then $l = 45$; if $w = 45$, then $l = 57$.

Since we want w less than l, we have (57, 45).

Now we check in the original problem: The perimeter is $2 \cdot 45 + 2 \cdot 57$, or 204. The area is $45 \cdot 57$, or 2565. The numbers check, so the answer is $l = 57$ m and $w = 45$ m.

EXERCISE SET 8.4

Solve.

1. $x^2 + y^2 = 25$
   $y - x = 1$

2. $x^2 + y^2 = 100$
   $y - x = 2$

3. $4x^2 + 9y^2 = 36$
   $3y + 2x = 6$

4. $9x^2 + 4y^2 = 36$
   $3x + 2y = 6$

5. $y^2 = x + 3$
   $2y = x + 4$

6. $y = x^2$
   $3x = y + 2$

7. $x^2 - xy + 3y^2 = 27$
   $x - y = 2$

8. $2y^2 + xy + x^2 = 7$
   $x - 2y = 5$

9. The sum of two numbers is 13 and the sum of their squares is 109. What are the numbers?

10. The sum of two numbers is 15 and the difference of their squares is also 15. What are the numbers?

11. A rectangle has perimeter 28 ft and the length of a diagonal is 10 ft. What are its dimensions?

12. A rectangle has perimeter 6 in. and the length of a diagonal is $\sqrt{5}$ in. What are its dimensions?

☆
13. Find an equation of the circle that passes through the points $(4, 6)$, $(-6, 2)$, and $(1, -3)$.

14. Find two numbers whose product is 2 and the sum of whose reciprocals is $\frac{33}{8}$.

15. Find an equation of a circle that passes through $(-2, 3)$ and $(-4, 1)$ and whose center is on the line $5x + 8y = -2$.

16. The sum of two numbers is 1, and their product is 1. Find the sum of their cubes. There is a method to solve this problem that is easier than solving a system of one first-degree equation and one second-degree equation. Can you discover it?

8.5 SYSTEMS OF SECOND-DEGREE EQUATIONS

ALGEBRAIC SOLUTIONS

Two second-degree equations can have common solutions in various ways. If the graphs happen to be a circle and a hyperbola, for example, there are five possibilities, as shown below.

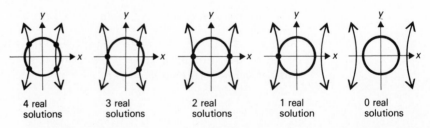

| 4 real solutions | 3 real solutions | 2 real solutions | 1 real solution | 0 real solutions |

To solve systems of two second-degree equations we can use either the substitution method or the addition method.

Example 1 Solve.

$$2x^2 + 5y^2 = 22 \qquad (1)$$
$$3x^2 - y^2 = -1 \qquad (2)$$

Here we use the addition method.

$$2x^2 + 5y^2 = 22$$
$$\underline{15x^2 - 5y^2 = -5} \qquad \text{Multiplying by 5 on both sides of (2)}$$
$$17x^2 = 17 \qquad \text{Adding}$$
$$x^2 = 1$$
$$x = \pm 1$$

If $x = 1$, $x^2 = 1$, and if $x = -1$, $x^2 = 1$, so substituting 1 or -1 for x in equation (2) we have

$$3 \cdot 1 - y^2 = -1$$
$$y^2 = 4$$
$$y = \pm 2.$$

Thus if $x = 1$, $y = 2$ or $y = -2$, and if $x = -1$, $y = 2$ or $y = -2$. The possible solutions are $(1, 2)$, $(1, -2)$, $(-1, 2)$, and $(-1, -2)$.

Check: Since $(2)^2 = 4$, $(-2)^2 = 4$, $(1)^2 = 1$, and $(-1)^2 = 1$, we can check all four pairs at one time.

$$\begin{array}{c|c} 2x^2 + 5y^2 = 22 & \\ \hline 2(\pm 1)^2 + 5(\pm 2)^2 & 22 \\ 2 + 20 & \\ 22 & \end{array}$$

$$\begin{array}{c|c} 3x^2 - y^2 = -1 & \\ \hline 3(\pm 1)^2 - (\pm 2)^2 & -1 \\ 3 - 4 & \\ -1 & \end{array}$$

The solutions are $(1, 2)$, $(1, -2)$, $(-1, 2)$, and $(-1, -2)$.

Example 2 Solve the system.

$$x^2 + 4y^2 = 20 \qquad (1)$$
$$xy = 4 \qquad (2)$$

Here we use the substitution method. First we solve equation (2) for y:

$$y = \frac{4}{x}.$$

Then we substitute $\dfrac{4}{x}$ for y in equation (1) and solve for x:

$$x^2 + 4\left(\frac{4}{x}\right)^2 = 20$$

$$x^2 + \frac{64}{x^2} = 20$$

$$x^4 + 64 = 20x^2 \qquad \text{Multiplying by } x^2$$
$$x^4 - 20x^2 + 64 = 0$$
$$u^2 - 20u + 64 = 0 \qquad \text{Letting } u = x^2$$
$$(u - 16)(u - 4) = 0 \qquad \text{Factoring}$$
$$u = 16 \quad \text{or} \quad u = 4 \qquad \text{Principle of zero products}$$

Then $x = 4$ or $x = -4$ or $x = 2$ or $x = -2$.

Since $y = 4/x$, if $x = 4$, $y = 1$; if $x = -4$, $y = -1$; if $x = 2$, $y = 2$; and if $x = -2$, $y = -2$.

The ordered pairs $(4, 1)$, $(-4, -1)$, $(2, 2)$, $(-2, -2)$ check. They are the solutions.

SOLVING PROBLEMS

Example 3   The area of a rectangle is 300 yd^2 and the length of a diagonal is 25 yd. Find the dimensions.

First draw a picture.

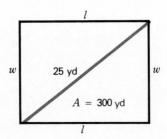

We use l for the length and w for the width and translate to equations.

From the Pythagorean theorem: $\quad l^2 + w^2 = 25^2$

Area: $\qquad lw = 300$

Now we solve the system
$$l^2 + w^2 = 625$$
$$lw = 300.$$

We get $(20, 15)$ and $(-20, -15)$. We check in the original problem: $15^2 + 20^2 = 25^2$ and $15 \cdot 20 = 300$, so $(20, 15)$ is a solution of the problem. Lengths of sides cannot be negative so $(-20, -15)$ is not a solution. The answer is $l = 20$ yd and $w = 15$ yd.

EXERCISE SET 8.5

Solve.

1. $x^2 + y^2 = 25$
$\quad\;\; y^2 = x + 5$

2. $y = x^2$
$\quad\; x = y^2$

3. $x^2 + y^2 = 9$
$\quad\; x^2 - y^2 = 9$

4. $y^2 - 4x^2 = 4$
$\quad 4x^2 + y^2 = 4$

5. $x^2 + y^2 = 5$
$\quad\;\;\; xy = 2$

6. $x^2 + y^2 = 20$
$\quad\quad\;\; xy = 8$

7. $\quad xy - y^2 = 2$
$\quad 2xy - 3y^2 = 0$

8. $2xy + 3y^2 = 7$
$\quad 3xy - 2y^2 = 4$

9. The product of two numbers is 156. The sum of their squares is 313. Find the numbers.

10. The product of two numbers is 60. The sum of their squares is 136. Find the numbers.

11. The area of a rectangle is $\sqrt{3}$ m^2 and the length of a diagonal is 2 m. Find the dimensions.

12. The area of a rectangle is $\sqrt{2}$ m^2 and the length of a diagonal is $\sqrt{3}$ m. Find the dimensions.

☆ _____

13. The square of a certain number exceeds twice the square of another number by $\frac{1}{8}$. The sum of their squares is $\frac{5}{16}$. Find the numbers.

14. Find an equation of a circle that passes through $(4, -20)$, $(10, -2)$, and $(-4, -4)$.

15. Four squares with sides 5 in. long are cut from the corners of a rectangular metal sheet that has an area of 340 in^2. The edges are bent up to form an open box with a volume of 350 in^3. Find the dimensions of the box.

TEST OR REVIEW—CHAPTER 8

1. The legs of a right triangle have lengths 3 and $\sqrt{5}$. Find the length of the hypotenuse.

2. Find the distance between $(-6, 2)$ and $(6, 8)$.

3. Write an equation of the circle with center $(2, -7)$ and having a radius of length $3\sqrt{2}$.

4. Find the center and radius of the circle $x^2 + y^2 + 8x - 10y - 8 = 0$.

5. Graph: $\dfrac{x^2}{16} + \dfrac{y^2}{4} = 1$.

6. Graph: $y = x^2 - 4x - 1$.

7. Graph: $\dfrac{x^2}{9} - \dfrac{y^2}{4} = 1$.

8. Graph: $xy = -5$.

9. Solve.

$$\dfrac{x^2}{16} + \dfrac{y^2}{9} = 1$$

$$3x + 4y = 12$$

10. Solve.

$$x^2 + y^2 = 16$$

$$\dfrac{x^2}{16} - \dfrac{y^2}{9} = 1$$

9

EXPONENTIAL AND LOGARITHMIC FUNCTIONS

9.1 EXPONENTIAL FUNCTIONS

In Chapter 6 we gave meaning to expressions with rational-number exponents such as

$$5^{1/4}, \quad 3^{-3/4}, \quad 7^{2.34}, \quad 8^{1.73}.$$

For example, $5^{1.73}$, or $5^{173/100}$, means to raise 5 to the 173rd power and then take the 100th root. We shall now give meaning to expressions with irrational exponents, such as

$$5^{\sqrt{2}}, \quad 7^{\pi}, \quad 9^{-\sqrt{3}}.$$

Consider $5^{\sqrt{3}}$. Let us think of rational numbers r close to $\sqrt{3}$ and look at 5^r. As r gets closer to $\sqrt{3}$, 5^r gets closer to some real number.

r closes in on $\sqrt{3}$.	5^r closes in on some real number p.
r	5^r
$1 < \sqrt{3} < 2$	$5^1 < p < 5^2$
$1.7 < \sqrt{3} < 1.8$	$5^{1.7} < p < 5^{1.8}$
$1.73 < \sqrt{3} < 1.74$	$5^{1.73} < p < 5^{1.74}$
$1.732 < \sqrt{3} < 1.733$	$5^{1.732} < p < 5^{1.733}$

> The symbol $<$ means "is less than," so $1 < \sqrt{3} < 2$ means that 1 is less than $\sqrt{3}$ and that $\sqrt{3}$ is less than 2.

As r closes in on $\sqrt{3}$, 5^r closes in on some real number p. We define $5^{\sqrt{3}}$ to be the number p.

Any positive irrational exponent can be defined in a similar way. Negative irrational exponents are then defined in the same way as negative integer exponents. Then the expression a^x has meaning for any real number x. The usual laws of exponents still hold, but we will not prove that here. We now define exponential functions.

The function $f(x) = a^x$, where a is a positive constant, is called *the exponential function*, **base a.**

Example 1   Graph the exponential function $y = 2^x$.

We compute some function values, thinking of y as $f(x)$.

$$f(1) = 2^1 = 2$$
$$f(3) = 2^3 = 8$$
$$f(-2) = 2^{-2} = \frac{1}{2^2} = \frac{1}{4}$$

We can keep a record of values in a table.

x	0	1	2	-1	-2	-3
y, or $f(x)$	1	2	4	$\frac{1}{2}$	$\frac{1}{4}$	$\frac{1}{8}$

Next, we plot these points and connect them with a smooth curve.

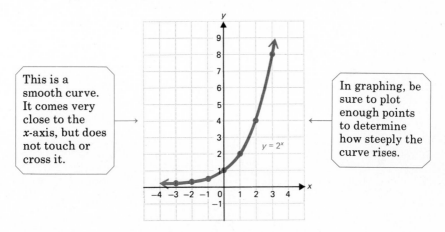

This is a smooth curve. It comes very close to the x-axis, but does not touch or cross it.

In graphing, be sure to plot enough points to determine how steeply the curve rises.

$y = 2^x$

Note that as x increases, the function values increase indefinitely. As x decreases, the function values decrease, becoming very close to 0. The x-axis is an *asymptote*, meaning that the curve comes very close to it.

Example 2 Graph the exponential function $y = (\frac{1}{2})^x$.

We compute some function values, thinking of y as $f(x)$, and keep the results in a table.

$$f(2) = \left(\frac{1}{2}\right)^2 = \frac{1}{4}$$

$$f(3) = \left(\frac{1}{2}\right)^3 = \frac{1}{8}$$

$$f(0) = \left(\frac{1}{2}\right)^0 = 1$$

$$f(-3) = \left(\frac{1}{2}\right)^{-3} = \frac{1}{\left(\frac{1}{2}\right)^3} = \frac{1}{\frac{1}{8}} = 8$$

x	-3	-2	-1	0	1	2	3
y, or $f(x)$	8	4	2	1	$\frac{1}{2}$	$\frac{1}{4}$	$\frac{1}{8}$

We plot these points and draw the curve. Note that this graph is a reflection of the one in Example 1.

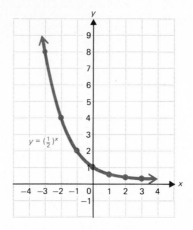

The preceding examples and exercises illustrate exponential functions of various bases. If $a = 1$, the graph is the line $y = 1$. For other positive values of a, the graphs are more interesting.

A When $a > 1$, the function $f(x) = a^x$ increases from left to right. See the colored curves below. The greater the value of a, the steeper the curve.

B When $a < 1$, the function $f(x) = a^x$ decreases from left to right. See the black curve below. As a approaches 1, the curve becomes less steep. All such functions go through the point (0, 1).

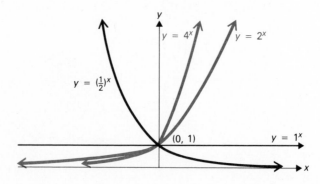

Example 3 Graph: $y = 2^{x-2}$.

We construct a table of values and then plot points and connect them. Be sure to note that $x - 2$ is the *exponent*.

x	0	-1	-2	1	2	3	4	5
y	$\frac{1}{4}$	$\frac{1}{8}$	$\frac{1}{16}$	$\frac{1}{2}$	1	2	4	8

The graph looks just like the graph of $y = 2^x$, but is moved to the right.

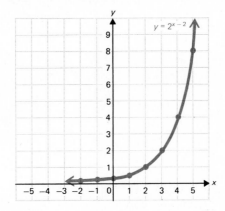

INVERSES

If in $y = a^x$ we interchange x and y, we get another kind of function, called the *inverse* of the exponential function.

Example 4 Graph: $x = 2^y$.

To construct a table of values, we can use the one we constructed in Example 1, interchanging x and y.

x	1	2	4	$\frac{1}{2}$	$\frac{1}{4}$	$\frac{1}{8}$
y, or $f(x)$	0	1	2	-1	-2	-3

We plot these points and connect them with a smooth curve.

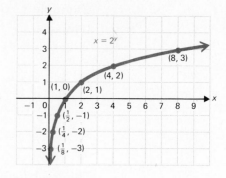

> This curve does not touch or cross the y-axis.

Note that this curve looks just like the graph of $y = 2^x$, except that it is reflected across the line $y = x$, as shown on the right.

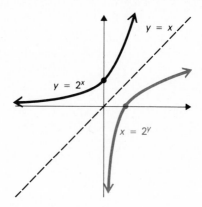

EXERCISE SET 9.1

Graph.

1. $y = 2^x$

2. $y = 3^x$

3. $y = 5^x$

4. $y = 6^x$

5. $y = 2^{x+1}$

6. $y = 2^{x-1}$

7. $y = \left(\dfrac{1}{2}\right)^x$

8. $y = \left(\dfrac{1}{3}\right)^x$

9. $y = \left(\dfrac{1}{4}\right)^x$

10. $y = \left(\dfrac{1}{5}\right)^x$

11. $x = 4^y$

12. $x = 3^y$

☆
Graph both equations using the same set of axes.

13. $y = \left(\dfrac{1}{4}\right)^x$, $x = \left(\dfrac{1}{4}\right)^y$

14. $y = 3^{-(x-1)}$, $x = {}^{-(y-1)}$

Graph.

15. $y = 2^{2x-1}$

16. $y = \left(\dfrac{1}{2}\right)^x - 1$

17. $y = 3^x + 3^{-x}$

18. $y = 2^{-(x-1)^2}$

19. $y = |2^{x^2} - 1|$

20. $y = |2^x - 2|$

9.2 LOGARITHMIC FUNCTIONS

GRAPHS OF LOGARITHMIC FUNCTIONS

The inverse of an exponential function is called a *logarithmic function*, or *logarithm function*. Given an exponential function $y = a^x$, if we reflect its graph across the line $y = x$, we get the graph of a logarithm function, $x = a^y$. We use the symbol $\log_a (x)$ or $\log_a x$ to name the function values* of a logarithmic function $x = a^y$. Thus we write $y = \log_a x$ rather than $x = a^y$. Almost always, we use a number a that is greater than 1 for a logarithm base.

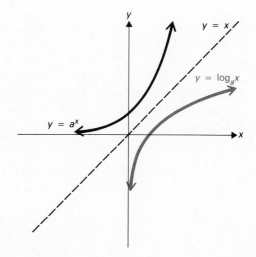

---
* The parentheses in $\log_a (x)$ are like those in $f(x)$. For logarithm functions we usually omit the parentheses and write $\log_a x$.

> $y = \log_a x$ **is defined to mean** $x = a^y$, **where** $x > 0$ **and** a **is a positive constant different from 1.**

We read "$\log_a x$" as "the logarithm, base a, of x." Note that $\log_a x$ represents the exponent in the equation $x = a^y$. Thus the logarithm, base a, of a number x is the power to which a is raised to get x.

> Remember: A logarithm is an exponent!

Example 1   Graph: $y = \log_5 x$.

This is equivalent to $5^y = x$. The graph is a reflection of $y = 5^x$ across the line $y = x$. We make a table of values for $y = 5^x$ and then interchange x and y.

For $y = 5^x$

x	0	1	2	-1	-2
y	1	5	25	$\frac{1}{5}$	$\frac{1}{25}$

For $y = \log_5 x$ (or $5^y = x$)

x	1	5	25	$\frac{1}{5}$	$\frac{1}{25}$
y	0	1	2	-1	-2

We plot the latter set of ordered pairs and connect them with a smooth curve. The graph of $y = 5^x$ has been shown only for reference.

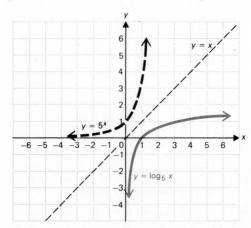

FROM EXPONENTIAL TO LOGARITHMIC EQUATIONS

We use the definition on the preceding page to convert from exponential to logarithmic equations.

Examples   Convert to logarithmic equations.

2. $8 = 2^x \rightarrow x = \log_2 8$ The exponent is the logarithm.
                                    The base remains the base.

3. $y^{-1} = 4 \rightarrow -1 = \log_y 4$

4. $a^b = c \rightarrow b = \log_a c$

FROM LOGARITHMIC TO EXPONENTIAL EQUATIONS

We also use the definition to convert from logarithmic to exponential equations.

Examples   Convert to exponential equations.

5. $y = \log_3 5 \rightarrow 3^y = 5$ The logarithm is the exponent.
                                    The base does not change.

6. $-2 = \log_a 7 \rightarrow a^{-2} = 7$

7. $a = \log_b d \rightarrow b^a = d$

SOLVING LOGARITHMIC EQUATIONS

Certain equations involving logarithms can be solved by first converting to exponential equations.

Example 8   Solve: $\log_2 x = -3$.

$$\log_2 x = -3$$
$$2^{-3} = x \qquad \text{Converting to an exponential equation}$$
$$\tfrac{1}{8} = x \qquad \text{Computing } 2^{-3}$$

This checks, so the solution is $\tfrac{1}{8}$.

Example 9   Solve: $\log_x 16 = 2$.

$$\log_x 16 = 2$$
$$x^2 = 16 \qquad \text{Converting to an exponential equation}$$
$$x = 4 \quad \text{or} \quad x = -4 \qquad \text{Solving}$$

Check: $\log_4 16 = 2$ because $4^2 = 16$. Thus 4 is a solution.

Since all logarithm bases must be positive, $\log_{-4} 16$ is not defined. Therefore, -4 is not a solution.

Solving an equation like $\log_b a = x$ amounts to finding the logarithm, base b, of the number a. You have done this before in graphing logarithmic functions. To think of finding logarithms as solving equations may help in some cases.

Example 10   Find $\log_{10} 1000$.

Let $\log_{10} 1000 = x$. Then,

$$10^x = 1000 \qquad \text{Converting to an exponential equation}$$
$$x = 3$$

Therefore, $\log_{10} 1000 = 3$.

Example 11   Find $\log_{10} 0.01$.

Let $\log_{10} 0.01 = x$. Then,

$$10^x = 0.01 \quad \text{or} \quad 10^x = \frac{1}{100}, \text{or } 10^x = 10^{-2}$$

$$x = -2$$

Therefore, $\log_{10} 0.01 = -2$.

Example 12   Find $\log_5 1$.

Let $\log_5 1 = x$. Then,

$$5^x = 1 \qquad \text{Converting to an exponential equation}$$
$$x = 0$$

Therefore, $\log_5 1 = 0$.

EXERCISE SET 9.2

Graph.

1. $y = \log_2 x$

2. $y = \log_6 x$

Convert to logarithmic equations.

3. $10^3 = 1000$ **4.** $10^2 = 100$ **5.** $5^{-3} = \dfrac{1}{125}$ **6.** $4^{-5} = \dfrac{1}{1024}$

7. $8^{1/3} = 2$ **8.** $16^{1/4} = 2$ **9.** $10^{0.3010} = 2$ **10.** $10^{0.4771} = 3$

Convert to exponential equations.

11. $t = \log_3 8$ **12.** $h = \log_7 10$ **13.** $\log_5 25 = 2$ **14.** $\log_6 6 = 1$

15. $\log_{10} 0.1 = -1$ **16.** $\log_{10} 0.01 = -2$ **17.** $\log_{10} 7 = 0.845$ **18.** $\log_{10} 3 = 0.4771$

Solve.

19. $\log_3 x = 2$ **20.** $\log_4 x = 3$ **21.** $\log_x 16 = 2$ **22.** $\log_x 64 = 3$

23. $\log_2 x = -1$ **24.** $\log_3 x = -2$ **25.** $\log_8 x = \dfrac{1}{3}$ **26.** $\log_{32} x = \dfrac{1}{5}$

Find the following.

27. $\log_{10} 100$ **28.** $\log_{10} 100{,}000$ **29.** $\log_{10} 0.1$ **30.** $\log_{10} 0.001$

31. $\log_{10} 1$ **32.** $\log_{10} 10$ **33.** $\log_5 625$ **34.** $\log_2 64$

35. $\log_5 \dfrac{1}{25}$ **36.** $\log_2 \dfrac{1}{16}$ **37.** $\log_3 1$ **38.** $\log_4 4$

☆ ───────────────────────────────

Graph both equations using the same set of axes.

39. $y = \left(\dfrac{3}{2}\right)^x$, $y = \log_{3/2} x$

Graph.

40. $y = \log_2 (x - 1)$ **41.** $y = \log_3 |x + 1|$

Solve.

42. $|\log_3 x| = 3$ **43.** $\log_{125} x = \dfrac{2}{3}$ **44.** $\log_\pi \pi^4 = x$

45. $\log_{\sqrt{5}} x = -3$ **46.** $\log_b b^{2x^2} = x$ **47.** $\log_4 (3x - 2) = 2$

48. $\log_8 (2x + 1) = -1$ **49.** $\log_x \sqrt[5]{36} = \dfrac{1}{10}$ **50.** $\log_{10} (x^2 + 21x) = 2$

Simplify.

51. $\log_{27} 9$ **52.** $\log_{1/4} \dfrac{1}{64}$ **53.** $\log_3 3^{x^2}$

54. $\log_{81} 3 \cdot \log_3 81$ **55.** $\log_{10} (\log_4 (\log_3 81))$ **56.** $8^{\log_8 (3x - 8)}$

57. $\log_2 (\log_2 (\log_4 256))$ **58.** $\log_{\sqrt{3}} \dfrac{1}{81}$ **59.** $\log_{1/5} 25$

9.3 PROPERTIES OF LOGARITHMIC FUNCTIONS

LOGARITHMS OF PRODUCTS

Logarithmic functions are important in many applications and in later mathematics. We now establish some basic properties.

> *Property 1.* **For any positive numbers M and N,**
>
> $$\log_a M \cdot N = \log_a M + \log_a N.$$
>
> **(The logarithm of a product is the sum of the logarithms of the factors. The number a can be any logarithm base.)**

Example 1   Express $\log_2 (4 \cdot 16)$ as a sum of logarithms.

$$\log_2 (4 \cdot 16) = \log_2 4 + \log_2 16 \qquad \text{By Property 1}$$

Example 2   Express as a single logarithm: $\log_{10} 0.01 + \log_{10} 1000$.

$$\log_{10} 0.01 + \log_{10} 1000 = \log_{10} (0.01 \times 1000) \qquad \text{By Property 1}$$
$$= \log_{10} 10$$

A PROOF OF PROPERTY 1 (Optional)

Let $\log_a M = x$ and $\log_a N = y$. Converting to exponential equations, we have $a^x = M$ and $a^y = N$.

Now we multiply the latter two equations, to obtain

$$M \cdot N = a^x \cdot a^y = a^{x+y}.$$

Converting back to a logarithmic equation, we get

$$\log_a M \cdot N = x + y.$$

Remembering what x and y represent, we get

$$\log_a M \cdot N = \log_a M + \log_a N,$$

which was to be shown.

LOGARITHMS OF POWERS

The second basic property is as follows.

Property 2.   **For any positive number M and any real number p,**

$$\log_a M^p = p \cdot \log_a M.$$

(The logarithm of a power of M is the exponent times the logarithm of M. The number a can be any logarithm base.)

Examples   Express as a product.

3. $\log_a 9^{-5} = -5 \log_a 9$ By Property 2

4. $\log_a \sqrt[4]{5} = \log_a 5^{1/4}$ Writing exponential notation

$\qquad\qquad = \frac{1}{4} \log_a 5$ By Property 2

A PROOF OF PROPERTY 2 (Optional)

Let $x = \log_a M$. Then convert to an exponential equation, to get $a^x = M$. Raising both sides to the pth power, we obtain

$$(a^x)^p = M^p, \quad \text{or} \quad a^{xp} = M^p.$$

Converting back to a logarithmic equation, we get

$$\log_a M^p = xp.$$

But $x = \log_a M$, so

$$\log_a M^p = (\log_a M)p,$$

or

$$p \cdot \log_a M,$$

which was to be shown.

LOGARITHMS OF QUOTIENTS

Here is the third basic property.

Property 3.   **For any positive numbers M and N,**

$$\log_a \frac{M}{N} = \log_a M - \log_a N.$$

(The logarithm of a quotient is the logarithm of the dividend minus the logarithm of the divisor. The number a can be any logarithm base.)

Example 5   Express as a difference of logarithms: $\log_t \dfrac{6}{U}$.

$$\log_t \frac{6}{U} = \log_t 6 - \log_t U \qquad \text{By Property 3}$$

Example 6   Express as a single logarithm: $\log_b 17 - \log_b 27$.

$$\log_b 17 - \log_b 27 = \log_b \frac{17}{27} \qquad \text{By Property 3}$$

Example 7   Express as a single logarithm: $\log_{10} 10{,}000 - \log_{10} 100$.

$$\log_{10} 10{,}000 - \log_{10} 100 = \log_{10} \frac{10{,}000}{100} = \log_{10} 100$$

USING THE PROPERTIES TOGETHER

Examples   Express in terms of logarithms of x, y and z.

8. $\log_a \dfrac{x^2 y^3}{z^4} = \log_a (x^2 y^3) - \log_a z^4 \qquad \text{Using Property 3}$

$\qquad\qquad\quad = \log_a x^2 + \log_a y^3 - \log_a z^4 \qquad \text{Using Property 1}$

$\qquad\qquad\quad = 2 \log_a x + 3 \log_a y - 4 \log_a z \qquad \text{Using Property 2}$

9. $\log_a \sqrt[4]{\dfrac{xy}{z^3}} = \log_a \left(\dfrac{xy}{z^3}\right)^{1/4} \qquad \text{Writing exponential notation}$

$\qquad\qquad\quad = \dfrac{1}{4} \cdot \log_a \dfrac{xy}{z^3} \qquad \text{Using Property 2}$

$\qquad\qquad\quad = \dfrac{1}{4}[\log_a xy - \log_a z^3] \qquad \text{Using Property 3}$

$\qquad\qquad\quad = \dfrac{1}{4}[\log_a x + \log_a y - 3 \log_a z] \qquad \text{Using Properties 1 and 2}$

Examples   Express as a single logarithm.

10. $\dfrac{1}{2} \log_a x - 7 \log_a y + \log_a z$

$\qquad = \log_a \sqrt{x} - \log_a y^7 + \log_a z \qquad \text{Using Property 2}$

$\qquad = \log_a \dfrac{\sqrt{x}}{y^7} + \log_a z \qquad \text{Using Property 3}$

$\qquad = \log_a \dfrac{z\sqrt{x}}{y^7} \qquad \text{Using Property 1}$

11. $\log_a \dfrac{b}{\sqrt{x}} + \log_a \sqrt{bx}$

$= \log_a b - \log_a \sqrt{x} + \log_a \sqrt{bx}$ Using Property 3

$= \log_a b - \dfrac{1}{2}\log_a x + \dfrac{1}{2}\log_a (bx)$ Property 2

$= \log_a b - \dfrac{1}{2}\log_a x + \dfrac{1}{2}[\log_a b + \log_a x]$ Property 1

$= \log_a b - \dfrac{1}{2}\log_a x + \dfrac{1}{2}\log_a b + \dfrac{1}{2}\log_a x$

$= \dfrac{3}{2}\log_a b$ Collecting like terms

$= \log_a b^{3/2}$ Using Property 2

Example 11 could also be done as follows:

$\log \dfrac{b}{\sqrt{x}} + \log_a \sqrt{bx}$

$= \log_a \dfrac{b}{\sqrt{x}} \sqrt{bx}$ Using Property 1

$= \log_a b\sqrt{b}$

$= \log_a b^{3/2}$

EXERCISE SET 9.3

Express as a sum of logarithms.

1. $\log_2 (32 \cdot 8)$ **2.** $\log_3 (27 \cdot 81)$ **3.** $\log_4 (64 \cdot 16)$

4. $\log_5 (25 \cdot 125)$ **5.** $\log_c Bx$ **6.** $\log_t 5Y$

Express as a single logarithm.

7. $\log_a 6 + \log_a 70$ **8.** $\log_b 65 + \log_b 2$

9. $\log_c K + \log_c y$ **10.** $\log_t H + \log_t M$

Express as a product.

11. $\log_a x^3$ **12.** $\log_b t^5$

Express as a difference of logarithms.

13. $\log_a \dfrac{67}{5}$

14. $\log_t \dfrac{T}{7}$

Express as a single logarithm.

15. $\log_a 15 - \log_a 7$

16. $\log_b 42 - \log_b 7$

Express in terms of logarithms of x, y, and z.

17. $\log_a x^2 y^3 z$

18. $\log_a 5xy^4z^3$

19. $\log_b \dfrac{xy^2}{z^3}$

20. $\log_c \sqrt[3]{\dfrac{x^4}{y^3z^2}}$

Express as a single logarithm and simplify if possible.

21. $\dfrac{2}{3}\log_a x - \dfrac{1}{2}\log_a y$

22. $\dfrac{1}{2}\log_a x + 3\log_a y - 2\log_a x$

23. $\log_a 2x + 3(\log_a x - \log_a y)$

24. $\log_a x^2 - 2\log_a \sqrt{x}$

25. $\log_a (x^2 - 4) - \log_a (x - 2)$

26. $\log_a \dfrac{a}{\sqrt{x}} - \log_a \sqrt{ax}$

☆
_____

True or False.

27. $\dfrac{\log_a M}{c} = \log_a M^{1/c}$

28. $\log_N (M \cdot N)^x = x \log_N M + x$

29. $\log_a \dfrac{1}{8}x = 8 \log_a x$

30. $\log_a (9x^5) = 45 \log_a x$

Express as a single logarithm and simplify if possible.

31. $\log_a (x^8 - y^8) - \log_a (x^2 + y^2)$

32. $\log_a (x + y) + \log_a (x^2 - xy + y^2)$

Express as a sum or difference of logarithms.

33. $\log_a \sqrt{1 - s^2}$

34. $\log_a \dfrac{c - d}{\sqrt{c^2 - d^2}}$

Given that $\log_a 2 = 0.301$, $\log_a 6 = 0.778$, and $\log_a 7 = 0.845$, find:

35. $\log_a \dfrac{2}{7}$ **36.** $\log_a \sqrt{24}$ **37.** $\log_a 21$

38. If $\log_a x = 2$, $\log_a y = 3$, and $\log_a z = 4$, what is $\log_a \dfrac{\sqrt[3]{x^2z}}{\sqrt[3]{y^2z^{-2}}}$?

9.4 COMMON LOGARITHMS

LOGARITHMS AND COMPUTATION

Base 10 logarithms are known as *common logarithms*. Table 2 at the back of the book contains common logarithms of numbers from 1 to 10, to four decimal places.

Before calculators and computers became so available, common logarithms were used extensively in calculating. Today, computations with logarithms are mainly of historical interest, but logarithm *functions* are of great importance.

The study of logarithms in computation does help you become more familiar with the important properties of logarithm functions.

The following is a short table of powers of 10, or *logarithms* base 10, accurate to four decimal places.

$$
\begin{aligned}
1 &= 10^{0.0000}, &\text{or}\quad \log_{10} 1 &= 0.0000 \\
2 &\approx 10^{0.3010}, &\text{or}\quad \log_{10} 2 &\approx 0.3010 \\
3 &\approx 10^{0.4771}, &\text{or}\quad \log_{10} 3 &\approx 0.4771 \\
4 &\approx 10^{0.6021}, &\text{or}\quad \log_{10} 4 &\approx 0.6021 \\
5 &\approx 10^{0.6990}, &\text{or}\quad \log_{10} 5 &\approx 0.6990 \\
6 &\approx 10^{0.7782}, &\text{or}\quad \log_{10} 6 &\approx 0.7782 \\
7 &\approx 10^{0.8451}, &\text{or}\quad \log_{10} 7 &\approx 0.8451 \\
8 &\approx 10^{0.9031}, &\text{or}\quad \log_{10} 8 &\approx 0.9031 \\
9 &\approx 10^{0.9542}, &\text{or}\quad \log_{10} 9 &\approx 0.9542 \\
10 &= 10^{1.0000}, &\text{or}\quad \log_{10} 10 &= 1.0000 \\
11 &\approx 10^{1.0414}, &\text{or}\quad \log_{10} 11 &\approx 1.0414 \\
12 &\approx 10^{1.0792}, &\text{or}\quad \log_{10} 12 &\approx 1.0792 \\
13 &\approx 10^{1.1139}, &\text{or}\quad \log_{10} 13 &\approx 1.1139 \\
14 &\approx 10^{1.1461}, &\text{or}\quad \log_{10} 14 &\approx 1.1461 \\
15 &\approx 10^{1.1761}, &\text{or}\quad \log_{10} 15 &\approx 1.1761 \\
16 &\approx 10^{1.2041}, &\text{or}\quad \log_{10} 16 &\approx 1.2041
\end{aligned}
$$

To illustrate how logarithms can be used for computation we will use the above table and do some easy calculations. The calculations are very simple, and we are doing them only to illustrate how logarithms work. As you read and work examples, remember that the logarithms in the table are only approximate because they have been rounded. Therefore, our answers will not always be exact.

Example 1   Find 3×4 using the table of exponents.

$$3 \times 4 \approx 10^{0.4771} \times 10^{0.6021} \qquad \text{From the table}$$
$$\approx 10^{1.0792} \qquad \text{Adding exponents}$$

From the table we see that $10^{1.0792} \approx 12$, so $3 \times 4 \approx 12$.

Note from Example 1 that we find a product by *adding* the logarithms of the factors. Then we find the number having that result as its logarithm. This is called finding the *antilogarithm*. Thus reading the table on p. 481 from right to left gives us antilogarithms. From that table we see that the antilogarithm of a number x is 10^x.

Example 2   Find $\frac{14}{2}$ using base 10 logarithms.

$$\log_{10} \frac{14}{2} = \log_{10} 14 - \log_{10} 2 \qquad \text{Using Property 3}$$

$$\approx 1.1461 - 0.3010 \qquad \text{Finding the logs from the table}$$

$$\log_{10} \frac{14}{2} \approx 0.8451 \qquad \text{Subtracting}$$

$$\frac{14}{2} \approx \text{antilog}_{10}\, 0.8451 \approx 7 \qquad \text{Using the table in reverse}$$

Example 3   Find $\sqrt[4]{16}$ using base 10 logarithms.

$$\log_{10} \sqrt[4]{16} = \log_{10} 16^{1/4}$$

$$= \frac{1}{4} \cdot \log_{10} 16 \qquad \text{Using Property 2}$$

$$\approx \frac{1}{4} \cdot 1.2041$$

$$\log_{10} \sqrt[4]{16} \approx 0.3010 \qquad \text{Rounding to four decimal places}$$
$$\sqrt[4]{16} \approx \text{antilog}_{10}\, 0.3010 \approx 2$$

Example 4   Find 2^3 using base 10 logarithms.

$$\log_{10} 2^3 = 3 \cdot \log_{10} 2 \qquad \text{Using Property 2}$$
$$\approx 3 \cdot 0.3010 \approx 0.9030$$
$$\log_{10} 2^3 \approx 0.9030$$
$$2^3 \approx \text{antilog}_{10}\, 0.9030 \approx 8$$

USING A LOGARITHM TABLE

We will often omit the base, 10, when working with common logarithms. That is,

$\log M$ will be agreed to mean $\log_{10} M$.

Table 2 contains logarithms of numbers from 1 to 10. Part of that table is as follows.

x	0	1	2	3	4	5	6	7	8	9
5.0	0.6990	0.6998	0.7007	0.7016	0.7024	0.7033	0.7042	0.7050	0.7059	0.7067
5.1	0.7076	0.7084	0.7093	0.7101	0.7110	0.7118	0.7126	0.7135	0.7143	0.7152
5.2	0.7160	0.7168	0.7177	0.7185	0.7193	0.7202	0.7210	0.7218	0.7226	0.7235
5.3	0.7243	0.7251	0.7259	0.7267	0.7275	0.7284	0.7292	0.7300	0.7308	0.7316
5.4	0.7324	0.7332	0.7340	0.7348	0.7356	0.7364	0.7372	0.7380	0.7388	0.7396

To illustrate the use of the table, let us find $\log 5.24$. We locate the row headed 5.2, then move across to the column headed 4. We find $\log 5.24$ (the colored entry in the table): $\log 5.24 = 0.7193$.

We can find antilogarithms by reversing this process. For example, antilog $0.7193 = 10^{0.7193} \approx 5.24$. Similarly, antilog $0.7292 \approx 5.36$.

Using Table 2 and scientific notation we can find logarithms of numbers that are not between 1 and 10. First recall that logarithms are exponents, so

$$\log_a a^k = k \quad \text{for any number } k.$$

Thus,

$$\log_{10} 10^k = k \quad \text{for any number } k.$$

Thus we know that

$\log 10 = 1$ because $10 = 10^1$,

$\log 100 = 2$ because $100 = 10^2$,

$\log 1000 = 3$ because $1000 = 10^3$,

$\log 0.001 = -3$ because $0.001 = 10^{-3}$.

Examples

5. $\log 52.4 = \log(5.24 \times 10^1)$ Writing scientific notation for 52.4
$= \log 5.24 + \log 10^1$ Using Property 1
$\approx 0.7193 + 1$ Using log table

6. $\log 52{,}400 = \log(5.24 \times 10^4)$
$= \log 5.24 + \log 10^4$
$\approx 0.7193 + 4$

7. $\log 0.00524 = \log(5.24 \times 10^{-3})$
$= \log 5.24 + \log 10^{-3}$
$\approx 0.7193 + (-3)$

In all of the examples above, 0.7193 is the fractional part of the logarithm. It is called a *mantissa*. The integer part varies and is known as the *characteristic*. Table 2 contains only mantissas. Characteristics must be supplied. The characteristic is the exponent used in writing scientific notation. Note that characteristics may be positive, negative, or zero, but mantissas are never negative.

$$\log 5240 \approx 0.7193 + 3$$

Mantissa: a   *Characteristic:*
number from 0 to 1   an integer

The preceding examples illustrate the importance of using the base 10. It allows great economy in the printing of tables. If we know the logarithms of numbers from 1 to 10 we can find the logarithm of any number. For any base other than 10 that would not be the case.

NEGATIVE CHARACTERISTICS

To find the logarithm of a number, we of course add the characteristic and the mantissa. In Example 6, we found that

$$\log 52{,}400 \approx 0.7193 + 4.$$

If we add the characteristic and mantissa, we get

4.7193.

In this case, we can still *see* the characteristic and mantissa. In Example 7, we found that

$$\log 0.00524 \approx 0.7193 + (-3).$$

Again, we can add the characteristic and mantissa to obtain standard notation for the logarithm.

$$
\begin{array}{r}
-3.0000 \\
0.7193 \\
\hline
-2.2807
\end{array}
$$

Caution! When we add, we do *not* get -3.7193.

Thus log $0.00524 \approx -2.2807$.

In this case the mantissa is no longer visible after we do the addition. Neither is the characteristic.

In some applications it does not matter that the characteristic and the mantissa are not preserved. In some cases it does.

> **If the characteristic and mantissa of a logarithm are to be preserved when the characteristic is negative, they must not be added.**

ANTILOGARITHMS

To find antilogarithms we reverse the procedure for finding logarithms. Recall that antilog $x = 10^x$ for any number x.

Example 8   Find antilog 2.6085.

$$
\text{antilog } 2.6085 = 10^{2.6085} = 10^{(2+0.6085)}
$$
$$
= 10^2 \cdot 10^{0.6085}
$$

Now $10^{0.6085} = $ antilog 0.6085 and we can find it from the table. It is 4.06. Thus we have

$$
\text{antilog } 2.6085 = 10^2 \times 4.06, \quad \text{or} \quad 406.
$$

In Example 8 we, in effect, separated the number 2.6085 into two parts, the characteristic (an integer) and a number between 0 and 1 (the mantissa). We found the antilogarithm of the mantissa from the table and the antilogarithm of the characteristic is 10^2. We then had scientific notation for our answer.

Example 9   Find antilog 3.7118.

From the table we find antilog $0.7118 \approx 5.15$. Thus

$$
\text{antilog } 3.7118 \approx 5.15 \times 10^3 \qquad \text{Note that 3 is the characteristic.}
$$
$$
\approx 5150.
$$

Example 10   Find antilog $0.7143 + (-3)$.

The characteristic is -3 and the mantissa is 0.7143. From the table we find that antilog $0.7143 \approx 5.18$. Thus

$$\text{antilog } [0.7143 + (-3)] \approx 5.18 \times 10^{-3}$$
$$\approx 0.00518.$$

LOGARITHMS ON A CALCULATOR (Optional)

Tables of common logarithms contain only mantissas. The reason for this is to save space. When we find a logarithm using a table, we must supply the characteristic ourselves. Scientific calculators that have $\boxed{\log}$ keys contain logarithms, *including* characteristic and mantissa. On a calculator it is not usually important to keep the characteristic and mantissa separate.

Example 11   Use a calculator to find log 475,000.

We enter 475,000 and then press the $\boxed{\log}$ key.

We find that log $475{,}000 \approx 5.6767$.

The characteristic is 5 and the mantissa is 0.6767

Example 12   Use a calculator to find log 0.00008952.

We enter 0.00008952 and then press the $\boxed{\log}$ key.

We find that log $0.00008952 \approx -4.0481$.

We do not know the characteristic or the mantissa. If we should need them we can add 5 and then add -5.

Using the calculator we find that $5 + (-4.0481) \approx 0.9519$. Now we add -5 and we have

$$\log 0.00008952 \approx \underset{\text{mantissa}}{0.9519} + \underset{\text{characteristic}}{(-5)}.$$

Calculators do not usually have an $\boxed{\text{antilog}}$ key. To find an antilogarithm, you must remember that

$$\text{antilog}_{10} X = 10^{X}.$$

To find the antilogarithm of a number, remember that the logarithm is an *exponent* and raise 10 to that power.

Example 13 Find antilog 3.2145.

We press the proper keys to find $10^{3.2145}$. (The key strokes are different on different calculators.)

We find that antilog $3.2145 \approx 1639$.

Example 14 Find antilog $[0.6646 + (-5)]$.

We first use the calculator to do the addition $0.6646 + (-5)$.

We get -4.3354.

Next, we raise 10 to this power (we find $10^{-4.3354}$).

Antilog $(-4.3354) \approx 0.00004620$.

EXERCISE SET 9.4

Find each of the following using the table of exponents. Show your work.

1. 7×2
2. 3×5
3. $\dfrac{12}{3}$
4. $\dfrac{16}{8}$

Find these logarithms using Table 2.

5. log 2.46
6. log 7.65
7. log 5.31
8. log 8.57
9. log 1.07
10. log 4.60
11. log 347
12. log 8720
13. log 52.5
14. log 20.6
15. log 3870
16. log 62,400
17. log 0.64
18. log 0.00216
19. log 0.0000404
20. log 0.00347

Find these antilogarithms using Table 2.

21. antilog 3.3674
22. antilog 4.9222
23. antilog 1.2553
24. antilog 2.6294
25. antilog $9.7875 - 10$
26. antilog $8.9881 - 10$
27. antilog $7.5391 - 10$
28. antilog $7.7774 - 10$
29. $10^{1.4014}$
30. $10^{2.5391}$
31. $10^{7.9881-10}$
32. $10^{8.5391-10}$
33. $10^{6.7875-10}$
34. $10^{4.6294-10}$

9.5 CALCULATIONS WITH LOGARITHMS (Optional)

Before calculators and computers were readily available, logarithms were a very important tool for doing computations. Today, they are not important for that purpose. Their importance is in working certain kinds of problems and in mathematical theory. Therefore, this section is mainly of historical interest. Doing calculations with logarithms can also serve to fix the ideas and properties of logarithms in one's mind.

In calculating with logarithms, it is necessary to preserve both mantissas and characteristics. When negative characteristics occur, we write the logarithm as a difference, in which 10 or a multiple of 10 is subtracted.

Example 1   Write $0.7196 + (-3)$ as a positive number minus a multiple of 10.

We add 7 and subtract 7, and we get

$7.7196 - 10.$ Note that the mantissa is still visible.

Suppose we wish to divide a logarithm by a number, say 5. We then use 50 for our multiple of 10, as in Example 2.

Example 2   Write $0.1425 + (-6)$ as a positive number minus 50.

We add 44 and subtract 44, and we get

$44.1425 - 50.$

We could now divide the number in Example 2 by 5 and still have a multiple of 10 subtracted.

Example 3   Divide $44.1425 - 50$ by 5. Find the characteristic and the mantissa of the result.

$$
\begin{array}{r}
8.8285 \quad -\ 10 \\
\hline
5\overline{\smash{)}44.1425 \quad -\ 50} \\
40 \\
\hline
4\ 1 \\
4\ 0 \\
\hline
14 \\
10 \\
\hline
42 \\
40 \\
\hline
25
\end{array}
$$

Mantissa 0.8285; characteristic (-2).

The kinds of calculations in which logarithms can be used are multiplication, division, raising to powers, and taking roots.

Example 4 Find: $\dfrac{0.0578 \times 32.7}{8460}$.

We write a *plan* for the use of logarithms, and then look up all of the mantissas at one time.

Let
$$N = \frac{0.0578 \times 32.7}{8460}.$$

Then, using Properties 1 and 3, we get

$\log N = \log 0.0578 + \log 32.7 - \log 8460.$

This gives us the plan. We use a straight line to indicate addition and a wavy line to indicate subtraction.

Completion of plan:

$$
\begin{aligned}
\log 0.0578 &\approx\ 8.7619 - 10 \\
\log 32.7 &\approx\ 1.5145 \\
\hline
\log \text{numerator} &\approx\ 10.2764 - 10 \qquad \text{Sum of the logs above} \\
\log 8460 &\approx\ 3.9274 \\
\hline
\log N &\approx\ 6.3490 - 10 \qquad \text{Subtracting logs} \\
N &\approx\ 0.0002234 \qquad \text{Taking the antilog}
\end{aligned}
$$

This answer was found from the table. Sometimes we had to estimate the fourth digit.

Example 5 Use logarithms to find $(92.8)^3 \times \sqrt[5]{0.986}$.

Let $N = (92.8)^3 \times \sqrt[5]{0.986}$.

Then

$$
\begin{aligned}
\log N &= \log (92.8)^3 + \log \sqrt[5]{0.986} \qquad \text{Using Property 1} \\
&= 3 \log 92.8 + \tfrac{1}{5} \log (0.986). \qquad \text{Using Property 2}
\end{aligned}
$$

Powers and roots are involved. In such cases it is easier to first find the following logs:

$\log 92.8 \approx 1.9675,$

$\log 0.986 \approx 9.9939 - 10.$

Consider $9.9939 - 10$. We are going to multiply this number by $\tfrac{1}{5}$, so we rename this number as follows:

$\log 0.986 \approx 49.9939 - 50.$

Completion of plan and work:

$$\log 92.8 \approx 1.9675$$
$$3 \log 92.8 \approx 5.9025 \quad \text{Multiplying by 3}$$
$$\log 0.986 \approx 49.9939 - 50$$
$$\tfrac{1}{5} \log 0.986 \approx 9.9988 - 10 \quad \text{Multiplying by } \tfrac{1}{5}$$
$$\overline{\log N \approx 15.9013 - 10} \quad \text{Adding logs}$$
$$N \approx 796{,}700 \quad \text{Taking the antilog (We estimated the fourth digit.)}$$

EXERCISE SET 9.5

Use logarithms to calculate.

1. 3.14×60.4

2. 541×0.0152

3. $286 \div 1.05$

4. $12.8 \div 81.6$

5. $\sqrt{76.9}$

6. $\sqrt{678}$

7. $\sqrt[3]{56.9}$

8. $\sqrt[4]{2600}$

9. $(1.36)^{4.2}$

10. $(0.727)^{3.6}$

11. $\sqrt[3]{\dfrac{3.24 \times (3.16)^2}{78.4 \times 24.6}}$

12. $\dfrac{70.7 \times (10.6)^2}{18.6 \times \sqrt{276}}$

9.6 INTERPOLATION

LOGARITHMS

In Section 9.4 we used Table 2 when three digits occur, as in 4.58 or 0.0458. We can also use the table when four digits occur, by a process called *interpolation*. The simplest kind of interpolation is called *linear interpolation*. What we say here applies to tables of many different functions.

Let us consider how a table of values for any function is made. We select numbers of the domain x_1, x_2, x_3, and so on. Then we compute or somehow determine the corresponding function values (outputs), $f(x_1)$, $f(x_2)$, $f(x_3)$, and so on. Then we tabulate the results. We might also graph the results.

x	x_1	x_2	x_3	x_4	$\cdots$
$f(x)$	$f(x_1)$	$f(x_2)$	$f(x_3)$	$f(x_4)$	$\cdots$

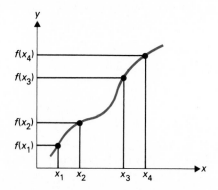

Suppose we want to find an output $f(x)$ for an x not in the table. If x is halfway between x_1 and x_2, then we can take the number halfway between $f(x_1)$ and $f(x_2)$ as an approximation to $f(x)$. If x is one-fifth of the way between x_2 and x_3, we can take the number that is one-fifth of the way between $f(x_2)$ and $f(x_3)$ as an approximation to $f(x)$. What we do is divide the length from x_2 to x_3 in a certain ratio, and then divide the length from $f(x_2)$ to $f(x_3)$ in the same ratio. This is *linear interpolation*.

We can show this geometrically. The length from x_1 to x_2 is divided in a certain ratio by x. The length from $f(x_1)$ to $f(x_2)$ is divided in the same ratio by y. The number y approximates $f(x)$ with the noted error.

Note the slanted line in the figure. The approximation y comes from this line. This explains the use of the term *linear interpolation*.

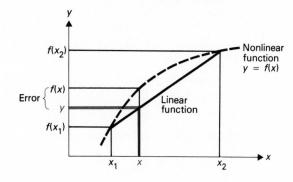

Example 1   Find: log 34870.

a) Find the characteristic. Since $34870 = 3.487 \times 10^4$, the characteristic is 4.

b) Find the mantissa. From Table 2 we have:

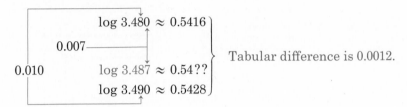

The tabular difference (the difference between consecutive values in the table) is 0.0012. Now 3.487 is $\frac{7}{10}$ of the way from 3.480 to 3.490. So we take 0.7 of 0.0012, which is 0.00084, and round it to 0.0008. We add this to 0.5416. The mantissa is 0.5424.

c) Add the characteristic and mantissa:

log 34870 $\approx$ 4.5424.

With practice you will take 0.7 of 12, forgetting the zeros, but adding in the same way.

Example 2   Find: log 0.009543.

a) Find the characteristic. Since $0.009543 = 9.543 \times 10^{-3}$, the characteristic is -3, or $7 - 10$.

b) Find the mantissa. From Table 2 we have:

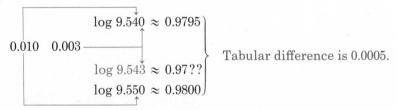

Now 9.543 is $\frac{3}{10}$ of the way from 9.540 to 9.550, so we take 0.3 of 0.0005, which is 0.00015, and round it to 0.0002. We add this to 0.9795. The mantissa that results is 0.9797.

c) Add the characteristic and the mantissa:

log 0.009543 $\approx$ 7.9797 $-$ 10.

ANTILOGARITHMS

We interpolate when finding antilogarithms, using the table in reverse.

Example 3 Find: antilog 4.9164.

a) The characteristic is 4. The mantissa is 0.9164.

b) Find the antilog of the mantissa, 0.9164. From Table 2 we have:

$$
0.0006 \quad 0.0005
\begin{cases}
\text{antilog } 0.9159 \approx 8.240 \\
\text{antilog } 0.9164 \approx 8.24? \\
\text{antilog } 0.9165 \approx 8.250
\end{cases}
\quad \text{The difference is } 0.010.
$$

The difference between 0.9159 and 0.9165 is 0.0006. Thus 0.9164 is $\frac{0.0005}{0.0006}$, or $\frac{5}{6}$, of the way between 0.9159 and 0.9165. Then antilog 0.9164 is $\frac{5}{6}$ of the way between 8.240 and 8.250, so we take $\frac{5}{6}(0.010)$, which is $0.00833\ldots$, and round it to 0.008. Thus the antilog of the mantissa is 8.248.

So antilog $4.9164 \approx 8.248 \times 10^4 = 82{,}480$.

Example 4 Find: antilog $(7.4122 - 10)$.

a) The characteristic is -3. The mantissa is 0.4122.

b) Find the antilog of the mantissa, 0.4122. From Table 2 we have:

$$
0.0017 \quad 0.0006
\begin{cases}
\text{antilog } 0.4116 \approx 2.580 \\
\text{antilog } 0.4122 \approx 2.58? \\
\text{antilog } 0.4133 \approx 2.590
\end{cases}
\quad \text{The difference is } 0.010.
$$

The difference between 0.4116 and 0.4133 is 0.0017. Thus 0.4122 is $\frac{0.0006}{0.0017}$, or $\frac{6}{17}$, of the way between 0.4116 and 0.4133. Then antilog 0.4122 is $\frac{6}{17}$ of the way between 2.580 and 2.590, so we take $\frac{6}{17}(0.010)$, which is 0.0035 to four places. We round it to 0.004. Thus the antilog of the mantissa is 2.584.

So antilog $(7.4122 - 10) \approx 2.584 \times 10^{-3} = 0.002584$.

EXERCISE SET 9.6

Find the following logarithms using interpolation.

1. log 41.63

2. log 472.1

3. log 2.944

4. log 21.76

5. log 650.2

6. log 37.37

7. log 0.1425

8. log 0.09045

9. log 0.004257

10. log 4518

11. log 0.1776

12. log 0.08356

13. log 600.6

14. log 800.1

Find the following antilogarithms using interpolation.

15. antilog 1.6350

16. antilog 2.3512

17. antilog 0.6478

18. antilog 1.1624

19. antilog 0.0342

20. antilog 4.8453

21. antilog 9.8564 − 10

22. antilog 8.9659 − 10

23. antilog 7.4128 − 10

24. antilog 9.7278 − 10

25. antilog 8.2010 − 10

26. antilog 7.8630 − 10

☆

Use Table 2 and interpolation to find the following.

27. $\sqrt{23.37}$

28. $\sqrt[3]{12.93}$

9.7 EXPONENTIAL AND LOGARITHMIC EQUATIONS

EXPONENTIAL EQUATIONS

Equations with variables in exponents, such as $3^{2x-1} = 4$, are called *exponential equations*. We solve such equations by taking the logarithm on both sides and then using Property 2.

Example 1 Solve: $2^{3x-5} = 16$.

$$\log 2^{3x-5} = \log 16 \qquad \text{Taking log on both sides}$$

$$(3x - 5) \log 2 = \log 16 \qquad \text{Using Property 2}$$

$$3x - 5 = \frac{\log 16}{\log 2}$$

$$x = \frac{\dfrac{\log 16}{\log 2} + 5}{3} \approx \frac{\dfrac{1.2041}{0.3010} + 5}{3} \qquad \begin{array}{l} \text{Solving for } x \text{ and} \\ \text{finding logarithms} \end{array}$$

$$x \approx 3.0001 \qquad \text{Calculating}$$

The answer is approximate because the logarithms are approximate.

Example 2   Solve: $2^{3x} \cdot 2^{-5} = 16$.

$$2^{3x-5} = 16 \qquad \text{Multiplying (adding exponents) on the left}$$

$$\log 2^{3x-5} = \log 16 \qquad \text{Taking log on both sides}$$

$$(3x - 5)\log 2 = \log 16 \qquad \text{Using Property 2}$$

$$3x - 5 = \frac{\log 16}{\log 2}$$

$$x = \frac{\dfrac{\log 16}{\log 2} + 5}{3} \approx \frac{\dfrac{1.2041}{0.3010} + 5}{3} \qquad \begin{array}{l}\text{Solving for } x \text{ and}\\ \text{finding logarithms}\end{array}$$

$$x \approx 3.0001 \qquad \text{Calculating}$$

LOGARITHMIC EQUATIONS

Equations containing logarithmic expressions are called *logarithmic equations*. To solve them, we obtain a single logarithmic expression on one side and then take the antilogarithm.

Example 3   Solve: $\log x + \log (x - 3) = 1$.

$$\log x(x - 3) = 1 \qquad \begin{array}{l}\text{Using Property 1 to obtain}\\ \text{a single logarithm}\end{array}$$

$$x(x - 3) = 10^1 \qquad \text{Taking the antilog on both sides}$$

$$x^2 - 3x - 10 = 0$$

$$(x + 2)(x - 5) = 0 \qquad \text{Factoring}$$

$$x = -2 \quad \text{or} \quad x = 5 \qquad \text{Principle of zero products}$$

Check:

$$\frac{\log x + \log (x - 3) = 1}{\log (-2) + \log (-2 - 3) \;\big|\; 1} \qquad \begin{array}{l}\dfrac{\log x + \log (x - 3) = 1}{\log 5 + \log (5 - 3) \;\big|\; 1}\\ \log 5 + \log 2 \;\big|\\ \log 10 \;\big|\\ 1 \;\big|\end{array}$$

The number -2 is not a solution because negative numbers do not have logarithms.

The solution is 5.

APPLICATIONS

There are many applications of exponential and logarithmic equations and functions.

Example 4 (*Compound interest*). The amount A that principal P will be worth after t years at interest rate r, compounded annually, is given by the formula

$$A = P(1 + r)^t.$$

Suppose $4000 principal is invested at 12% interest compounded annually and yields $10,706. How many years was it invested?

Using the formula $A = P(1 + r)^t$, we have

$$10{,}706 = 4000(1 + 0.12)^t \text{ or } 10{,}706 = 4000(1.12)^t.$$

We solve for t.

$$\log 10{,}706 = \log 4000(1.12)^t = \log 4000 + t \log 1.12$$

$$\frac{\log 10{,}706 - \log 4000}{\log 1.12} = t$$

$$\frac{4.0296 - 3.6021}{0.0492} \approx t$$

$$8.69 \approx t.$$

The money was invested for about $8\frac{2}{3}$ years.

Example 5 (*Earthquake magnitude*). The magnitude R (on the Richter scale) of an earthquake of intensity I is defined as follows:

$$R = \log \frac{I}{I_0},$$

where I_0 is a minimum intensity used for comparison.

An earthquake has an intensity $10^{8.6}$ times I_0. What is its magnitude on the Richter scale?

We substitute into the formula:

$$R = \log \frac{10^{8.6} \cdot I_0}{I_0} = \log 10^{8.6}$$

$$R = 8.6.$$

EXERCISE SET 9.7

Solve.

1. $2^x = 9$ **2.** $2^x = 30$ **3.** $2^x = 10$ **4.** $2^x = 33$

5. $5^{4x-7} = 125$

6. $4^{3x+5} = 16$

7. $3^{x^2} \cdot 3^{4x} = \dfrac{1}{27}$

8. $3^{5x} \cdot 9^{x^2} = 27$

9. $4^x = 7$

10. $8^x = 10$

11. $\log x + \log (x - 9) = 1$

12. $\log x + \log (x + 9) = 1$

13. $\log x - \log (x + 3) = -1$

14. $\log (x + 9) - \log x = 1$

15. $\log_2 (x + 1) + \log_2 (x - 1) = 3$

16. $\log_4 (x + 3) - \log_4 (x - 5) = 2$

17. (*Doubling time*). How many years will it take an investment of $1000 to double itself when interest is compounded annually at 10%?

18. (*Tripling time*). How many years will it take an investment of $1000 to triple itself when interest is compounded annually at 10%?

19. The Los Angeles earthquake of 1971 had an intensity $10^{6.7}$ times I_0. What was its magnitude on the Richter scale?

20. The San Francisco earthquake of 1906 had an intensity $10^{8.25}$ times I_0. What was its magnitude on the Richter scale?

☆ ───────────────────────────────

Solve.

21. $8^x = 16^{3x+9}$

22. $27^x = 81^{2x-3}$

23. $\log_2 x + \log_2 (x - 2) = 3$

24. $\log_4 (x + 6) - \log_4 x = 2$

25. $\log_6 (\log_2 x) = 0$

26. $\log_x (\log_3 27) = 3$

27. $\log_5 \sqrt{x^2 - 9} = 1$

28. $x \log \frac{1}{8} = \log 8$

29. $\log (\log x) = 5$

30. $2^{x^2+4x} = \frac{1}{8}$

31. $3^{\log_3(8x-4)} = 5$

32. $\log x^2 = (\log x)^2$

33. $10^{\log 14} = 3x - 19$

34. $\log_5 |x| = 4$

35. $\log x^{\log x} = 25$

36. $\log \sqrt{2x} = \sqrt{\log 2x}$

37. $\log_a a^{x^2+4x} = 21$

38. $(\log_a x)^{-1} = \log_a x$

39. $x^{\log_{10} x} = \dfrac{x^{-4}}{1000}$

40. $3^{2x} - 8 \cdot 3^x + 15 = 0$

41. $(81^{x-2})(27^{x+1}) = 9^{2x-3}$

42. $25^{x+2} = 3120 + 25^x$

43. $3^{2x} - 3^{2x-1} = 18$

44. True or false: If $r^a = s^d$ and $s^b = r^c$, then $ab = cd$.

45. Given that $2^y = 16^{x-3}$ and $3^{y+2} = 27^x$, find the value of $x + y$.

46. If $x = (\log_{125} 5)^{\log_5 125}$, what is the value of $\log_3 x$?

47. Solve this system of equations: $w^{2u} = v^u$, $5^u = 5 \cdot 25^w$, $u + v + w = 131$.

48. 🖩 Suppose an amount P_0 is invested in a savings account at interest rate k, compounded continuously. The balance P, after t years, is given by

$$P = P_0 e^{kt}, \text{ where } e \approx 2.718282.$$

You will need a calculator with $\boxed{e^x}$ and $\boxed{\ln x}$ keys, where $\log_e x = \ln x$.

Suppose $1000 is invested at 14%.

a) What is the balance after 1 year? 2 years? 10 years?
b) After what amount of time will the $1000 double itself?
c) Suppose P_0 dollars are invested at interest rate k and that amount grows to $2P_0$ dollars in time T. Find a formula relating T and k.

TEST OR REVIEW—CHAPTER 9

1. Graph: $y = 2^{x+3}$.   **2.** Graph: $x = 2^y$.

3. Graph: $y = \log_3 x$.

4. Convert to a logarithmic equation: $4^{-3} = x$.

5. Convert to an exponential equation: $\log_4 16 = 2$.

6. Express in terms of logarithms of a, b, and c: $\log \dfrac{a^3 b^{1/2}}{c^2}$.

7. Express as a single logarithm: $\dfrac{1}{3} \log_a x - 3 \log_a y + 2 \log_a z$.

8. Use logarithms to compute:

$$\sqrt[3]{\frac{4.76 \times 1.82^2}{72.1^4}}.$$

9. Use Table 2 to find log 0.0123.

10. Use Table 2 to find antilog 5.6484.

11. Use Table 2 to find antilog $0.2614 + (-7)$.

12. Use Table 2 to find log 0.01234.

13. Solve: $7^x = 1.2$.

14. Solve: $\log (x^2 - 1) - \log (x - 1) = 1$.

10

INEQUALITIES AND SETS

10.1 THE ADDITION PRINCIPLE FOR SOLVING INEQUALITIES

ORDER AND INEQUALITIES

The order of the real numbers is often pictured on a number line.

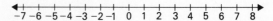

If a number occurs to the left of another, we say the first number *is less than* the second, or the second *is greater than* the first. We use the symbol $<$ to mean *is less than* and $>$ to mean *is greater than*.

Examples   Insert the proper symbol, $<$, $>$, or $=$.

1. 1   5   Since 1 occurs to the left of 5, $1 < 5$.

2. 5   1   Since $1 < 5$, we also know that $5 > 1$.

3. -3 7   Since -3 occurs to the left of 7, $-3 < 7$.

4. 7 -3 Since $-3 < 7$, we also know that $7 > -3$.

5. -9 -2 Since -9 occurs to the left of -2, $-9 < -2$.

6. -2 -9 Since $-9 < -2$, we also know that $-2 > -9$.

The symbol $\leqslant$ means *is less than or equal to*, and the symbol $\geqslant$ means *is greater than or equal to*.

Examples   Which of the following are true?

7. $-1 \leqslant 7$ True, because -1 is less than 7 ($-1 < 7$ is true).

8. $3 \leqslant 3$ True, because $3 = 3$ is true.

9. $-5 \geqslant 7$ False, because $-5 > 7$ and $-5 = 7$ are not true.

An *inequality* is any sentence having one of the verbs $<$, $>$, $\leqslant$, or $\geqslant$. As for equations, a *solution* of an inequality is any number making it a true sentence.

Examples   Determine whether the given number is a solution of the inequality.

10. $x + 3 < 6$,   5

We substitute and get $5 + 3 < 6$, or $8 < 6$, a false sentence. Therefore 5 is not a solution.

11. $2x - 3 > -3,$ 1

We substitute and get $2(1) - 3 > -3$, or $-1 > -3$, a true sentence. Therefore 1 is a solution.

12. $4x - 1 \leqslant 3x + 2,$ 3

We substitute and get $4(3) - 1 \leqslant 3(3) + 2$, or $11 \leqslant 11$, a true sentence. Therefore 3 is a solution.

SOLVING INEQUALITIES

There is an addition principle for inequalities, similar to the one for solving equations.

> *The Addition Principle for Inequalities.* **If any number is added on both sides of a true inequality, another true inequality is obtained.**

To solve inequalities using the addition principle, we do almost exactly what we would do to solve it if it were an equation.

Example 13   Solve $x + 3 > 6$. Then graph.

$$x + 3 + (\boxed{-3}) > 6 + (\boxed{-3}) \qquad \text{Using the addition principle; adding } -3$$

$$x > 3$$

Any number greater than 3 makes the last sentence true. Hence there are many solutions.

> To *solve* an inequality or equation means to find all of its solutions. That is, to find all of the numbers that make it true. When you have found all of the solutions of an equation or inequality, then you have *solved it, no matter how* you found the solutions.

We graph the solutions as follows.

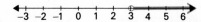

Note that there is an open circle on the graph at 3. This indicates that the number 3 is not a solution.

> Remember that the graph of any equation or inequality is a picture of its solution set.

CHECKING

In Example 13 there was an unlimited number of solutions. We cannot check them by substituting into the original inequality, as we do for equations. There are too many of them. We do not really need to check, however. Let us see why. Consider the first and last inequalities in Example 13:

$$x + 3 > 6, \qquad x > 3.$$

Any number that makes the first one true must make the last one true. We know this by the addition principle. Now the question is, will any number that makes the last one true also be a solution of the first one? Let us use the addition principle again.

$$x > 3$$

$$x + \boxed{3} > 3 + \boxed{3} \qquad \text{Using the addition principle; adding 3}$$

$$x + 3 > 6$$

Now we know that any number that makes $x > 3$ true will also make $x + 3 > 6$ true. Whenever we use the addition principle, the first and last inequalities will have the same solutions.

Example 14   Solve $5x - 2 \leqslant 4x - 1$. Then graph.

$$5x - 2 + (\boxed{-4x}) \leqslant 4x - 1 + (\boxed{-4x}) \qquad \text{Adding } -4x$$

$$x - 2 \leqslant -1$$

$$x - 2 + 2 \leqslant -1 + 2 \qquad \text{Adding 2}$$

$$x \leqslant 1$$

The graph consists of all points to the left of 1, and also the point for 1. The solid dot at 1 indicates that the number 1 is one of the solutions.

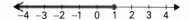

COMPOUND INEQUALITIES

A compound inequality is made up of two or more inequalities. For example, the sentence $-2 < x < 3$ is an abbreviation for $-2 < x$ *and* $x < 3$. In order for a number to be a solution, it must satisfy both of the simple inequalities.

Example 15   Graph: $-2 < x < 1$.

The inequality as stated means $-2 < x$ *and* $x < 1$. The graph consists of all points that are to the right of -2 *and also* to the left of 1.

Example 16   Graph: $-2 \leqslant x < 3$.

The inequality as stated means $-2 \leqslant x$ *and* $x < 3$. The graph consists of all points between -2 and 3, and also the point -2.

Example 17   Solve: $-3 < x + 5 < 7$.

Method 1.   We first rewrite the sentence using the word *and*. Then we solve each inequality separately.

$$-3 < x + 5 \qquad\qquad and \qquad\qquad x + 5 < 7$$

$$-3 + (\,-5\,) < x + 5 + (\,-5\,) \quad and \quad x + 5 + (\,-5\,) < 7 + (\,-5\,)$$

$$-8 < x \qquad\qquad and \qquad\qquad x < 2$$

$$-8 < x < 2 \qquad \text{Translating back}$$

Method 2.   Note that in Method 1 we did the same thing to each inequality. We can abbreviate as follows.

$$-3 + (\,-5\,) < x + 5 + (\,-5\,) < 7 + (\,-5\,) \qquad \text{Adding } -5$$

$$-8 < x < 2$$

Consider the sentence $x < -3$ *or* $x \geqslant 3$. Any sentence with *or* like this will be true when one or both parts are true. In this case both parts cannot be true at the same time. The solution set consists of those numbers that are less than -3, together with 3 and the numbers greater than 3.

> An inequality with the word *or* in it *cannot* be shortened as in Example 17.

Example 18   Graph: $x < -3$ *or* $x \geqslant 3$.

The graph consists of all points to the left of -3 and also all points to the right of 3. The point for 3 is also included.

Example 19   Solve: $x - 5 < -2$ or $x - 5 > 2$.

We solve each inequality separately but keep writing the word "or."

$$x - 5 + \boxed{5} < -2 + \boxed{5} \quad or \quad x - 5 + \boxed{5} > 2 + \boxed{5} \qquad \text{Adding 5}$$

$$x < 3 \qquad\qquad or \qquad\qquad x > 7$$

$$\underset{\text{—We leave } or \text{ in the answer.}}{\uparrow}$$

EXERCISE SET 10.1

Solve. Then graph.

1. $x - 7 > 5$ **2.** $x + 3 < 2$

3. $x - 3 \geqslant -1$ **4.** $x + 2 \leqslant 3$

Solve. You need not graph.

5. $x - \dfrac{3}{4} \geqslant \dfrac{11}{8}$ **6.** $\dfrac{3}{8} \leqslant x - \dfrac{5}{8}$

7. $3x - 2 \leqslant 2x + 1$ **8.** $4x + 4 \geqslant 3x - 1$

Graph.

9. $-2 \leqslant x < 5$ **10.** $3 > x \geqslant 0$

11. $x > 5$ *or* $x \leqslant 1$ **12.** $x \leqslant -2$ *or* $x > 2$

Solve. You need not graph.

13. $2 < x + 3 \leqslant 9$ **14.** $-6 \leqslant x + 1 < 9$

15. $x + 7 \leqslant -2$ *or* $x + 7 > 2$ **16.** $x + 9 < -4$ *or* $x + 9 \geqslant 4$

☆
 Solve.

17. $x + 6.922 \leqslant -1.998$ *or* $x + 7.0123 > 2.212$

18. $-5.888 \leqslant x + 1.003 < 9.354$

19. $-\dfrac{1}{3} < x - \dfrac{1}{6} \leqslant 6$ **20.** $0.9x > 1.8 - 0.1x$

21. $3x - \dfrac{1}{8} \leqslant \dfrac{3}{8} + 2x$ **22.** $2x + 3 \leqslant x - 6$ or $3x - 2 \leqslant 4x + 5$

23. $x + 4 < 2x - 6 \leqslant x + 12$

24. $2x - 3 < \dfrac{13}{4}x + 10 - 1.25x$

25. $7 - 3x \geqslant 5.7x + 8 - 8.7x$

26. $x + \dfrac{1}{10} \leqslant -\dfrac{1}{10} \text{ or } x + \dfrac{1}{10} \geqslant \dfrac{1}{10}$

10.2 SETS

NOTATION FOR SETS

To name a set, we can write the names of the members, separated by commas. Then we enclose the names within braces. The names can be written in any order.

Example 1 Write set notation for the set of whole numbers less than 5.

$\{0, 1, 2, 3, 4\}$

For some infinite (unending) sets we can write names for some of the members and then write dots to indicate the rest of the members.

Example 2 Write set notation for the set of even whole numbers.

$\{0, 2, 4, 6, 8, \ldots\}$

Letters are also used to name sets. The set of natural numbers, $\{1, 2, 3, \ldots\}$, is often named N and the set of integers, $\{\ldots, -3, -2, -1, 0, 1, 2, 3, \ldots\}$, is often named Z. The symbol $\in$ is used to denote membership in a set.

Examples Determine whether the following are true or false.

3. $5 \in N$

This is true, because 5 is a member of the set of natural numbers.

4. $\frac{1}{2} \in \{0, 3, 7, 9\}$

This is false, because $\frac{1}{2}$ is not a member of the set named.

For some sets, it is necessary to use a different kind of notation. Consider, for example, the set of all real numbers less than $\sqrt{2}$. We again use braces, but instead of trying to list members, we state a defining property, as follows.

$$\{x \mid x < \sqrt{2}\}$$

The set of — such that x is less
all x than $\sqrt{2}$.

Example 5   Write set notation for the set of all real numbers greater than $-\pi$.

$$\{x \mid x > -\pi\}$$

SOLUTION SETS

Given an equation or inequality, the set of numbers that make it true is called its *solution set*.

Example 6   Find the solution set of the equation $(x - 1)(x + 2) = 0$.

By the principle of zero products we know that the numbers 1 and -2 make the equation true. Thus the solution set is

$$\{1, -2\}.$$

Example 7   Find the solution set of the inequality $3x - 2 \leqslant 2x + 1$.

We solve the inequality.

$$3x - 2 \leqslant 2x + 1$$

$$3x + (-2x) - 2 \leqslant 2x + (-2x) + 1 \quad \text{Adding } -2x$$

$$x - 2 + 2 \leqslant 1 + 2 \quad \text{Adding } 2$$

$$x \leqslant 3$$

All numbers less than or equal to 3 are solutions. Thus the solution set is

$$\{x \mid x \leqslant 3\}.$$

SUBSETS AND THE EMPTY SET

When all the members of one set are also members of another, we say that the first set is a *subset* of the second. We use the symbol $\subset$ to indicate that one set is a subset of another. For example,

$A \subset B$

means that A is a subset of B.

Examples   Determine whether the following are true or false.

 8. $\{1, 2, 3, 4\} \subset \{0, 1, 2, 3, 4, 5\}$

This is true, because every member of the first set is also a member of the second.

 9. $\{x \mid x \leqslant 3\} \subset \{x \mid x < 3\}$

This is false, because there is one member of the first set that is not a member of the second. It is the number 3.

10. $\{y \mid y > -6\} \subset \{y \mid y > -10\}$

This is true, because any number greater than -6 is also greater than -10.

The set without any members is called the *empty set*, and is named $\emptyset$. The empty set is a subset of every set.

Example 11   List all the subsets of $\{1, 2, 3\}$.

$\{1, 2, 3\}$ Every set is a subset of itself.
$\{1, 2\}$
$\{1, 3\}$ There are three subsets with two members each.
$\{2, 3\}$
$\{1\}$
$\{2\}$ There are three subsets with one member each.
$\{3\}$
$\emptyset$ The empty set is a subset of any set.

INTERSECTIONS AND UNIONS

The *intersection* of two sets A and B is the set of all members that are common to A and B. We denote the intersection of sets A and B as

$A \cap B$.

The intersection of two sets is often pictured as follows.

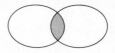

Example 12   Find: $\{1, 2, 3, 4, 5\} \cap \{-2, -1, 0, 1, 2, 3\}$.

The numbers 1, 2, and 3 are common to the two sets, so the intersection is $\{1, 2, 3\}$.

The *union* of two sets is formed by putting them together. We denote the union of sets A and B as

   $A \cup B$.

The union of two sets is often pictured as follows.

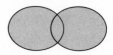

Example 13   Find: $\{2, 3, 4\} \cup \{3, 5, 7\}$.

The numbers in either or both sets are 2, 3, 4, 5, and 7, so the union is $\{2, 3, 4, 5, 7\}$.

In many situations, the word *and* corresponds to intersection and the word *or* to union.

Example 14   Graph: $-3 < x < 1$.

The inequality as stated means $-3 < x$ *and* $x < 1$. We graph $-3 < x$ and also graph $x < 1$.

Then we take the intersection of the graphs.

The last graph is the intersection of the two preceding graphs.

Example 15 Graph: $x < -3$ *or* $x \geqslant 2$.

We graph $x < -3$. We also graph $x \geqslant 2$. Then we take the union of the two graphs.

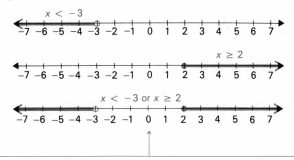

> A compound inequality using the word *or*, as in this example, *cannot* be shrunk to look like the one in Example 14. Remember, when the word *or* occurs, you must keep that word.

EXERCISE SET 10.2

Write set notation for the following.

1. The even whole numbers less than 10

2. The odd whole numbers less than 10

3. The odd whole numbers greater than 10

4. The even whole numbers greater than 10

5. The real numbers greater than $\sqrt{3}$

6. The real numbers less than $-\sqrt{5}$

Find the solution sets.

7. $(x + 5)(x - 5) = 0$

8. $(x + 4)(x + 7) = 0$

9. $x^2 - x - 6 = 0$

10. $x^2 + x - 12 = 0$

11. $5x + 3 > 4x - 2$

12. $7x - 7 < 6x + 5$

Which of the following are true and which are false?

13. $\{1, 5, 8\} \subset \{1, 2, 5, 7, 8\}$

14. $\{2, 4\} \subset \{0, 2, 3, 4, 7\}$

15. $\{1, 2, 3, 4, 5\} \subset \{2, 3, 4\}$

16. $\{8, 9, 13, 17\} \subset \{4, 5, 17\}$

17. $\{3, 8, 13\} \subset \{6, 8, 25, 38\}$

18. $\{9, 14, 27\} \subset \{9, 27, 38, 50\}$

19. $\emptyset \subset \{1, 2\}$

20. $\emptyset \subset \emptyset$

21. $\{x \mid x > 3\} \subset \{x \mid x > 2\}$

22. $\{x \mid x > 3\} \subset \{x \mid x > 4\}$

23. $\{x \mid x < 5\} \subset \{x \mid x \leqslant 5\}$

24. $\{x \mid x < 5\} \subset \{x \mid x < 5\}$

List all the subsets.

25. $\{-4, 7\}$

26. $\{a, b, c\}$

Find these unions and intersections.

27. $\{5, 6, 7, 8\} \cap \{4, 6, 8, 10\}$

28. $\{9, 10, 27\} \cap \{8, 10, 38\}$

29. $\{4, 5, 6, 7, 8\} \cup \{1, 4, 6, 11\}$

30. $\{8, 9, 27\} \cup \{2, 8, 27\}$

31. $\{2, 4, 6, 8\} \cap \{1, 3, 5\}$

32. $\emptyset \cap \emptyset$

33. $\{1, 2, 3, 4\} \cap \{1, 2, 3, 4\}$

34. $\{8, 9, 10\} \cap \emptyset$

35. $\{4, 8, 11\} \cup \emptyset$

36. $\emptyset \cup \emptyset$

Graph.

37. $-5 < x \leqslant 2$

38. $3 \leqslant x < 7$

39. $x < -3 \quad or \quad x \geqslant 5$

40. $x \leqslant 5 \quad or \quad x > 10$

☆ _____

41. If a set has n members, how many subsets does it have?

Use the following sets when answering questions 42–65.

$A = \{x | -4 < x \leqslant 2\}$

$B = \emptyset$

$C = \{\ldots, -4, -2, 0, 2, 4, \ldots\}$

$D = \{0, 1, 2, 3, 4, 5, 6, 7, 8, 9\}$

$E = \{\ldots, -5, -3, -1, 1, 3, 5, \ldots\}$

$F = \{x | x \leqslant 3\}$

$G = \{0, 1\}$

$H = \{x | x < -10 \text{ or } x \geqslant 5\}$

$J = \{\{0\}, \{1\}, \{0, 1\}\}$

$K = \{x | x \leqslant -9 \text{ or } x \geqslant 9\}$

$L = \left\{ x | x > -\dfrac{10}{3} \right\}$

$M = \{0\}$

$N = \{1, 2, 3, \ldots\}$

$P = \{1\}$

$S = \{1, 2, 3\}$

$T = \{x | -1 \leqslant x < 1\}$

$W = \{0, 1, 2, 3, \ldots\}$

$Z = \{\ldots, -3, -2, -1, 0, 1, 2, 3, \ldots\}$

Determine whether the following are true or false. (The symbol $\notin$ means "is not a member of.")

42. $W \subset D$

43. $B \subset J$

44. $1 \in J$

45. $\{1\} \in J$

46. $-\dfrac{2}{3} \notin L$

47. $A \subset F$

48. $-\dfrac{11}{2} \in E$

49. $S \subset N \subset W \subset Z$

50. $2^7 \in C$

51. $G \subset T$

52. $1 \subset G$

53. $L \subset A$

Find these unions and intersections.

54. $C \cup E$

55. $E \cap C$

56. $M \cap G \cap P$

57. $M \cap B$ **58.** $T \cup P$ **59.** $A \cap C$

60. $A \cap F$ **61.** $K \cup H$ **62.** $K \cap H$

63. $H \cap Z$ **64.** $M \cup B$ **65.** $F \cap L$

10.3 THE MULTIPLICATION PRINCIPLE FOR SOLVING INEQUALITIES

THE MULTIPLICATION PRINCIPLE

Consider the true inequality

$$4 < 9.$$

If we multiply both numbers by 2 we get another true inequality:

$$8 < 18.$$

If we multiply both numbers by -3 we get a false inequality:

$$-12 < -27.$$

However, if we now reverse the inequality symbol we get a true inequality:

$$-12 > -27.$$

> *The Multiplication Principle for Inequalities.* **If we multiply on both sides of a true inequality by a positive number, we get another true inequality. If we multiply by a negative number and reverse the inequality symbol, we get another true inequality.**

When we solve an inequality using the multiplication principle, we do not need to check, provided that we do not multiply by an expression containing a variable.

Examples Solve.

1. $3y < \frac{3}{4}$

$\frac{1}{3} \cdot 3y < \frac{1}{3} \cdot \frac{3}{4}$ Multiplying by $\frac{1}{3}$

$y < \frac{1}{4}$

Any number less than $\frac{1}{4}$ is a solution. The solution set is $\{y \mid y < \frac{1}{4}\}$.

2. $-4x < \frac{4}{5}$

$\boxed{-\frac{1}{4}} \cdot (-4x) > \boxed{-\frac{1}{4}} \cdot \frac{4}{5}$ Multiplying by $-\frac{1}{4}$ and reversing
the inequality sign

$x > -\frac{1}{5}$

Any number greater than $-\frac{1}{5}$ is a solution. The solution set is $\{x \mid x > -\frac{1}{5}\}$.

Example 3   Solve: $-5x \geqslant -80$.

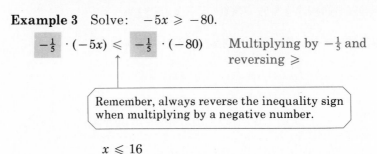

$\boxed{-\frac{1}{5}} \cdot (-5x) \leqslant \boxed{-\frac{1}{5}} \cdot (-80)$ Multiplying by $-\frac{1}{5}$ and
reversing $\geqslant$

Remember, always reverse the inequality sign
when multiplying by a negative number.

$x \leqslant 16$

Any number less than 16 is a solution, and 16 is also a solution. The solution set is

$\{x \mid x \leqslant 16\}$.

USING THE PRINCIPLES TOGETHER

We use the addition and multiplication principles together in solving inequalities in much the same way as for equations.

Example 4   Solve: $16 - 7y \geqslant 10y - 4$.

$\boxed{-16} + 16 - 7y \geqslant \boxed{-16} + 10y - 4$ Adding -16

$-7y \geqslant 10y - 20$

$\boxed{-10y} - 7y \geqslant \boxed{-10y} + 10y - 20$ Adding $-10y$

$-17y \geqslant -20$

$\boxed{-\frac{1}{17}} \cdot (-17y) \leqslant \boxed{-\frac{1}{17}} \cdot (-20)$ Multiplying by $-\frac{1}{17}$ and
reversing the inequality
sign

$y \leqslant \frac{20}{17}$

The solution set is $\{y \mid y \leqslant \frac{20}{17}\}$.

In Example 4, we could have begun by adding $7y$, as follows.

$$16 - 7y + \boxed{7y} \geqslant 10y - 4 + \boxed{7y} \qquad \text{Adding } 7y$$

$$16 \geqslant 17y - 4$$

$$16 + \boxed{4} \geqslant 17y - 4 + \boxed{4} \qquad \text{Adding } 4$$

$$20 \geqslant 17y$$

$$\tfrac{20}{17} \geqslant y \qquad \text{Multiplying by } \tfrac{1}{17}$$

The solution set can be named $\{y \mid \tfrac{20}{17} \geqslant y\}$, or $\{y \mid y \leqslant \tfrac{20}{17}\}$.

Examples   Solve.

5. $-1 < 2x + 3 \leqslant 11$

$$-1 + \boxed{(-3)} < 2x + 3 + \boxed{(-3)} \leqslant 11 + \boxed{(-3)} \qquad \text{Adding } -3$$

$$-4 < 2x \leqslant 8$$

$$\boxed{\tfrac{1}{2}} \cdot (-4) < \boxed{\tfrac{1}{2}} \cdot 2x \leqslant \boxed{\tfrac{1}{2}} \cdot 8 \qquad \text{Multiplying by } \tfrac{1}{2}$$

$$-2 < x \leqslant 4$$

The solution set is $\{x \mid -2 < x \leqslant 4\}$.

6. $3 \leqslant 5 - 2x < 7$

$$3 - 5 \leqslant -2x < 7 - 5 \qquad \text{Adding } -5$$

$$-2 \leqslant -2x < 2$$

$$-\tfrac{1}{2} \cdot (-2) \geqslant -\tfrac{1}{2} \cdot (-2x) > -\tfrac{1}{2} \cdot 2 \qquad \text{Multiplying by } -\tfrac{1}{2} \text{ and reversing the inequality signs}$$

$$1 \geqslant x > -1$$

The solution set is $\{x \mid 1 \geqslant x > -1\}$.

SOLVING PROBLEMS

Certain problems translate quite naturally to inequalities, rather than equations.

Example 7   In a business course, there will be three tests. You must get a total score of 270 for an A. You get 91 and 86 on the first two tests. What scores on the last test will give you an A?

We translate. For an A,

Total score $\geqslant$ 270

$91 + 86 + x \geqslant 270.$

We have used x for your score on the last test. Solving the inequality, we get

$x \geqslant 93.$

Thus a score of 93 or higher will give you an A.

Example 8   On your new job, you can be paid in one of two different ways.

Plan A: A salary of $600 per month, plus a commission of 4% of gross sales.

Plan B: A salary of $800 per month, plus a commission of 6% of gross sales over $10,000.

For what gross sales is Plan A better than Plan B, assuming that gross sales are always more than $10,000?

To translate, we will write an inequality that states that the income from Plan A is greater than that from Plan B.

Income from A > Income from B

$600 + 4\% \cdot x > 800 + (x - 10{,}000) \cdot 6\%.$

We have used x to represent the gross sales for the month. Solving this inequality, we get

$x < 20{,}000.$

Thus for gross sales under $20,000, Plan A is better.

Example 9   A mason can be paid in two ways.

Plan A: $300 plus $3 per hour.

Plan B: Straight $8.50 per hour.

Suppose that the job takes n hours. For what values of n is Plan B better for the mason?

We translate:

Income from B > Income from A

$$8.50n > 300 + 3 \cdot n$$

Solving this inequality, we get

$$n > 54.55.$$

Thus if the job takes over 54.55 hours (54 hr, 33 min), Plan B is better.

EXERCISE SET 10.3

Solve.

1. $3x \leqslant 18$ **2.** $4x > 24$ **3.** $-3y > \dfrac{3}{8}$ **4.** $-8y \leqslant -\dfrac{32}{5}$

5. $\dfrac{3}{4} + 7x < \dfrac{5}{8}$ **6.** $6.2 + 4.2x \geqslant 11.1$ **7.** $2x - 7 \geqslant 5x - 5$

8. $4 - 5x > 2x + 3$ **9.** $8 < 3x + 2 \leqslant 14$ **10.** $-7 \leqslant 5x - 2 < 12$

11. A car rents for $13.95 per day, plus 10¢ per mile. You are on a daily budget of $76. What mileages will allow you to stay within the budget?

12. You are taking a history course. There will be 4 tests. You have scores of 89, 92, and 95 on the first three. You must make a total of 360 to get an A. What scores on the last test will give you an A?

13. A person is going to invest $25,000, part at 7% and part at 8%. What is the most that can be invested at 7% in order to make at least $1800 interest per year?

14. A person is going to invest $20,000, part at 6% and part at 8%. What is the most that can be invested at 6% in order to make at least $1500 interest per year?

15. On your new job, you can be paid in one of two ways.

Plan A: A salary of $500 per month, plus a commission of 4% of gross sales.

Plan B: A salary of $750 per month plus a commission of 5% of gross sales over $8000.

For what gross sales is Plan B better than Plan A, assuming that gross sales are always more than $8000?

16. A mason can be paid in two ways.

Plan A: $500 plus $3 per hour.

Plan B: Straight $8.00 per hour.

Suppose the job takes n hours. For what values of n is Plan A better for the mason than Plan B?

☆

Solve.

17. $\dfrac{5m - 3}{4} \leqslant m + 1$

18. $\dfrac{3}{7}x - \dfrac{3}{14} < \dfrac{73}{70} - \dfrac{4}{35}x$

19. $x - 10 < 5x + 6 \leqslant x + 10$

20. $\dfrac{2}{3} \leqslant -\dfrac{4}{5}(u - 2) < 1$

21. $2 - (3 - 4a) < a - (1 - 3a)$

22. $2[5(3 - y) - 2(y - 2)] > y + 4$

23. $-\dfrac{2}{15} \leqslant \dfrac{2}{3}x - \dfrac{2}{5} \leqslant \dfrac{2}{15}$

24. $2x - \dfrac{3}{4} < -\dfrac{1}{10}$ or $2x - \dfrac{3}{4} > \dfrac{1}{10}$

25. $3x < 4 - 5x < 5 + 3x$

26. $(x + 6)(x - 4) > (x + 1)(x - 3)$

True or False. Let a, b, and c represent real numbers.

27. If $b > c$, then $b \not< c$.

28. If $-b < -a$, then $a < b$.

29. If $c \neq a$, then $a \not< c$.

30. If $a < c$ and $c < b$, then $b \not> a$.

31. If $a < c$ and $b < c$, then $a < b$.

32. If $-a < c$ and $-c > b$, then $a < b$.

10.4 ABSOLUTE VALUE

DISTANCE

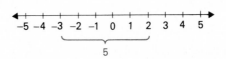

The number line above shows that the distance between -3 and 2 is 5. Another way to find the distance between two numbers on a number line is to subtract and take the absolute value, as follows.

$$|-3 - 2| = |-5| = 5, \quad \text{or} \quad |2 - (-3)| = |5| = 5$$

Note that the order in which we subtract does not matter since we are taking the absolute value afterward.

For any numbers a and b, the distance between them is $|a - b|$.

Example 1   Find the distance between -8 and -92.

$$|-8 - (-92)| = |84| = 84$$

Example 2 Find the distance between x and 0.

$$|x - 0| = |x|$$

EQUATIONS AND INEQUALITIES WITH ABSOLUTE VALUE

Example 3 Solve $|x| = 4$. Then graph, using a number line.

Note that $|x| = |x - 0|$, so that $|x - 0|$ is the distance from x to 0. The solutions of the equation are those numbers x whose distance from 0 is 4. Those numbers are 4 and -4. The solution set is $\{4, -4\}$. The graph consists of just two points, as shown.

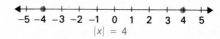

Remember, to *solve* an equation or inequality means to find all of its solutions (its solution set). There may be many ways of doing it, including guessing. To *graph* an equation or inequality means to make a picture of its solution set. For inequalities in one variable we make that picture on a number line.

Example 4 Solve $|x| < 4$. Then graph.

Again, note that $|x| = |x - 0|$, so $|x - 0|$ is the distance from x to 0. Thus the solutions of $|x - 0| < 4$ are those numbers x whose distance from 0 is less than 4. The solutions are those numbers x such that $-4 < x < 4$. The solution set is $\{x \mid -4 < x < 4\}$. The graph is as follows.

Example 5 Solve $|x| \geqslant 4$. Then graph.

Since $|x| = |x - 0|$, the solutions of $|x - 0| \geqslant 4$ are those numbers x whose distance from 0 is greater than or equal to 4, in other words, those numbers x such that $x \leqslant -4$ *or* $x \geqslant 4$. The solution set is $\{x \mid x \leqslant -4 \text{ or } x \geqslant 4\}$. The graph is as follows.

Examples 3 through 5 illustrate three cases of solving inequalities with absolute value. The expression inside of the absolute value signs can be something besides a single letter. We may wish to solve $|5x - 2| < 4$, for example. The following statement gives the general principles.

For any positive number b, and any expression X:

1. The solutions of $|X| = b$ are those numbers that satisfy

$$X = b \quad \text{or} \quad X = -b.$$

For example, the solutions of $|5x - 1| = 8$ are those numbers x for which

$$5x - 1 = 8 \quad \text{or} \quad 5x - 1 = -8.$$

2. The solutions of $|X| < b$ are those numbers that satisfy

$$-b < X < b.$$

For example, the solutions of $|6x + 7| < 5$ are those numbers x for which

$$-5 < 6x + 7 < 5.$$

3. The solutions of $|X| > b$ are those numbers that satisfy

$$X < -b \quad \text{or} \quad X > b.$$

For example, the solutions of $|2x - 9| > 4$ are those numbers x for which

$$2x - 9 < -4 \quad \text{or} \quad 2x - 9 > 4.$$

The above still hold if $<$ is replaced by $\leqslant$ and $>$ is replaced by $\geqslant$.

Example 6   Solve $|5x - 4| = 11$. Then graph.

We use Property 1. In this case, b is 11 and X is $5x - 4$.

$$|X| = b$$
$$|5x - 4| = 11 \qquad \text{Replacing } X \text{ by } 5x - 4 \text{ and } b \text{ by } 11$$
$$5x - 4 = 11 \quad \text{or} \quad 5x - 4 = -11 \qquad \text{Property 1}$$
$$5x = 15 \quad \text{or} \qquad 5x = -7 \qquad \text{Adding 4}$$
$$x = 3 \quad \text{or} \qquad x = -\frac{7}{5} \qquad \text{Multiplying by } \frac{1}{5}$$

The solution set is $\{3, -\frac{7}{5}\}$.

The graph is as follows.

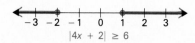

$$|5x - 4| = 11$$

Example 7   Solve $|3x - 2| < 4$. Then graph.

We use Property 2. In this case, b is 4 and X is $3x - 2$.

$$|X| < b$$
$$|3x - 2| < 4 \qquad \text{Replacing } X \text{ by } 3x - 2 \text{ and } b \text{ by } 4$$
$$-4 < 3x - 2 < 4 \qquad \text{Property 2}$$
$$-2 < 3x < 6 \qquad \text{Adding 2}$$
$$-\frac{2}{3} < x < 2 \qquad \text{Multiplying by } \frac{1}{3}$$

The graph is as follows

$$|3x - 2| < 4$$

Example 8   Solve $|4x + 2| \geqslant 6$. Then graph.

We use Property 3. In this case, b is 6 and X is $4x + 2$.

$$|X| \geqslant b$$
$$|4x + 2| \geqslant 6 \qquad \text{Replacing } X \text{ by } 4x + 2 \text{ and } b \text{ by } 6$$
$$4x + 2 \leqslant -6 \quad or \quad 4x + 2 \geqslant 6 \qquad \text{Property 3}$$
$$4x \leqslant -8 \quad or \qquad 4x \geqslant 4 \qquad \text{Adding } -2$$
$$x \leqslant -2 \quad or \qquad x \geqslant 1 \qquad \text{Multiplying by } \frac{1}{4}$$

The graph is as follows.

$$|4x + 2| \geq 6$$

In Example 8, we have the symbol $\geqslant$. Whenever $>$ or $\geqslant$ occurs, we must use the word *or* in solving. A common error is to try to construct a compound inequality without the word *or*.

EXERCISE SET 10.4

Find the distance between each pair of numbers.

1. $-8, \quad -42$ **2.** $-9, \quad -36$ **3.** $26, \quad 15$ **4.** $54, \quad 18$

5. $-9, \quad 24$ **6.** $-18, \quad 37$ **7.** $-2, \quad 0$ **8.** $15, \quad 0$

Solve. Then graph.

9. $|x| = 3$ **10.** $|x| = 5$

11. $|x| < 3$ **12.** $|x| \leqslant 5$

13. $|x| \geqslant 2$ **14.** $|y| > 8$

15. $|x - 3| < 12$ **16.** $|x - 2| < 6$

17. $|2x + 3| \leqslant 4$ **18.** $|5x + 2| \leqslant 3$

19. $|2y - 7| > 5$ **20.** $|3y - 4| > 8$

21. $|4x - 9| \geqslant 14$ **22.** $|9y - 2| \geqslant 17$

☆ ───────────────────────────

Solve.

23. $\left| \dfrac{2x - 1}{3} \right| \leqslant 1$ **24.** $\left| \dfrac{3x - 2}{5} \right| \geqslant 1$

25. $|3x| = 7$ **26.** $\left| 4x - \dfrac{1}{2} \right| \leqslant \dfrac{1}{12}$

27. $|3x - 4| > -2$ **28.** $|x - 6| \leqslant -8$

29. $\left| \dfrac{5}{9} + 3x \right| < \dfrac{1}{6}$ **30.** $1 - \left| \dfrac{1}{4}x + 8 \right| > \dfrac{3}{4}$

31. $|x + 5| > x$ **32.** $2 \leqslant |x - 1| \leqslant 5$

33. $|7x - 2| = x + 4$ **34.** $|x + 1| \leqslant |x - 3|$

10.5 INEQUALITIES IN TWO VARIABLES

SOLUTIONS

The solutions of inequalities in two variables are ordered pairs, as is the case with equations in two variables.

Example 1 Determine whether $(-3, 2)$ is a solution of $5x - 4y > 13$.

We replace x by -3 and y by 2, using alphabetical order.

$$5x - 4y > 13$$

$5(-3) \quad -4 \cdot 2$	13
$-15 - 8$	
-23	

Since $-23 > 13$ is false, $(-3, 2)$ is not a solution.

GRAPHS

Inequalities in two variables are graphed on a plane.

Example 2 Graph: $y < x$.

For comparison, we first graph the line $y = x$. We draw it dashed. Every solution of $y = x$ is an ordered pair having the same first and second coordinates.

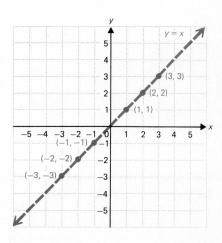

Now look at a vertical line and some ordered pairs on it. The one shown goes through (2, 2).

For any point here, the y-value is greater than the x-value.

For any point here, the y-value is less than the x-value.

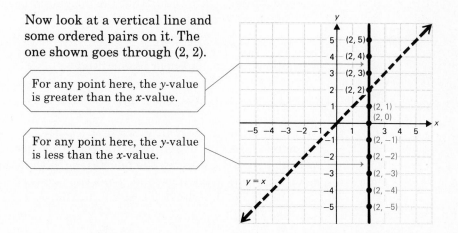

For all points below $y = x$, the second coordinate is less than the first, that is, $y < x$. For points above the line, $y > x$. The same thing happens for any vertical line. Thus all points below the line $y = x$ are solutions. We shade the half-plane below $y = x$. This is the graph of the inequality $y < x$. The points on the line $y = x$ are not included. That is why we drew the line dashed.

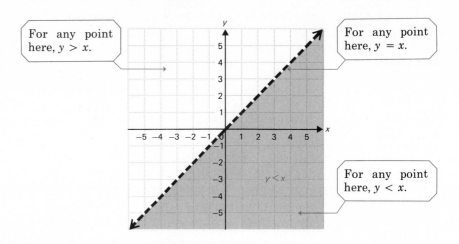

RECOGNIZING AND GRAPHING LINEAR INEQUALITIES

In Chapter 2 we learned to recognize linear equations. If the inequality sign in a sentence is replaced by an equals sign and the result is a linear equation, then we say that the inequality is a *linear inequality*. The following are linear inequalities:

$$y < 3x + 2, \qquad 5y - 4x - 2 \geqslant 0, \qquad 6x + 3 \leqslant 8y - 2.$$

The following are not linear inequalities:

$$y < x^2 + 2, \qquad xy^3 \geqslant 5 - 2y, \qquad \frac{1}{y} \leqslant x + 2y.$$

To graph a linear inequality, we first replace the inequality sign by an equals sign. Then we graph the resulting line. If the inequality symbol is $<$ or $>$, we draw the line dashed. If the inequality symbol is $\leqslant$ or $\geqslant$, we draw the line solid. The graph consists of a half-plane as in Example 2 and, if the line is solid, the line as well.

Example 3   Graph: $6x - 2y < 12$.

a) We first graph the line $6x - 2y = 12$. Since the inequality symbol is $<$, we draw it dashed. The line is most easily graphed using intercepts.

b) We determine which half-plane to shade by trying some point off the line. The point $(0, 0)$ is easy to check unless the line goes through the origin.

$$\frac{6x - 2y < 12}{\begin{array}{c|c} 6 \cdot 0 - 2 \cdot 0 & 12 \\ 0 & \end{array}}$$

Any point not on the line can be used as a check. The origin is an *easy* one to use.

Since $0 < 12$, the point is in the graph. We shade that half-plane. The graph consists of the shaded half-plane, but not the dashed line.

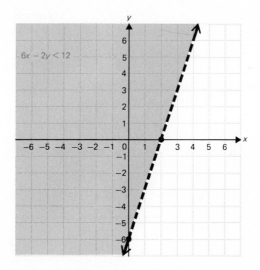

Example 4   Graph: $8x + 3y \geqslant 24$.

a) We graph the line $8x + 3y = 24$. Since the inequality symbol is $\geqslant$, we draw the line solid.

b) The point $(0, 0)$ is not in the graph, because $8 \cdot 0 + 3 \cdot 0 = 0$, which is less than 24. We shade the other half-plane. The graph consists of the shaded half-plane and the solid line as well.

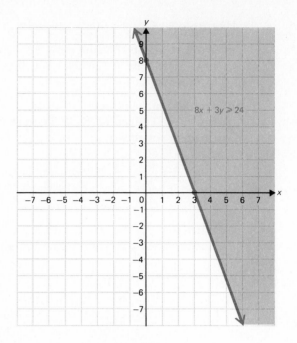

Example 5   Graph: $y > 3$.

a) We first graph $y = 3$, using a dashed line.

b) Checking any point above this line, we see that it satisfies the inequality.

c) We shade the upper half-plane.

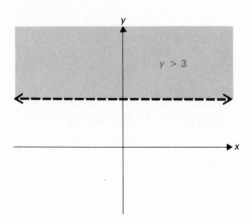

Example 6 Graph: $-2 < x \leqslant 5$.

a) We think of this as a system of inequalities:

$$-2 < x$$
$$x \leqslant 5.$$

b) We graph the equation $-2 = x$, and see that the graph of the first inequality is the half-plane to the right of the line $-2 = x$ (see the graph on the left below).

c) We graph the second inequality, starting with the line $x = 5$, and find that its graph is the line and also the half-plane to the left of it (see the graph on the right below).

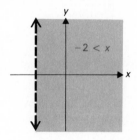 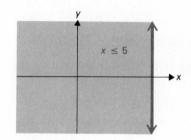

d) We shade the intersection of these graphs.

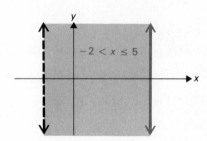

EXERCISE SET 10.5

Determine whether the ordered pair is a solution of the inequality.

1. $5y + 4x < 1$, $(-2, 1)$

2. $5y + 4x < 1$, $(2, -1)$

3. $y \geqslant 3x - 4$, $(2, 2)$

4. $y \geqslant 3x - 4$, $(2, 3)$

Graph the following on a plane.

5. $y > 2x$ **6.** $y < 3x$ **7.** $y \leqslant x + 1$

8. $y \geqslant x + 3$ **9.** $x + y < 4$ **10.** $x + y \geqslant 5$

11. $3x + 4y \geqslant 12$ **12.** $2x + 3y \leqslant 6$ **13.** $7x + 2y \leqslant 21$

14. $5x + 4y > 20$ **15.** $x < -4$ **16.** $y \geqslant 2$

17. $-4 < y < -1$ **18.** $-2 \leqslant x < 3$

☆

Graph.

19. $4x - 6 > 2x + 4$ **20.** $y \leqslant |x|$

21. $y > x^2 - 3$ **22.** $y \leqslant \dfrac{1}{4}x^2 - 1$

23. Solve $2x^2 + x < 6$ by graphing.

Graph each system of inequalities.

To graph the solution set of a system of inequalities, graph each inequality separately, and find their intersection.)

24. $x + y \leqslant 3$ **25.** $3x + 4y \geqslant 12$ **26.** $2x + y \leqslant 5$ **27.** $y - x \leqslant 2$

     $x - y \leqslant 1$ $5x + 6y \leqslant 30$ $x + 2y \leqslant 4$ $x - y \leqslant 2$

                                                       $x \geqslant 0$ $2 \leqslant y \leqslant 3$

                                                       $y \geqslant 0$

TEST OR REVIEW—CHAPTER 10

1. Solve and graph: $6 < 3x + 2 \leqslant 12$.

Find the solution set and write set notation for it.

2. $(x + 2)(x - 3) = 0$

3. $4x + 3 > 3x - 2$

4. $-16x < 48$

5. Which of the following are true?

    a) $\{2, 3\} \subset \{2, 3, 4\}$
    b) $\emptyset \subset \{1, 2, 3\}$
    c) $\{x \mid x > 3\} \subset \{x \mid x > 0\}$

6. Find the intersection: $\{1, 3, 5, 7, 9\} \cap \{3, 5, 11, 13\}$.

7. Find the union: $\{1, 3, 5, 7, 9\} \cup \{3, 5, 11, 13\}$.

8. Graph: $2 < x \leqslant 5$.   **9.** Graph: $x \leqslant 2$ *or* $x > 5$.

Solve. Then graph.

10. $|3x - 2| < 7$ **11.** $|2x - 3| \geqslant 6$

12. Graph: $2x + 3y < 12$

11

SOME ADVANCED TOPICS

11.1 MORE ABOUT FUNCTIONS

REVIEW OF FUNCTIONS

Before studying this chapter you may wish to review the ideas of functions presented earlier in the book. They are to be found on the following pages.

178–181
320–327
328–330
368–372
373–374

FINDING FUNCTION VALUES

When a function is given by a formula or equation, function values are found by substituting into the formula. That is true even if the inputs are not specific numbers. We may use such things as $2m$ or $a + h$ for inputs, but we still proceed by substituting.

Example 1   The function g given by $g(x) = x^2 - 1$ takes an input x and squares it. Then it subtracts 1 from the result. Find the following function values.

a) $g(-3) = (-3)^2 - 1$ Substituting the input, -3
$\qquad\quad = 9 - 1$ Squaring -3
$\qquad\quad = 8$ Simplifying

b) $g(3m) = (3m)^2 - 1$ Substituting the input, $3m$
$\qquad\qquad = 9m^2 - 1$

c) $g(a + h) = (a + h)^2 - 1$ Substituting the input, $a + h$
$\qquad\qquad\ = a^2 + 2ah + h^2 - 1$ Squaring $a + h$

d) $g(a) + g(a + h)$

$\qquad = [a^2 - 1] + [(a + h)^2 - 1]$ Substituting the inputs a
$\qquad\qquad\qquad\qquad\qquad\qquad\qquad$ and $a + h$
$\qquad = [a^2 - 1] + [a^2 + 2ah + h^2 - 1]$ Squaring $a + h$
$\qquad = a^2 - 1 + a^2 + 2ah + h^2 - 1$ Removing parentheses
$\qquad = 2a^2 + 2ah + h^2 - 2$ Collecting like terms

INVERSES

On page 371 we considered the inverse of

$$y = 2^x.$$

The inverse can be obtained by interchanging x and y, to obtain

$$x = 2^y.$$

If for any equation with two variables, such as x and y, we interchange the variables, we obtain an equation of the inverse relation. That inverse may or may not be a function. If the inverse of a function f is also a function, it can be named f^{-1} (read "f-inverse").

> The -1 in f^{-1} is not an exponent.

Example 2   Given $f(x) = x + 1$, find a formula for $f^{-1}(x)$.

a) Let us think of this as $y = x + 1$.

b) To find the inverse we interchange x and y: $x = y + 1$.

c) Now we solve for y: $y = x - 1$.

d) Thus $f^{-1}(x) = x - 1$.

Note in Example 2 that f maps any x onto $x + 1$ (this function adds 1 to each number of the domain). Its inverse, f^{-1}, maps any number x onto $x - 1$ (this inverse function subtracts 1 from each member of its domain). Thus the function and its inverse do opposite things.

Example 3   Let $g(x) = 2x + 3$. Find a formula for $g^{-1}(x)$.

a) Let us think of this as $y = 2x + 3$.

b) To find the inverse we interchange x and y: $x = 2y + 3$.

c) Now we solve for y: $y = (x - 3)/2$.

d) Thus $g^{-1}(x) = (x - 3)/2$.

Note in Example 3 that g maps any x onto $2x + 3$ (this function doubles each input and then adds 3). Its inverse, g^{-1}, maps any input onto $(x - 3)/2$ (this inverse function subtracts 3 from each input and then divides it by 2). Thus the function and its inverse do opposite things.

To find a formula for the inverse of a function, if the inverse is also a function,

a) Write y instead of $f(x)$;
b) Interchange x and y;
c) Solve for y;
d) Write $f^{-1}(x)$ instead of y.

Example 4   Consider $f(x) = x^2$. Let us think of this as $y = x^2$.

To find the inverse we interchange x and y: $x = y^2$. Now we solve for y: $y = \pm\sqrt{x}$. Note that for each positive x we get two y's. For example, if y is 4, x can be either 2 or -2, as shown in the graph. Note also that the inverse fails the vertical line test.

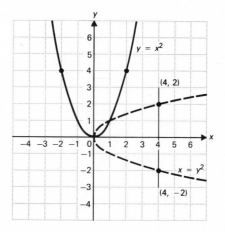

If we restrict the domain of $f(x) = x^2$ to nonnegative numbers, then its inverse is a function, $f^{-1}(x) = \sqrt{x}$.

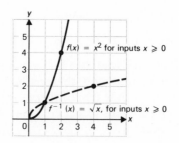

Let us consider inverses of functions in terms of function machines. Suppose that the function f programmed into a machine has an inverse that is also a function. Suppose then that the function machine has a reverse switch. When the switch is thrown, the machine is then programmed to do the inverse mapping f^{-1}. Inputs then enter at the opposite end, and the entire process is reversed.

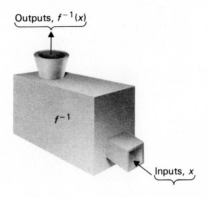

TRANSLATIONS

In Section 7.8 we considered graphs of quadratic functions. We compared graphs of

$$f(x) = ax^2 \quad \text{and} \quad f(x) = a(x - h)^2,$$

where h is some constant. We noted that replacing x by $x - h$ moved the graph to the right or left. If h is positive, as in $x - 2$ or $x - 5$ (a positive number being *subtracted* from x), then the graph is moved to the right (in the positive direction). We compare, for example, the graphs of $f(x) = x^2$ and $f(x) = (x - 3)^2$.

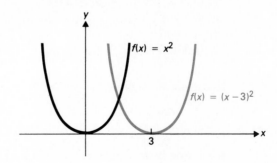

If the constant h is negative, as in $x + 3$ or $x + 7$ [a negative number being *subtracted* from x, as in $x - (-3)$], then the graph is moved in the negative direction.

A similar thing happens with the graph of any function.

> **If in $y = f(x)$ we replace x by $x - h$, where h is a constant, the graph will be moved according to h. If h is positive, the graph is moved in the positive direction, a distance h. If h is negative, the graph is moved in the negative direction.**

Example 5   Graph $f(x) = 2^{x-1}$. We know what the graph of $f(x) = 2^x$ looks like. (See page 369.) The graph of $f(x) = 2^{x-1}$ will look just like it, except that it will be moved one unit to the right.

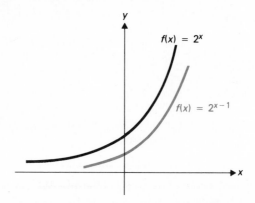

What we have said about replacing x by $x - h$ also holds for replacing y by $y - k$, or $f(x)$ by $f(x) - k$, where k is some constant.

> **If in $y = f(x)$ we replace y by $y - k$, where k is a constant, the graph will be moved according to k. If k is positive, the graph is moved in the positive direction (up) a distance k. If k is negative, the graph is moved in the negative direction.**

Example 6   Graph $y = x^2 + 3$.

$$y = x^2 + 3$$
$$y - 3 = x^2 \qquad \text{Subtracting 3 to get an expression } y - k$$

We know how to graph $y = x^2$. To graph $y - 3 = x^2$, we move that graph up (in the positive y-direction) a distance of 3.

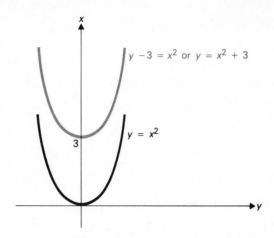

For functions we more often write $y = x^2 + 3$ than $y - 3 = x^2$. If you wish, you can remember the rule stated accordingly, as follows.

> **If in $y = f(x)$, we replace $f(x)$ by $f(x) + k$, where k is some constant, the graph will be moved according to k. If k is positive, the graph is moved upward a distance k. If k is negative, the graph is moved downward.**

EXERCISE SET 11.1

Find the following function values.

1. $f(x) = 5x^2 + 4x$
  a) $f(0)$ b) $f(-1)$
  c) $f(3)$ d) $f(t)$
  e) $f(t - 1)$ f) $\dfrac{f(a + h) - f(a)}{h}$

2. $g(x) = 3x^2 - 2x + 1$.
  a) $g(0)$ b) $g(-1)$
  c) $g(3)$ d) $g(t)$
  e) $g(a + h)$ f) $\dfrac{g(a + h) - g(a)}{h}$

3. $f(x) = 2|x| + 3x$
  a) $f(1)$ b) $f(-2)$
  c) $f(-4)$ d) $f(2y)$
  e) $f(a + h)$ f) $\dfrac{f(a + h) - f(a)}{h}$

4. $g(x) = x^3 - 2x$
  a) $g(1)$ b) $g(-2)$
  c) $g(-4)$ d) $g(3y)$
  e) $g(2 + h)$ f) $\dfrac{g(2 + h) - g(2)}{h}$

5. 🔢 $f(x) = 4.3x^2 - 1.4x$
  a) $f(1.034)$ b) $f(-3.441)$
  c) $f(27.35)$ d) $f(-16.31)$

6. 🔢 $g(x) = \sqrt{2.2|x| + 3.5}$
  a) $g(17.3)$ b) $g(-64.2)$
  c) $g(0.095)$ d) $g(-6.33)$

7. $f(x) = \dfrac{x^2 - x - 2}{2x^2 - 5x - 3}$

    a) $f(0)$ b) $f(4)$

    c) $f(-1)$ d) $f(3)$

8. $s(x) = \sqrt{\dfrac{3x - 4}{2x + 5}}$

    a) $s(10)$ b) $s(2)$

    c) $s(1)$ d) $s(-1)$

Find inverses of the following.

9. $f(x) = 3x + 2$

10. $g(x) = 5x - 3$

11. $g(x) = \dfrac{2x + 1}{3}$

12. $f(x) = \dfrac{3x + 2}{5}$

13. $g(x) = 2x^2 + 3$

(Restrict x to being greater than 3.)

14. $f(x) = 3x^2 - 2$

(Restrict x to being greater than -2.)

15. $f(x) = 10^x$

16. $g(x) = \log_{10} x$

Graph the following.

17. a) $f(x) = 3^x$
    b) $f(x) = 3^{x+2}$
    c) $f(x) = 3^{x-1}$

18. a) $g(x) = 2x^2$
    b) $g(x) = 2x^2 + 3$
    c) $g(x) = 2x^2 - 4$

19. a) $y = x^2$
    b) $y = (x - 3)^2$
    c) $y = (x + 4)^2$

20. a) $y = 2x + 5$
    b) $y = 2(x + 4) + 5$
    c) $y = 2(x - 3) + 5$

21. a) $f(x) = \log_{10} x$
    b) $f(x) = \log_{10} (x - 5)$
    c) $f(x) = \log_{10} (x + 3)$

22. a) $g(x) = \log_2 x$
    b) $g(x) = \log_2 (x - 4)$
    c) $g(x) = \log_2 (x + 6)$

23. a) $y = 2x^2$
    b) $y = 2(x - 4)^2 + 5$
    c) $y = 2(x + 3)^2 - 4$

24. a) $y = \log_{10} x$
    b) $y = \log_{10} (x + 2) + 4$
    c) $y = \log_{10} (x - 5) - 7$

☆

For each function find $\dfrac{f(x + h) - f(x)}{h}$.

25. $f(x) = \dfrac{1}{x}$ 26. $f(x) = \dfrac{1}{x^2}$ 27. $f(x) = \sqrt{x}$ (Rationalize the numerator.)

11.2 SEQUENCES AND SERIES

When the elements of a set are considered in order, we say we have a *sequence*. We specify which element is first, which is second, and so on. That amounts to assigning elements of the set to certain natural numbers. Here is an example:

red, white, blue, purple

is a sequence in which red is assigned to the number 1, white is assigned to the number 2, and so on. A sequence is thus a *function*, where the domain is some set of natural numbers.

> *Definition:* **A** *sequence* **is a function having for its domain some set of natural numbers.**

As another example, consider the sequence given by*

$$a(n) = 2^n, \quad \text{or} \quad a_n = 2^n.$$

Some of the function values (also known as *terms* of the sequence) are as follows:

$$a_1 = 2^1 = 2, \qquad a_3 = 2^3 = 8,$$
$$a_2 = 2^2 = 4, \qquad a_6 = 2^6 = 64.$$

The first term of the sequence is a_1, the fifth term is a_5, and the nth term or *general* term is a_n. This sequence can also be denoted in the following ways:

a) $2, 4, 8, \ldots$;

b) $2, 4, 8, \ldots, 2^n$.

Example 1   Find the first four terms and the 57th term of the sequence whose general term is given by $a_n = (-1)^n/(n + 1)$.

$$a_1 = \frac{(-1)^1}{1 + 1} = -\frac{1}{2}, \qquad a_2 = \frac{(-1)^2}{2 + 1} = \frac{1}{3},$$

$$a_3 = \frac{(-1)^3}{3 + 1} = -\frac{1}{4}, \qquad a_4 = \frac{(-1)^4}{4 + 1} = \frac{1}{5},$$

$$a_{57} = \frac{(-1)^{57}}{57 + 1} = \frac{-1}{58}.$$

When a sequence is described merely by naming the first few terms, we do not know for sure what the general term is, but the reader is expected to make a guess by looking for a pattern.

Examples   For each sequence, make a guess at the general term.

2. $1, 4, 9, 16, 25, \ldots$

These are squares of numbers, so the general term may be n^2.

---

* The notation a_n means the same as $a(n)$ but is more commonly used with sequences.

3. $\sqrt{1}, \sqrt{2}, \sqrt{3}, \sqrt{4}, \ldots$

These are square roots of numbers, so the general term may be $\sqrt{n}$.

4. $-1, 2, -4, 8, -16, \ldots$

These are powers of 2 with alternating signs, so the general term may be $(-1)^n [2^{n-1}]$.

5. $2, 4, 8, \ldots$

If we see the pattern of powers of 2, we will see 16 as the next term and guess 2^n for the general term. If we see that we can get the second term by adding 2, the third term by adding 4, and the next term by adding 6, and so on, we will see 14 as the next term. A general term for the sequence is then $n^2 - n + 2$.

SUMS, OR SERIES

With any sequence there is associated another sequence, obtained by taking sums of terms of the given sequence. The general term of this sequence of "partial sums" is denoted S_n. Consider the sequence

$$3, 5, 7, 9, \ldots, 2n + 1.$$

We construct some terms of the sequence of partial sums.

$S_1 = 3,$ This is the first term of the given sequence.
$S_2 = 3 + 5 = 8,$ This is the sum of the first two terms.
$S_3 = 3 + 5 + 7 = 15,$ The sum of the first three terms
$S_4 = 3 + 5 + 7 + 9 = 24.$ The sum of the first four terms

A sum like any of the above is called a *series*.

EXERCISE SET 11.2

In each of the following the general, or nth, term of a sequence is given. In each case find the first four terms, the 10th term, and the 15th term.

1. $a_n = 3n + 1$ 2. $a_n = (n-1)(n-2)(n-3)$ 3. $a_n = \dfrac{1}{n}$

4. $a_n = \dfrac{(-1)^n}{n}$ 5. $a_n = \dfrac{n}{n+1}$ 6. $a_n = \left(-\dfrac{1}{2}\right)^{n-1}$

For each of the following sequences find the general term or nth term. Answers may vary.

7. $1, 3, 5, 7, 9, \ldots$ 8. $3, 9, 27, 81, 243, \ldots$

9. $\dfrac{2}{3}, \dfrac{3}{4}, \dfrac{4}{5}, \dfrac{5}{6}, \dfrac{6}{7}, \ldots$

10. $\sqrt{2}, \sqrt{4}, \sqrt{6}, \sqrt{8}, \sqrt{10}, \ldots$

11. $\sqrt{3}, 3, 3\sqrt{3}, 9, 9\sqrt{3}, \ldots$

12. $1 \cdot 2, 2 \cdot 3, 3 \cdot 4, 4 \cdot 5, \ldots$

13. $\log 1, \log 10, \log 100, \log 1000, \ldots$

14. $2, 4, 6, 8, 10, 12, 14, 16, \ldots$

For each sequence find the associated sum.

15. $1, 2, 3, 4, 5, 6, 7, \ldots$ Find S_7.

16. $1, -3, 5, -7, 9, -11, \ldots$ Find S_8.

17. $2, 4, 6, 8, \ldots$ Find S_5.

18. $1, \dfrac{1}{4}, \dfrac{1}{9}, \dfrac{1}{16}, \dfrac{1}{25}, \ldots$ Find S_5.

☆ ───

19. Find the first five terms of the sequence given by

$$a_n = \frac{1}{2n} \log 1000^n.$$

20. a) Find the first few terms of the sequence $a_n = n^2 - n + 41$.

   b) What pattern do you observe?

   c) Find the 41st term. Does the pattern you found in (b) still hold?

Find decimal notation, rounded to six decimal places, for the first six terms of each sequence.

21. ▦ $a_n = \left(1 + \dfrac{1}{n}\right)^n$

22. ▦ $a_n = \sqrt{n + 1} - \sqrt{n}$

──

11.3 ARITHMETIC SEQUENCES AND SERIES

ARITHMETIC SEQUENCES

Consider this sequence:

$$2, 5, 8, 11, 14, 17, \ldots$$

Note that adding 3 to any term produces the following term. Or, the difference between any term and the preceding one is 3. Sequences in which the difference between successive terms is constant are called *arithmetic sequences*.

We now find a formula for the general, or nth, term of any arithmetic sequence. Let us denote the difference between successive terms (called the *common difference*) by d, and write out the first few terms.

$$a_1$$
$$a_2 = a_1 + d$$
$$a_3 = a_2 + d = (a_1 + d) + d = a_1 + 2d$$
$$a_4 = a_3 + d = (a_1 + 2d) + d = a_1 + 3d$$

Generalizing, we obtain the following.

> 1. **The nth term of an arithmetic sequence is given by**
>
> $$a_n = a_1 + (n - 1)d,$$
>
> **where d is the common difference between successive terms.**

Example 1   Find the 14th term of the arithmetic sequence 4, 7, 10, 13, ...

First note that $a_1 = 4$, $d = 3$, and $n = 14$. Then using formula 1, we obtain

$$
\begin{aligned}
a_{14} &= 4 + (14 - 1) \cdot 3 \\
&= 4 + 39 \\
&= 43.
\end{aligned}
$$

Example 2   In the sequence of Example 1, which term is 301? That is, what is n if $a_n = 301$?

We substitute into formula 1 and solve for n:

$$
\begin{aligned}
a_n &= a_1 + (n - 1)d \\
301 &= 4 + (n - 1) \cdot 3 \\
301 &= 4 + 3n - 3 \\
300 &= 3n \\
100 &= n.
\end{aligned}
$$

The 100th term is 301.

Given two terms and their places in an arithmetic sequence, we can construct the sequence.

Example 3   The third term of an arithmetic sequence is 8, and the sixteenth term is 47. Find a_1 and d and construct the sequence.

We know that $a_3 = 8$ and $a_{16} = 47$. Thus we would have to add d thirteen times to get from 8 to 47. That is,

$$13d = 47 - 8 = 39.$$

Solving, we obtain

$$d = 3.$$

Since $a_3 = 8$, we subtract d twice to get to a_1. Thus

$$a_1 = 2.$$

The sequence is $2, 5, 8, 11, \ldots$.

SUMS

We now look for a formula for the sum of the first n terms of an arithmetic sequence. Let us denote the terms as follows:

$$a_1, \quad (a_1 + d), \quad (a_1 + 2d), \ldots, (a_n - 2d), \quad (a_n - d), \quad a_n.$$

Then the sum of the first n terms is given by

$$S_n = a_1 + (a_1 + d) + (a_1 + 2d) + \cdots + (a_n - 2d) + (a_n - d) + a_n.$$

Reversing the right side, we obtain

$$S_n = a_n + (a_n - d) + (a_n - 2d) + \cdots + (a_1 + 2d) + (a_1 + d) + a_1.$$

Adding these two equations, we obtain

$$2S_n = (a_1 + a_n) + (a_1 + a_n) + \cdots + (a_1 + a_n),$$

a total of n terms on the right. Thus $2S_n = n(a_1 + a_n)$. Dividing by 2 gives us the formula we seek.

> **2.** **The sum of the first n terms of an arithmetic sequence is given by**
>
> $$S_n = \frac{n}{2}(a_1 + a_n).$$

We now derive a variant of this formula, which we can use in case a_n is not known. We substitute the expression for a_n given in formula 1, $a_n = a_1 + (n - 1)d$, into formula 2:

$$S_n = \frac{n}{2}(a_1 + [a_1 + (n - 1)d]).$$

We now have the following.

> **3.** **The sum of the first n terms of an arithmetic sequence is given by**
>
> $$S_n = \frac{n}{2}[2a_1 + (n - 1)d].$$

Example 4   Find the sum of the first 100 natural numbers.

The series is $1 + 2 + 3 + \cdots + 100$, so $a_1 = 1$, $a_n = 100$, and $n = 100$. We substitute into formula 2:

$$S_n = \frac{n}{2}(a_1 + a_n)$$

$$S_{100} = \frac{100}{2}(1 + 100) = 5050.$$

Example 5   Find the sum of the first 14 terms of the arithmetic sequence 2, 5, 8, 11, 14, ...

We note that $a_1 = 2$, $d = 3$, and $n = 14$. We use formula 3:

$$S_n = \frac{n}{2}[2a_1 + (n - 1)d],$$

$$S_{14} = \frac{14}{2} \cdot [2 \cdot 2 + (14 - 1)3] = 7 \cdot [4 + 13 \cdot 3] = 7 \cdot 43$$

$$S_{14} = 301.$$

Example 6   A family saves money in an arithmetic sequence. They save $500 one year, $600 the next, $700 the next and so on, for 10 years. How much do they save in all, exclusive of interest?

The amount saved is given by the series

$$\$500 + \$600 + \$700 + \cdots.$$

We have $a_1 = 500$, $n = 10$, and $d = 100$. We use formula 3:

$$S_n = \frac{n}{2}[2a_1 + (n - 1)d],$$

$$S_{10} = \frac{10}{2}[2 \cdot 500 + (10 - 1)100]$$

$$S_{10} = 9500.$$

EXERCISE SET 11.3

Exercises 1–10 refer to arithmetic sequences.

1. Find the 12th term of 2, 6, 10, ...

2. Find the 11th term of 0.07, 0.12, 0.17, ...

3. In the sequence in Exercise 1, which term is 106?

4. In the sequence in Exercise 2, which term is 1.67?

5. If $a_1 = 5$ and $d = 6$, what is a_{17}?

6. If $a_1 = 14$ and $d = -3$, what is a_{20}?

7. If $d = 4$ and $a_8 = 33$, what is a_1?

8. If $a_1 = 8$ and $a_{11} = 26$, what is d?

9. If $a_1 = 5$, $d = -3$ and $a_n = -76$, what is n?

10. If $a_1 = 25$, $d = -14$, and $a_n = -507$, what is n?

11. In an arithmetic sequence $a_{17} = -40$ and $a_{28} = -73$. Find a_1 and d. Find the first 5 terms of the sequence.

12. A bomb drops from an airplane and falls 16 ft the first sec, 48 ft the second sec, and so on, forming an arithmetic sequence. How many feet will the bomb fall during the 20th sec?

13. Find the sum of the first 20 terms of the sequence 5, 8, 11, 14, ...

14. Find the sum of the first 15 terms of the sequence $5, \dfrac{55}{7}, \dfrac{75}{7}, \dfrac{95}{7}, \ldots$.

15. Find the sum of the odd numbers from 1 to 99, inclusive.

16. Find the sum of the multiples of 7, from 7 to 98 inclusive.

17. An arithmetic sequence has $a_1 = 2$, $d = 5$, and $n = 20$. What is S_n?

18. Find the sum of all multiples of 4 that are between 14 and 523.

19. How many poles will be in a pile of telephone poles if there are 30 in the first layer, 29 in the second, and so on until there is one in the last layer?

20. If a student saved 10¢ on October 1, 20¢ on October 2, 30¢ on October 3, etc., how much would be saved during October? (October has 31 days.)

☆

21. Find a formula for the sum of the first n odd natural numbers.

22. Find three numbers in an arithmetic sequence such that the sum of the first and third is 10, and the product of the first and second is 15.

11.4 GEOMETRIC SEQUENCES AND SERIES

GEOMETRIC SEQUENCES

Consider this sequence:

3, 6, 12, 24, 48, 96, ...

Note that multiplying any term by 2 produces the following term. That is, the ratio of any term and the preceding one is 2. Sequences in which the ratio of successive terms is constant (but different from 1) are called *geometric sequences*.

We now find a formula for the general, or nth, term of any geometric sequence. Let us denote the ratio of successive terms (called the *common ratio*) by r, and write out the first few terms.

$$a_1$$
$$a_2 = a_1 r$$
$$a_3 = a_2 r = (a_1 r)r = a_1 r^2$$
$$a_4 = a_3 r = (a_1 r^2)r = a_1 r^3$$

Generalizing, we obtain the following.

4.   **The nth term of a geometric sequence is given by**

$$a_n = a_1 r^{n-1},$$

where r is the common ratio of successive terms.

Example 1   Find the 6th term of the geometric sequence $4, 20, 100, \ldots$.

First note that $a_1 = 4$, $n = 6$, and $r = 5$. We use formula 4:

$$a_n = a_1 r^{n-1}$$
$$a_6 = 4 \cdot 5^{6-1} = 4 \cdot 5^5$$
$$= 12{,}500.$$

Example 2   Find the 11th term of the geometric sequence $64,\ -32,$ $16,\ -8, \ldots$.

Note that $a_1 = 64$, $n = 11$, and $r = -\frac{1}{2}$. We use formula 4:

$$a_n = a_1 r^{n-1}$$
$$a_{11} = 64 \cdot \left(-\frac{1}{2}\right)^{11-1}$$
$$= 64 \cdot \left(-\frac{1}{2}\right)^{10}$$
$$= 2^6 \cdot \frac{1}{2^{10}}$$
$$= \frac{1}{2^4} \quad \text{or} \quad \frac{1}{16}.$$

Example 3   A college student borrows \$600 at 12% interest compounded annually. The loan is paid off at the end of 3 years. How much is paid back?

For principal P, at 12% interest, $P + 0.12P$ will be owed at the end of a year. This is also $1.12P$, and is the principal for the second year, so at the end of that year $1.12(1.12P)$ is owed. Thus the principal at the beginnings of successive years is

$$P, \quad 1.12P, \quad 1.12^2 P, \quad 1.12^3 P, \quad \text{and so on.}$$

We now have a geometric sequence with $a_1 = 600$, $n = 4$, and $r = 1.12$. We use formula 4:

$$a_n = a_1 r^{n-1}$$
$$a_4 = 600 \cdot (1.12)^{4-1}$$
$$= 600(1.12)^3$$
$$= 600 \cdot 1.404928$$
$$= 842.96.$$

The amount paid back is $842.96

SUMS

We now look for a formula for the sum of the first n terms of a geometric sequence. Let us denote the terms as follows:

$$a_1, \quad a_1 r, \quad a_1 r^2, \quad a_1 r^3, \ldots, \quad a_1 r^{n-1}.$$

Then the sum of the first n terms is given by

$$S_n = a_1 + a_1 r + a_1 r^2 + \cdots + a_1 r^{n-2} + a_1 r^{n-1}.$$

We next multiply on both sides of this equation by $-r$, obtaining

$$-r S_n = -a_1 r - a_1 r^2 - \cdots - a_1 r^{n-1} - a_1 r^n.$$

We now add the two equations, obtaining

$$S_n - r S_n = a_1 - a_1 r^n.$$

Solving for S_n gives us the formula we seek.

> **5.   The sum of the first n terms of a geometric sequence is given by**
>
> $$S_n = \frac{a_1 - a_1 r^n}{1 - r}, \quad r \neq 1.$$

We now derive a variant of this formula, which we can use when a_n is known. Using formula 4, we substitute a_n for $a_1 r^{n-1}$ into formula 5.

We then have the following.

> **6. The sum of the first n terms of a geometric sequence is given by**
>
> $$S_n = \frac{a_1 - ra_n}{1 - r}, \qquad r \neq 1.$$

Example 4 Find the sum of the first 6 terms of the geometric sequence 3, 6, 12, 24, ...

Note that $a_1 = 3$, $n = 6$, and $r = 2$. We use formula 5:

$$S_n = \frac{a_1 - a_1 r^n}{1 - r}$$

$$S_6 = \frac{3 - 3 \cdot 2^6}{1 - 2} = \frac{3 - 192}{-1}, \quad \text{or} \quad 189.$$

Recall the definition of a *sequence* as a function having for its domain some set of natural numbers. If that domain is an infinite (unending) set of natural numbers, then the sequence is called an *infinite sequence*. In the examples that follow, the domain will be the set of *all* natural numbers, even though that is not a requirement of the definition.

Let us look at two examples of infinite geometric sequences and their sequences of partial sums.

A. 2, 4, 8, 16, . . . , 2^n, . . .

$$S_1 = 2 \qquad\qquad S_4 = 30$$
$$S_2 = 6 \qquad\qquad S_5 = 62$$
$$S_3 = 14$$

The terms of the sequence S_n get larger and larger, without bound. Now let us look at the next sequence.

B. $\dfrac{1}{2}, \dfrac{1}{4}, \dfrac{1}{8}, \dfrac{1}{16}, \ldots, \dfrac{1}{2^n}, \ldots$

$$S_1 = \frac{1}{2} \qquad\qquad\qquad\qquad S_4 = S_3 + \frac{1}{16} = \frac{15}{16}$$

$$S_2 = \frac{1}{2} + \frac{1}{4} = \frac{3}{4} \qquad\qquad S_5 = S_4 + \frac{1}{32} = \frac{31}{32}$$

$$S_3 = S_2 + \frac{1}{8} = \frac{3}{4} + \frac{1}{8} = \frac{7}{8}$$

It appears that we can write a formula for S_n as follows:

$$S_n = \frac{2^n - 1}{2^n}.$$

Not all infinite sequences have sums, for example, sequence A above. An infinite sum is also called an *infinite series*.

We now find a formula for the "sum" of an infinite geometric sequence by looking at values of S_n as n becomes very large. Recall that

$$S_n = \frac{a_1 - a_1 r^n}{1 - r} = \frac{a_1}{1 - r} - \frac{a_1 r^n}{1 - r}.$$

As n becomes large, r^n will also become large if $r > 1$. If r is negative and $|r| > 1$, then the values of r^n will alternate between positive and negative numbers having large absolute value. In either case, S_n will not approach a limit. If $r = 1$, the formula does not hold, and the sequence will not be geometric. However, if $|r| < 1$, then the values of r^n approach 0 as n becomes large, and we have the following.

> **7.   An infinite geometric sequence has a sum if and only if $|r| < 1$. That sum is given by**
>
> $$S_\infty = \frac{a_1}{1 - r}.$$

Example 5   Find the sum of the infinite geometric sequence 5, $\frac{5}{2}$, $\frac{5}{4}$, $\frac{5}{8}$, ... , if it exists.

Note that $a_1 = 5$. The sum exists because $r = \frac{1}{2}$, which is less than 1. We use formula 7:

$$S_\infty = \frac{a_1}{1 - r}$$

$$= \frac{5}{1 - \frac{1}{2}} = 10.$$

EXERCISE SET 11.4

1. Find the 10th term of the geometric sequence $\frac{8}{243}$, $\frac{4}{81}$, $\frac{2}{27}$, ...

2. Find the 5th term of the geometric sequence 2, -10, 50, ...

3. Find the sum of the first 8 terms of the geometric sequence 5, 10, 20, ...

4. Find the sum of the first 6 terms of the geometric sequence 16, -8, 4, ...

5. Find the sum of the first 7 terms of the geometric sequence $\frac{1}{18}$, $-\frac{1}{6}$, $\frac{1}{2}$, ...

6. Find the sum of the following geometric sequence: -8, 4, -2, . . . , $-\frac{1}{32}$.

7. A college student borrows $800 at 8% interest compounded annually. She pays off the loan in full at the end of 2 yr. How much does she pay?

8. A college student borrows $1000 at 8% interest compounded annually. He pays off the loan in full at the end of 4 yr. How much does he pay?

9. A ping-pong ball is dropped from a height of 16 ft and always rebounds $\frac{1}{4}$ of the distance of the previous fall. What distance does it rebound the 6th time?

10. Gaintown has a population of 100,000 now, and the population is increasing 10% every year. What will the population be in 5 yr?

11. 🔳 Find the sum of the first 5 terms of the geometric sequence

$1000, $1000(1.08), $1000(1.08)^2$, ...

12. 🔳 Find the sum of the first 6 terms of the geometric sequence

$200, $200(1.13), $200(1.13)^2$, ...

Round to the nearest cent.

Round to the nearest cent.

Find these infinite sums. (The series are geometric.)

13. $4 + 2 + 1 + \cdots$

14. $7 + 3 + \frac{9}{7} + \cdots$

15. $25 + 20 + 16 + \cdots$

16. $12 + 9 + \frac{27}{4} + \cdots$

☆

17. 🔳 A piece of paper is 0.01 in. thick. It is folded in such a way that its thickness is doubled each time for 20 times. How thick is the result?

18. 🔳 A superball dropped from the top of the Washington Monument (556 ft high) always rebounds $\frac{3}{4}$ of the distance of the previous fall. How far up and down has it traveled when it hits the ground for the sixth time?

11.5 BINOMIAL SERIES

Consider the following expanded powers of $(a + b)^n$, where $a + b$ is any binomial. Look for patterns.

$$(a + b)^0 = \qquad\qquad 1$$
$$(a + b)^1 = \qquad\qquad a + b$$
$$(a + b)^2 = \qquad\quad a^2 + 2ab + b^2$$
$$(a + b)^3 = \qquad a^3 + 3a^2b + 3ab^2 + b^3$$
$$(a + b)^4 = \quad a^4 + 4a^3b + 6a^2b^2 + 4ab^3 + b^4$$
$$(a + b)^5 = a^5 + 5a^4b + 10a^3b^2 + 10a^2b^3 + 5ab^4 + b^5$$

Each expansion is a polynomial that can be regarded as a series. There are some patterns to be noted in the expansions.

1. In each term, the sum of the exponents is n.

2. The exponents of a start with n and decrease to 0. The last term has no factor of a. The first term has no factor of b. The exponents of b start in the second term with 1 and increase to n. The sum of the exponents in each term is n.

3. There is one more term than the degree of the polynomial. The expansion of $(a + b)^n$ has $n + 1$ terms.

4. Now we consider the coefficients. Their pattern is not so easy to see, but the following will help.

$$(a + b)^5 = a^5 + \frac{5}{1} a^4 b + \frac{5 \cdot 4}{1 \cdot 2} a^3 b^2 + \frac{5 \cdot 4 \cdot 3}{1 \cdot 2 \cdot 3} a^2 b^3$$

$$+ \frac{5 \cdot 4 \cdot 3 \cdot 2}{1 \cdot 2 \cdot 3 \cdot 4} ab^4 + \frac{5 \cdot 4 \cdot 3 \cdot 2 \cdot 1}{1 \cdot 2 \cdot 3 \cdot 4 \cdot 5} b^5$$

The numerators of the coefficients, beginning with the second term, are 5, $5 \cdot 4$, $5 \cdot 4 \cdot 3$, and so on. The denominators are 1, $1 \cdot 2$, $1 \cdot 2 \cdot 3$, and so on. When the numerator becomes $5 \cdot 4 \cdot 3 \cdot 2 \cdot 1$ and the denominator becomes $1 \cdot 2 \cdot 3 \cdot 4 \cdot 5$, the coefficient is actually 1, and we have reached the last term.

The Binomial Theorem. **For any binomial $a + b$ and any natural number n,**

$$(a + b)^n = a^n + \frac{n}{1} a^{n-1} b + \frac{n(n - 1)}{1 \cdot 2} a^{n-2} b^2 + \frac{n(n - 1)(n - 2)}{1 \cdot 2 \cdot 3}$$

$$\times a^{n-3} b^3 + \cdots + \frac{n(n - 1)(n - 2) \cdots (1)}{1 \cdot 2 \cdot 3 \cdots n} b^n$$

Example 1 Expand $(2x + 3)^4$.

We think $(a + b)^4$, where $a = 2x$ and $b = 3$. Using the binomial theorem, we get

$$(2x + 3)^4 = (2x)^4 + \frac{4}{1} (2x)^3 \cdot 3 + \frac{4 \cdot 3}{1 \cdot 2} (2x)^2 \cdot 3^2$$

$$+ \frac{4 \cdot 3 \cdot 2}{1 \cdot 2 \cdot 3} (2x) \cdot 3^3 + \frac{4 \cdot 3 \cdot 2 \cdot 1}{1 \cdot 2 \cdot 3 \cdot 4} \cdot 3^4.$$

Now, simplifying, we get

$$16x^4 + 96x^3 + 216x^2 + 216x + 81.$$

Example 2   Expand $(x^2 - 2y)^5$.

We think of $(a + b)^5$, where $a = x^2$ and $b = -2y$. Using the binomial theorem, we get

$$[x^2 + (-2y)]^5 = (x^2)^5 + \frac{5}{1}(x^2)^4(-2y) + \frac{5 \cdot 4}{1 \cdot 2}(x^2)^3(-2y)^2$$

$$+ \frac{5 \cdot 4 \cdot 3}{1 \cdot 2 \cdot 3}(x^2)^2(-2y)^3$$

$$+ \frac{5 \cdot 4 \cdot 3 \cdot 2}{1 \cdot 2 \cdot 3 \cdot 4}(x^2)(-2y)^4$$

$$+ \frac{5 \cdot 4 \cdot 3 \cdot 2 \cdot 1}{1 \cdot 2 \cdot 3 \cdot 4 \cdot 5}(-2y)^5.$$

Simplifying, we get

$$x^{10} - 10x^8y + 40x^6y^2 - 80x^4y^3 + 80x^2y^4 - 32y^5.$$

Example 3   Find the fifth term in the expansion of $(2x - 5y)^6$.

We think of $(a + b)^6$, where $a = 2x$ and $b = -5y$. From the preceding pattern notice that in the fifth term the coefficient has four factors in the numerator and in the denominator. The exponent of b is the same as that number of factors. We can write the term without doing the entire expansion.

Fifth term: $\dfrac{6 \cdot 5 \cdot 4 \cdot 3}{1 \cdot 2 \cdot 3 \cdot 4}(2x)^2(-5y)^4$

4 factors   Exponent is also 4.

Exponents must add up to 6.

Simplifying, we get $37{,}500\ x^2y^4$.

Example 4   Find the eighth term in the expansion of $(3x - 2)^{10}$.

Eighth term: $\dfrac{10 \cdot 9 \cdot 8 \cdot 7 \cdot 6 \cdot 5 \cdot 4}{1 \cdot 2 \cdot 3 \cdot 4 \cdot 5 \cdot 6 \cdot 7}(3x)^3(-2)^7$

Simplifying, we get $-414{,}720\ x^3$.

EXERCISE SET 11.5

Find the indicated term of the binomial expansion.

1. 3rd, $(a + b)^6$

2. 6th, $(x + y)^7$

3. 12th, $(a - 2)^{14}$

4. 11th, $(x - 3)^{12}$

5. 5th, $(2x^3 - \sqrt{y})^8$

6. 4th, $\left(\dfrac{1}{b^2} + \dfrac{b}{3}\right)^7$

Expand.

7. $(m + n)^5$

8. $(a - b)^4$

9. $(x^2 - 3y)^5$

10. $(3c - d)^6$

11. $(1 - 1)^n$

12. $(1 + 3)^n$

13. $(\sqrt{3} - t)^4$

14. $(\sqrt{5} + t)^6$

☆

15. Find the middle term of the expansion of $(2u - 3v^2)^{10}$.

16. Find the middle two terms in the expansion of $(\sqrt{x} + \sqrt{3})^5$.

Expand.

17. $(x^{-2} + x^2)^4$

18. $\left(\dfrac{1}{\sqrt{x}} - \sqrt{x}\right)^6$

19. $(\sqrt{2} + 1)^6 - (\sqrt{2} - 1)^6$

20. $(1 - \sqrt{2})^4 + (1 + \sqrt{2})^4$

11.6 PROBABILITY

We say that when a coin is tossed the chances that it will fall heads are 1 out of 2, or that the *probability* that it will fall heads is $\frac{1}{2}$. Of course this does not mean that if a coin is tossed 10 times it will necessarily fall heads exactly 5 times. If the coin is tossed a great number of times, however, it will fall heads very nearly half of them.

EXPERIMENTAL AND THEORETICAL PROBABILITY

If we toss a coin a large number of times, say 1000, and count the number of heads, we can determine the probability. If there are 503 heads, we would calculate the probability to be

$$\frac{503}{1000} \quad \text{or} \quad 0.503.$$

This is an *experimental* determination of probability.

If we consider a coin and reason that it is just as likely to fall heads as tails we would calculate the probability to be $\frac{1}{2}$. This is a *theoretical* determination of probability. You might ask "What is the *true* probability?" In fact there is none. Experimentally we can determine probabilities within certain limits. These may or may not agree with what we obtain theoretically.

We need some terminology before we continue. Suppose we perform an experiment such as flipping a coin, throwing a dart, drawing a card from a deck, or checking an item off an assembly line for quality. The results of an experiment are called the *outcomes*. The set of all possible outcomes is called the *sample space*. An *event* is a set of outcomes; that is, a subset of the sample space. For example, for the experiment "throwing a dart," suppose the board was A in the figure below. Then one event is "hitting red," which is a subset of the sample space "red, white, gray," assuming the dart hits the target somewhere.

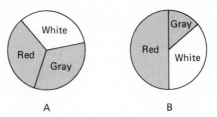

A   B

We denote the probability that an event E occurs as $P(E)$. For example, "getting a head" may be denoted by H. Then $P(H)$ represents the probability of getting a head. When the outcomes of an experiment all have the same probability of occurring, we say that they are *equally likely*. To see the distinction between events that are equally likely and those that are not, consider the dartboards shown here. For dartboard A, the events, *hitting red*, *hitting white*, and *hitting gray* are equally likely, but for board B they are not. A sample space that can be expressed as a union of equally likely events allows us to calculate probabilities of other events.

Principle P. **If an event E can occur m ways out of n possible equally likely outcomes of a sample space S, the probability of that event is given by**

$$P(E) = \frac{m}{n}.$$

A die (pl., dice) is a cube, with six faces, each containing a number of dots from 1 to 6, as shown.

Example 1 What is the probability of rolling a 3 on a die? On a fair die there are 6 equally likely outcomes, so the sample space contains the 6 items, getting a 1, 2, 3, 4, 5, or 6. There is just one way to get a 3. By Principle P, $P(3) = \frac{1}{6}$.

Example 2 What is the probability of rolling an even number on a die? The sample space is the same as that of Example 1. The *event* in this case is getting an even number. That can occur in 3 ways. Or, to say it another way, the event consists of the three elements of the sample space, 2, 4, and 6. By Principle P, $P(\text{even}) = \frac{3}{6}$, or $\frac{1}{2}$.

Example 3 What is the probability of getting a total of 8 on a roll of a pair of dice? (Assume that the dice are different, say one red and one black.)

On each die there are 6 outcomes. The outcomes are paired, so there are $6 \cdot 6$ or 36 elements in the sample space, as shown.

Red die

6	(1,6)	(2,6)	(3,6)	(4,6)	(5,6)	(6,6)
5	(1,5)	(2,5)	(3,5)	(4,5)	(5,5)	(6,5)
4	(1,4)	(2,4)	(3,4)	(4,4)	(5,4)	(6,4)
3	(1,3)	(2,3)	(3,3)	(4,3)	(5,3)	(6,3)
2	(1,2)	(2,2)	(3,2)	(4,2)	(5,2)	(6,2)
1	(1,1)	(2,1)	(3,1)	(4,1)	(5,1)	(6,1)
	1	2	3	4	5	6

Black die

The pairs that total 8 are shaded. Those elements taken together constitute the *event* of rolling an 8. There are 5 of them. By Principle P, $P(8) = \frac{5}{36}$.

The following are some results that follow from Principle P.

> **If an event E cannot ever occur, then $P(E) = 0$.**
>
> **If an event E is certain to occur (every trial is a success), then $P(E) = 1$.**
>
> **The probability of any event is a number from 0 to 1.**
>
> $$0 \leqslant P(E) \leqslant 1$$

THE FUNDAMENTAL COUNTING PRINCIPLE

A basic principle that helps determine the number of ways a compound event may occur is as follows.

> *The Fundamental Counting Principle.* **In a compound event in which the first event can occur in m ways and a second can occur in n ways, the number of ways in which the first and then the second can occur is $m \cdot n$.**

This principle can be extended to events consisting of more than two separate events.

Example 4 There are 11 names in a hat. You draw one name and then another. How many ways can that be done?

The first name can be drawn in 11 ways. After that, there are only 10 names left, so the second can be drawn in 10 ways. By the Fundamental Counting Principle, the total number of ways of drawing a first and a second name is $11 \cdot 10$, or 110.

Example 5 There are 11 names in a hat. You draw one name and then put it back in the hat. Then you draw a second name. How many ways can that be done?

The first name can be drawn in 11 ways. The second can also be drawn in 11 ways since the first name was replaced. By the Fundamental Counting Principle, the total number of ways is $11 \cdot 11$, or 121.

Example 6 There are 11 names in a hat. You draw two names, not caring which name is drawn first. In how many ways can that be done?

Suppose you draw name A first and then name B. By Example 4 there are 110 ways of doing that.

Suppose you draw name B first and then name A. By Example 4 there are 110 ways of doing that. Thus the total number of ways of drawing names A and B, not considering order, is 220.

Example 7   Two people are to be selected at random from a group consisting of 6 men and 4 women. What is the probability of selecting two women?

The total number of ways of selecting two people is $10 \cdot 9$, or 90. Thus the sample space contains 90 elements. The number of ways of selecting two women is $4 \cdot 3$, or 12. By Principle P the probability is $\frac{12}{90}$, or $\frac{2}{15}$.

Example 8   Suppose you draw 2 cards from a well-shuffled deck of 52 playing cards. What is the probability that you draw two spades?

There are 13 spades in the deck. The number of ways of drawing 2 spades from 13 spades is $13 \cdot 12$, or 156.

The number of ways of drawing two cards from the deck is $52 \cdot 51$, or 2652. The probability is $\frac{156}{2652}$, or $\frac{1}{17}$.

ORIGIN AND USE OF PROBABILITY

A desire to calculate odds in games of chance gave rise to the theory of probability. Today the theory of probability and its closely related field, mathematical statistics, have many applications, most of them not related to games of chance. Opinion polls, with such uses as predicting elections, are a familiar example. Quality control, in which a prediction about the percentage of faulty items manufactured is made without testing them all, is an important application, among many, in business. Still other applications are in the areas of genetics, medicine, and the kinetic theory of gases.

EXERCISE SET 11.6

Suppose we select, without looking, one marble from a bag containing 4 red marbles and 10 green marbles.

1. What is the probability of selecting a red marble?

2. What is the probability of selecting a green marble?

3. What is the probability of selecting a purple marble?

4. What is the probability of selecting a white marble?

Suppose 4 cards are drawn from a well-shuffled deck of 52 cards.

5. What is the probability that all 4 are spades?

6. What is the probability that all 4 are hearts?

7. If 4 marbles are drawn at random all at once from a bag containing 8 white marbles and 6 black marbles, what is the probability that 2 will be white and 2 will be black?

8. From a group of 8 men and 7 women, a committee of 4 is chosen. What is the probability that 2 men and 2 women will be chosen?

9. What is the probability of getting a total of 6 on a roll of a pair of dice?

10. What is the probability of getting a total of 3 on a roll of a pair of dice?

11. What is the probability of getting snake eyes (a total of 2) on a roll of a pair of dice?

12. What is the probability of getting box-cars (a total of 12) on a roll of a pair of dice?

13. From a bag containing 5 nickels, 8 dimes, and 7 quarters, 5 coins are drawn at random. What is the probability of getting 2 nickels, 2 dimes, and 1 quarter?

14. From a bag containing 6 nickels, 10 dimes, and 4 quarters, 6 coins are drawn at random. What is the probability of getting 3 nickels, 2 dimes, and 1 quarter?

15. Suppose you draw 2 cards from a well-shuffled deck of 52 cards. What is the probability that you will get 2 aces?

16. Suppose you draw 2 cards from a well-shuffled deck of 52 cards. What is the probability that they will both be red?

Roulette. A roulette wheel contains slots numbered 00, 0, 1, 2, 3, . . . , 35, 36. Eighteen of the slots numbered 1 through 36 are colored red and eighteen are colored black. The 00 and 0 slots are uncolored. The wheel is spun, and a ball is rolled around the rim until it falls into a slot. What is the probability that the ball falls in:

17. a black slot?

18. a red slot?

19. a red or black slot?

20. the 00 slot?

21. the 0 slot?

22. either the 00 or 0 slot? (Here the house always wins.)

23. an odd-numbered slot?

11.7 MATRICES AND SYSTEMS OF EQUATIONS

In solving systems of equations, we compute with the constants. The variables play no important role. We can simplify writing by omitting the variables. For example, the system

$$3x + 4y = 5,$$
$$x - 2y = 1$$

can be simplified as follows:

$$\begin{matrix} 3 & 4 & 5 \\ 1 & -2 & 1 \end{matrix}$$

We have written a rectangular array of numbers. Such an array is called a *matrix* (plural, *matrices*). We ordinarily write brackets around matrices, although that is not necessary when calculating.

The *rows* of a matrix are horizontal, and the *columns* are vertical.

$$\begin{bmatrix} 5 & -2 & 2 \\ 1 & 0 & 1 \\ 0 & 1 & 2 \end{bmatrix} \begin{matrix} \longleftarrow \text{Row 1} \\ \longleftarrow \text{Row 2} \\ \longleftarrow \text{Row 3} \end{matrix}$$

Column 1 Column 2 Column 3

A matrix can have various numbers of rows and columns. Let us now use matrices to solve systems of linear equations.

Example 1 Solve using matrices.

$$2x - y + 4z = -3,$$
$$x \qquad - 4z = 5,$$
$$6x - y + 2z = 10.$$

We write a matrix, using only the constants. Where there are missing terms we must write 0's:

$$\begin{bmatrix} 2 & -1 & 4 & -3 \\ 1 & 0 & -4 & 5 \\ 6 & -1 & 2 & 10 \end{bmatrix}.$$

We do exactly the same calculations as though we wrote the entire equations.

The first thing to do, if possible, is to interchange the rows so that each number below the first number in the first row is a multiple of that number. We can do that by interchanging rows 1 and 2:

$$\begin{bmatrix} 1 & 0 & -4 & 5 \\ 2 & -1 & 4 & -3 \\ 6 & -1 & 2 & 10 \end{bmatrix}.$$

This corresponds to interchanging equation (1) with equation (2).

Next we multiply the first row by -2 and add it to the second row.

Note that we get a 0 in the second row.

$$\begin{bmatrix} 1 & 0 & -4 & 5 \\ 0 & -1 & 12 & -13 \\ 6 & -1 & 2 & 10 \end{bmatrix}.$$

This corresponds to multiplying equation (1) by -2 and adding it to equation (2).

Now we multiply the first row by -6 and add it to the third row, giving us a 0 in the third row.

$$\begin{bmatrix} 1 & 0 & -4 & 5 \\ 0 & -1 & 12 & -13 \\ 0 & -1 & 26 & -20 \end{bmatrix}.$$

This corresponds to multiplying equation (1) by -6 and adding it to equation (3).

Next we multiply row 2 by -1 and add it to the third row, giving us a 0 in the third row.

$$\begin{bmatrix} 1 & 0 & -4 & 5 \\ 0 & -1 & 12 & -13 \\ 0 & 0 & 14 & -7 \end{bmatrix}.$$

This corresponds to multiplying equation (2) by -1 and adding it to equation (3).

If we now put the variables back, we have the following:

$$x \quad - \quad 4z = 5,$$
$$- y + 12z = -13,$$
$$14z = -7.$$

Now we proceed as follows. We solve (3) for z and get $z = -\frac{1}{2}$. Next we substitute $-\frac{1}{2}$ for z in (2) and solve for y: $-y + 12(-\frac{1}{2}) = -13$, so $y = 7$. Since there is no y-term in (1), we need only substitute $-\frac{1}{2}$ for z in (1) and solve for x: $x - 4(-\frac{1}{2}) = 5$, so $x = 3$. Therefore, the solution is $(3, 7, -\frac{1}{2})$.

Note in the preceding discussion that our goal was to get the matrix in the following form, where there are 0's below the *main diagonal*.

$$\begin{bmatrix} a & b & c & d \\ 0 & e & f & g \\ 0 & 0 & h & k \end{bmatrix}$$

Then we put the variables back and complete the solution.

Each of the operations used in the preceding steps corresponds to operations with the equations and produces equivalent systems of equations. We call the matrices *row equivalent*, and the operations row equivalent.

> Each of the following row-equivalent operations produces row equivalent matrices:
>
> a) Interchanging any two rows of a matrix;
> b) Multiplying each element of a row by the same nonzero constant;
> c) Multiplying each element of a row by a nonzero number and adding the result to another row.

EXERCISE SET 11.7

Solve, using matrices.

1. $4x + 2y = 11,$
$3x - y = 2$

2. $3x - 3y = 11,$
$9x - 2y = 5$

3. $x + 2y - 3z = 9,$
$2x - y + 2z = -8,$
$3x - y - 4z = 3$

4. $x - y + 2z = 0,$
$x - 2y + 3z = -1,$
$2x - 2y + z = -3$

5. $5x - 3y = -2,$
$4x + 2y = 5$

6. $3x + 4y = 7,$
$-5x + 2y = 10$

7. $4x - y - 3z = 1,$
$8x + y - z = 5,$
$2x + y + 2z = 5$

8. $3x + 2y + 2z = 3,$
$x + 2y - z = 5,$
$2x - 4y + z = 0$

9. $p + q + r = 1,$
$p + 2q + 3r = 4,$
$4p + 5q + 6r = 7$

10. $m + n + t = 9,$
$m - n - t = -15,$
$m + n + t = 3$

11. $2x + 2y - 2z - 2w = -10,$
$x + y + z + w = -5,$
$x - y + 4z + 3w = -2,$
$3x - 2y + 2z + w = -6$

12. $2x - 3y + z - w = -8,$
$x + y - z - w = -4,$
$x + y + z + w = 22,$
$x - y - z - w = -14$

☆
13. A collection of 34 coins consists of dimes and nickels. The total value is $1.90. How many dimes and how many nickels are there?

14. A collection of 43 coins consists of dimes and quarters. The total value is $7.60. How many dimes and how many quarters are there?

15. A collection of 22 coins consists of nickels, dimes, and quarters. The total value is $2.90. There are six more nickels than there are dimes. How many of each type of coin are there?

16. A collection of 18 coins consists of nickels, dimes, and quarters. The total value is $2.55. There are two more quarters than there are dimes. How many of each type of coin are there?

17. A tobacco dealer has two kinds of tobacco. One is worth $4.05 per lb and the other is worth $2.70 per lb. The dealer wants to blend the two tobaccos to get a 15-lb mixture worth $3.15 per lb. How much of each kind of tobacco should be used?

18. A grocer mixes candy worth $0.80 per lb with nuts worth $0.70 per lb to get a 20-lb mixture worth $0.77 per lb. How many pounds of candy and how many pounds of nuts should be used?

11.8 OPERATIONS ON MATRICES

A matrix of m rows and n columns is called a matrix with *dimensions* $m \times n$ (read "m by n").

A matrix with the same number of rows as columns is called a *square* matrix.

MATRIX ADDITION

To add matrices, we add the corresponding members. In order to do this the matrices must have the same dimensions.

Example 1 Add.

$$\begin{bmatrix} -5 & 0 \\ 4 & \frac{1}{2} \end{bmatrix} + \begin{bmatrix} 6 & -3 \\ 2 & 3 \end{bmatrix} = \begin{bmatrix} -5 + 6 & 0 - 3 \\ 4 + 2 & \frac{1}{2} + 3 \end{bmatrix}$$
$$= \begin{bmatrix} 1 & -3 \\ 6 & 3\frac{1}{2} \end{bmatrix}$$

Example 2 Add.

$$\begin{bmatrix} 1 & 3 & 2 \\ -1 & 5 & 4 \\ 6 & 0 & 1 \end{bmatrix} + \begin{bmatrix} -1 & -2 & 1 \\ 1 & -2 & 2 \\ -3 & 1 & 0 \end{bmatrix} = \begin{bmatrix} 0 & 1 & 3 \\ 0 & 3 & 6 \\ 3 & 1 & 1 \end{bmatrix}$$

Addition of matrices is both commutative and associative.

ZERO MATRICES

A matrix having zeros for all of its entries is called a zero *matrix* and is often denoted by *0*. When a zero matrix is added to another matrix of the same dimensions, that same matrix is obtained. Thus a zero matrix is an *additive identity*.

Example 3 Add.

$$\begin{bmatrix} 2 & -1 & 3 \\ 1 & 0 & -1 \end{bmatrix} + \begin{bmatrix} 0 & 0 & 0 \\ 0 & 0 & 0 \end{bmatrix} = \begin{bmatrix} 2 & -1 & 3 \\ 1 & 0 & -1 \end{bmatrix}$$

INVERSES AND SUBTRACTION

To subtract matrices, we subtract the corresponding members. Of course, the matrices must have the same dimensions for that to be possible.

Example 4 Subtract.

$$\begin{bmatrix} 1 & 2 \\ -2 & 0 \\ -3 & -1 \end{bmatrix} - \begin{bmatrix} 1 & -1 \\ 1 & 3 \\ 2 & 3 \end{bmatrix} = \begin{bmatrix} 0 & 3 \\ -3 & -3 \\ -5 & -4 \end{bmatrix}$$

The additive inverse of a matrix can be obtained by replacing each member by its additive inverse. Of course, when two matrices that are inverses of each other are added, a zero matrix is obtained.

Example 5 Let *A* represent

$$\begin{bmatrix} 1 & 0 & 2 \\ 3 & -1 & 5 \end{bmatrix}$$

and $-A$ represent

$$\begin{bmatrix} -1 & 0 & -2 \\ -3 & 1 & -5 \end{bmatrix}.$$

Show that $A + (-A) = 0$.

$$\underset{A}{\begin{bmatrix} 1 & 0 & 2 \\ 3 & -1 & 5 \end{bmatrix}} + \underset{(-A)}{\begin{bmatrix} -1 & 0 & -2 \\ -3 & 1 & -5 \end{bmatrix}} = \underset{0.}{\begin{bmatrix} 0 & 0 & 0 \\ 0 & 0 & 0 \end{bmatrix}}.$$

With numbers we can subtract by adding an inverse. The same is true of matrices. If we denote matrices by A and B, this fact can be stated as follows:

> **For matrices A and B of the same dimensions,**
> $$A - B = A + (-B).$$

Example 6   Subtract directly and also by adding an inverse.

a) $\underset{A}{\begin{bmatrix} 3 & -1 \\ -2 & 4 \end{bmatrix}} - \underset{B}{\begin{bmatrix} 2 & 1 \\ 3 & -2 \end{bmatrix}} = \begin{bmatrix} 1 & -2 \\ -5 & 6 \end{bmatrix}$

b) $\underset{A}{\begin{bmatrix} 3 & -1 \\ -2 & 4 \end{bmatrix}} + \underset{(-B)}{\begin{bmatrix} -2 & -1 \\ -3 & 2 \end{bmatrix}} = \begin{bmatrix} 1 & -2 \\ -5 & 6 \end{bmatrix}.$

MULTIPLYING MATRICES AND NUMBERS

We define a product of a matrix and a number, and call that product a *scalar* product.

> **The (scalar) product of a number k and a matrix A is the matrix, denoted kA, obtained by multiplying each number in A by the number k.**

Examples

Let $A = \begin{bmatrix} -3 & 0 \\ 4 & 5 \end{bmatrix}$.

7. Find $3A$.

$$3A = 3 \begin{bmatrix} -3 & 0 \\ 4 & 5 \end{bmatrix} = \begin{bmatrix} -9 & 0 \\ 12 & 15 \end{bmatrix},$$

8. Find $(-1)A$.

$$(-1)A = -1 \begin{bmatrix} -3 & 0 \\ 4 & 5 \end{bmatrix} = \begin{bmatrix} 3 & 0 \\ -4 & -5 \end{bmatrix}.$$

PRODUCTS OF MATRICES

We do not multiply two matrices by multiplying their corresponding members. The motivation for defining matrix products comes from systems of equations.

Let us begin by considering one equation:

$$3x + 2y - 2z = 4.$$

We will write the coefficients on the left side in a 1×3 matrix (a *row* matrix) and the variables in a 3×1 matrix (a *column* matrix). The 4 on the right is written in a 1×1 matrix:

$$[3 \quad 2 \quad -2] \begin{bmatrix} x \\ y \\ z \end{bmatrix} = [4].$$

We can return to our original equation by multiplying the members of the row matrix by those of the column matrix, and adding:

$$[3 \quad 2 \quad -2] \begin{bmatrix} x \\ y \\ z \end{bmatrix} = [3x + 2y - 2z].$$

We define multiplication accordingly. In this special case, we have a row matrix A and a column matrix B. Their product AB is a 1×1 matrix, having the single member 4 (also called $3x + 2y - 2z$).

Example 9 Find the product of these matrices.

$$[3 \quad 2 \quad -1] \begin{bmatrix} 1 \\ -2 \\ 3 \end{bmatrix} = [3 \cdot 1 + 2(-2) + (-1) \cdot 3] = [-4]$$

Let us continue by considering a system of equations:

$$3x + 2y - 2z = 4,$$
$$2x - y + 5z = 3,$$
$$-x + y + 4z = 7.$$

Consider the following matrices:

$$\underbrace{\begin{bmatrix} 3 & 2 & -2 \\ 2 & -1 & 5 \\ -1 & 1 & 4 \end{bmatrix}}_{A} \underbrace{\begin{bmatrix} x \\ y \\ z \end{bmatrix}}_{X} \underbrace{\begin{bmatrix} 4 \\ 3 \\ 7 \end{bmatrix}}_{B}.$$

If we multiply the first row of A by the (only) column of X, as we did above, we get $3x + 2y - 2z$. If we multiply the second row of A by the column in X, in the same way we get the following:

$$\begin{bmatrix} 3 & 2 & -2 \\ 2 & -1 & 5 \\ -1 & 1 & 4 \end{bmatrix} \begin{bmatrix} x \\ y \\ z \end{bmatrix} = 2x - y + 5z$$

Note that the first members are multiplied, the second members are multiplied, the third members are multiplied, and the results are added, to get the single number $2x - y + 5z$. What do we get when we multiply the third row of A by the column in X?

$$\begin{bmatrix} 3 & 2 & -2 \\ 2 & -1 & 5 \\ -1 & 1 & 4 \end{bmatrix} \begin{bmatrix} x \\ y \\ z \end{bmatrix} = -x + y + 4z$$

We define the product AX to be the column matrix

$$\begin{bmatrix} 3x + 2y - 2z \\ 2x - y + 5z \\ -x + y + 4z \end{bmatrix}.$$

Now consider this matrix equation:

$$\begin{bmatrix} 3x + 2y - 2z \\ 2x - y + 5z \\ -x + y + 4z \end{bmatrix} = \begin{bmatrix} 4 \\ 3 \\ 7 \end{bmatrix}.$$

Equality for matrices is the same as for numbers, that is, a sentence such as $a = b$ says that a and b are two names for the same thing. Thus if the above matrix equation is true, the "two" matrices are really the same one. This means that $3x + 2y - 2z$ is 4, $2x - y + 5z$ is 3, and $-x + y + 4z$ is 7, or that

$$3x + 2y - 2z = 4,$$
$$2x - y + 5z = 3,$$

and

$$-x + y + 4z = 7.$$

Thus the matrix equation $AX = B$ is equivalent to the original system of equations.

Example 10 Write a matrix equation equivalent to this system of equations:

$$4x + 2y - z = 3,$$
$$9x + z = 5,$$
$$4x + 5y - 2z = 1,$$
$$x + y + z = 0.$$

We write the coefficients on the left in a matrix. We multiply that matrix by the column matrix containing the variables, and set the product equal to the column matrix containing the constants on the right:

$$\begin{bmatrix} 4 & 2 & -1 \\ 9 & 0 & 1 \\ 4 & 5 & -2 \\ 1 & 1 & 1 \end{bmatrix} \begin{bmatrix} x \\ y \\ z \end{bmatrix} = \begin{bmatrix} 3 \\ 5 \\ 1 \\ 0 \end{bmatrix}.$$

Example 11 Multiply.

$$\begin{bmatrix} 3 & 1 & -1 \\ 1 & 2 & 2 \\ -1 & 0 & 5 \\ 4 & 1 & 2 \end{bmatrix} \begin{bmatrix} 1 \\ 2 \\ 1 \end{bmatrix} = \begin{bmatrix} 3 \cdot 1 + 1 \cdot 2 - 1 \cdot 1 \\ 1 \cdot 1 + 2 \cdot 2 + 2 \cdot 1 \\ -1 \cdot 1 + 0 \cdot 2 + 5 \cdot 1 \\ 4 \cdot 1 + 1 \cdot 2 + 2 \cdot 1 \end{bmatrix} = \begin{bmatrix} 4 \\ 7 \\ 4 \\ 8 \end{bmatrix}$$

In all the examples so far, the second matrix had only one column. If the second matrix has more than one column, we treat it in the same way when multiplying that we treated the single column. The product matrix will have as many columns as the second matrix.

Example 12 Multiply (compare with Example 11).

$$\underset{A}{\begin{bmatrix} 3 & 1 & -1 \\ 1 & 2 & 2 \\ -1 & 0 & 5 \\ 4 & 1 & 2 \end{bmatrix}} \underset{B}{\begin{bmatrix} 1 & 0 \\ 2 & 1 \\ 1 & 3 \end{bmatrix}} = \begin{bmatrix} 4 & 3 \cdot 0 + 1 \cdot 1 + (-1)3 \\ 7 & 1 \cdot 0 + 2 \cdot 1 + 2 \cdot 3 \\ 4 & -1 \cdot 0 + 0 \cdot 1 + 5 \cdot 3 \\ 8 & 4 \cdot 0 + 1 \cdot 1 + 2 \cdot 3 \end{bmatrix} = \begin{bmatrix} 4 & -2 \\ 7 & 8 \\ 4 & 15 \\ 8 & 7 \end{bmatrix}$$

Same as in Example 11

The rows of A multiplied by the second column of B

Example 13 Multiply.

$$\begin{bmatrix} 3 & 1 & -1 \\ 2 & 0 & 3 \end{bmatrix}\begin{bmatrix} 1 & 4 & 6 \\ 3 & -1 & 9 \\ 2 & 5 & 1 \end{bmatrix}$$

$$=\begin{bmatrix} 3\cdot 1+1\cdot 3-1\cdot 2 & 3\cdot 4+\ 1(-1)-1\cdot 5 & 3\cdot 6+1\cdot 9-1\cdot 1 \\ 2\cdot 1+0\cdot 3+3\cdot 2 & 2\cdot 4+0\cdot (-1)+3\cdot 5 & 2\cdot 6+0\cdot 9+3\cdot 1 \end{bmatrix}$$

$$=\begin{bmatrix} 4 & 6 & 26 \\ 8 & 23 & 15 \end{bmatrix}$$

If matrix A has n columns and matrix B has n rows, then we can compute the product AB, regardless of the other dimensions. The product will have as many rows as A and as many columns as B.

EXERCISE SET 11.8

For Exercises 1–16, let

$$A = \begin{bmatrix} 1 & 2 \\ 4 & 3 \end{bmatrix}, \quad B = \begin{bmatrix} -3 & 5 \\ 2 & -1 \end{bmatrix}, \quad C = \begin{bmatrix} 1 & -1 \\ -1 & 1 \end{bmatrix}, \quad D = \begin{bmatrix} 1 & 1 \\ 1 & 1 \end{bmatrix},$$

$$E = \begin{bmatrix} 1 & 3 \\ 2 & 6 \end{bmatrix}, \quad F = \begin{bmatrix} 3 & 3 \\ -1 & -1 \end{bmatrix}, \quad O = \begin{bmatrix} 0 & 0 \\ 0 & 0 \end{bmatrix}, \quad \text{and} \quad I = \begin{bmatrix} 1 & 0 \\ 0 & 1 \end{bmatrix}.$$

Find.

1. $A + B$ **2.** $B + A$ **3.** $E + O$ **4.** $2A$

5. $3F$ **6.** $(-1)D$ **7.** $3F + 2A$ **8.** $A - B$

9. $B - A$ **10.** AB **11.** BA **12.** OF

13. CD **14.** EF **15.** AI **16.** IA

In Exercises 17–20, let

$$A = \begin{bmatrix} 1 & 0 & -2 \\ 0 & -1 & 3 \\ 3 & 2 & 4 \end{bmatrix}, \quad B = \begin{bmatrix} -1 & -2 & 5 \\ 1 & 0 & -1 \\ 2 & -3 & 1 \end{bmatrix}, \quad C = \begin{bmatrix} -2 & 9 & 6 \\ -3 & 3 & 4 \\ 2 & -2 & 1 \end{bmatrix}, \quad \text{and} \quad I = \begin{bmatrix} 1 & 0 & 0 \\ 0 & 1 & 0 \\ 0 & 0 & 1 \end{bmatrix}.$$

Find.

17. AB **18.** BA **19.** CI **20.** IC

Multiply.

21. $[-3 \quad 2] \begin{bmatrix} 4 \\ -2 \end{bmatrix}$

22. $[-2 \quad 0 \quad 4] \begin{bmatrix} 8 \\ -6 \\ \frac{1}{2} \end{bmatrix}$

23. $[-5 \quad 1 \quad 2] \begin{bmatrix} 1 & 3 \\ -1 & 0 \\ 4 & -2 \end{bmatrix}$

24. $\begin{bmatrix} -3 & 2 \\ 0 & 1 \\ -4 & 5 \end{bmatrix} \begin{bmatrix} 4 \\ 2 \end{bmatrix}$

Write a matrix equation equivalent to each of the following systems of equations.

25. $3x - 2y + 4z = 17,$
    $2x + \ y - 5z = 13$

26. $3x + 2y + 5z = 9,$
    $4x - 3y + 2z = 10$

27. $\ x - \ y + 2z - 4w = 12,$
    $2x - \ y - \ z + \ w = 0,$
    $\ x + 4y - 3z - \ w = 1,$
    $3x + 5y - 7z + 2w = 9$

28. $2x + \ 4y - \ 5z + 12w = 2,$
    $4x - \ \ y + 12z - \ \ w = 5,$
    $-x + \ 4y \ \ \ \ \ \ \ \ + 2w = 13,$
    $2x + 10y + \ \ z \ \ \ \ \ \ \ = 5$

Compute.

29. ▦ $\begin{bmatrix} 3.61 & -2.14 & 16.7 \\ -4.33 & 7.03 & 12.9 \\ 5.82 & -6.95 & 2.34 \end{bmatrix} \begin{bmatrix} 3.05 & 0.402 & -1.34 \\ 1.84 & -1.13 & 0.024 \\ -2.83 & 2.04 & 8.81 \end{bmatrix}$

30. ▦ $\begin{bmatrix} -1.23 & 4.51 & -17.4 \\ 61.2 & -8.81 & 0.123 \\ 14.14 & 6.92 & -14.4 \end{bmatrix} \begin{bmatrix} 4.24 & 16.1 & 41.3 \\ 0.146 & -6.06 & -18.9 \\ -8.43 & 1.12 & 0.0245 \end{bmatrix}$

TEST OR REVIEW—CHAPTER 11

1. Let $f(x) = 3x^2 - 2x$. Find

    a) $f(3a)$,   b) $f(a + h)$,   c) $\dfrac{f(a + h) - f(a)}{h}$.

2. Find the inverse of g, where $g(x) = \dfrac{5x - 2}{3}$.

3. Find the tenth term in the arithmetic sequence $\frac{3}{4}, \frac{13}{12}, \frac{17}{12}, \ldots$.

4. Find the sum of the first 18 terms of the arithmetic sequence $4, 7, 10, \ldots$.

5. The first term of an arithmetic sequence is 5. The seventeenth term is 53. Find the third term.

6. Find the sum of the infinite geometric sequence $2 + 1 + \frac{1}{2} + \cdots$.

7. The Greek alphabet contains 24 letters. How many different fraternity or sorority names can be formed using three different Greek letters? How many can be formed if repetitions of letters are allowed?

8. What is the probability of rolling a 10 on one roll of a pair of dice?

9. There are three red marbles and five blue marbles in a bag. Two marbles are drawn. What is the probability that they are both red?

10. Solve using matrices.

$$3x + 2y - z = 4$$
$$3x - 2y + z = 5$$
$$4x - 5y - z = -1$$

11. Find the fourth term in the expansion of $(a + x)^{12}$.

12. Expand $(x^2 - 3y)^5$.

$$\text{Let } A = \begin{bmatrix} 1 & -1 & 0 \\ 2 & 3 & -2 \\ -2 & 0 & 1 \end{bmatrix}, \qquad B = \begin{bmatrix} -1 & 0 & 6 \\ 1 & -2 & 0 \\ 0 & 1 & -3 \end{bmatrix}.$$

Find each of the following.

13. $A + B$ **14.** $-3A$ **15.** $-A$ **16.** AB **17.** $A - B$

FINAL EXAMINATION

Chapter 1

1. Simplify: $|-18|$.

2. Evaluate $\dfrac{ab - ac}{bc}$ when $a = -2$, $b = 3$, and $c = -4$.

3. Simplify:

$2y - [3 - 4(5 - 2y) - 3y]$.

4. Simplify:

$(-9x^2y^3) \cdot (5x^4y^{-7})$.

Chapter 2

5. Solve:

$9(x - 1) - 3(x - 2) = 1$.

6. Find three consecutive even integers whose sum in 198.

7. Solve $A = P + Prt$, for r.

8. Find an equation of the line containing the point $(-1, 4)$ and perpendicular to the line whose equation is $x - y = 6$.

9. Graph: $3x - y = 6$.

Chapter 3

10. Evaluate:

$\begin{vmatrix} -5 & -7 \\ 4 & 6 \end{vmatrix}$.

11. Solve:

$5x + 3y = 2$
$3x + 5y = -2$.

12. Solve:

$x + y - z = 0$
$3x + y + z = 6$
$x - y + 2z = 5$.

13. The difference between two numbers is 6. Three times the smaller minus twice the larger is 7. Find the numbers.

14. A bartender has two kinds of beer. Beer A is 2% alcohol and beer B is 6% alcohol. He wants to mix the two beers to get an 80-liter mixture that is 3.2% alcohol. How many liters of each should he use?

Chapter 4

15. Add:

$-3x^2 + 4x^3 - 5x - 1$ and
$9x^3 - 4x^2 + 7 - x$.

16. Subtract:

$(5a^2 - 3ab - 7b^2) -$
$(2a^2 + 5ab + 8b^2)$.

Multiply.

17. $(2a - 1)(3a + 5)$

18. $(3x^2 - 5y)^2$

Factor.

19. $4x^2 - 12x + 9$

20. $27a^3 - 8$

21. $a^2 + 2ab + b^2 - 25$

22. Solve: $x^2 - 2x = 48$.

23. Solve for x: $ax - bx = c^2$.

24. For the function described by $f(x) = 3x^2 - 4x$, find $f(-2)$.

Chapter 5

25. Solve:

$$\frac{6}{x} + \frac{6}{x + 2} = \frac{5}{2}.$$

26. Solve $I = \dfrac{E}{R + r}$, for R.

Perform the following operations and simplify.

27. $\dfrac{1}{x - 2} - \dfrac{4}{x^2 - 4} + \dfrac{3}{x + 2}$

28. $\dfrac{x^2 - 6x + 8}{3x + 9} \cdot \dfrac{x + 3}{x^2 - 4}$

29. $\dfrac{3x + 3y}{5x - 5y} \div \dfrac{3x^2 + 3y^2}{5x^2 - 5y^2}$

30. $\dfrac{x - \dfrac{a^2}{x}}{1 + \dfrac{a}{x}}$

31. Use synthetic division to find the quotient and remainder:

$(7x^4 - 5x^3 + x^2 - 4) \div (x - 2)$.

32. An airplane can fly 190 miles with the wind in the same time it takes to fly 160 miles against the wind. The speed of the wind is 30 mph. How fast can the plane fly in still air?

33. A can do a certain job in 21 min. B can do the same job in 14 min. How long would it take to do the job if the two worked together?

Chapter 6

34. Multiply. Write scientific notation for the answer.

$(8.9 \times 10^{-17})(7.6 \times 10^4)$

35. Multiply and simplify:

$\sqrt{8x} \cdot \sqrt{8x^3y}.$

36. Divide and simplify:

$\dfrac{\sqrt[3]{15x}}{\sqrt[3]{3y^2}}.$

37. Rationalize the denominator:

$\dfrac{1 - \sqrt{x}}{1 + \sqrt{x}}.$

38. Write a single radical expression:

$\dfrac{\sqrt[3]{(x + 1)^5}}{\sqrt{(x + 1)^3}}.$

39. Solve:

$\sqrt{x - 5} = 5 - \sqrt{x}.$

40. Multiply these complex numbers:

$(3 + 2i)(4 - 7i).$

Chapter 7

Solve.

41. $7x^2 + 14 = 0$

42. $x^2 + 4x = 3$

43. Write a quadratic equation whose solutions are $5\sqrt{2}$ and $-5\sqrt{2}$.

44. Solve:

$x^4 - 29x^2 + 100 = 0.$

45. Graph $f(x) = -2(x - 3)^2 + 1.$

a) Label the vertex.
b) Draw the line of symmetry.
c) Find the maximum or minimum value.

46. The centripetal force F of an object moving in a circle varies directly as the square of the velocity v and inversely as the radius r of the circle. If $F = 8$ when $v = 1$ and $r = 10$, what is F when $v = 2$ and $r = 16$?

47. A farmer wants to fence in a rectangular area next to a river. (Note that no fence will be needed along the river.) What is the area of the largest region that can be fenced in with 100 ft of fencing?

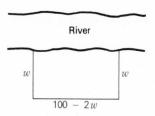

River

w w

$100 - 2w$

Chapter 8

48. Solve:

$$x^2 + y^2 = 8$$
$$x^2 - y^2 = 2.$$

49. The perimeter of a rectangle is 34 ft. The length of a diagonal is 13 ft. Find its dimensions.

50. Find the center and radius of the circle:

$$x^2 + y^2 - 4x + 6y - 23 = 0.$$

51. Graph:

$$\frac{x^2}{25} + \frac{y^2}{4} = 1.$$

Chapter 9

52. Graph: $y = \log_2 x$.

53. Express as a single logarithm:

$$\frac{2}{3} \log_a x - \frac{1}{2} \log_a y + 5 \log_a z.$$

54. Write an exponential equation with the same meaning as $\log_a c = 5$.

55. Solve:
$$\log (x^2 - 1) - \log (x + 1) = 2.$$

Find each of the following, using Table 2.

56. log 14.1

57. antilog 7.7808 − 10

58. Solve:

$5^x = 8.$

59. Solve:

$\log (x^2 - 25) - \log (x + 5) = 3.$

Chapter 10

60. Find the intersection: $\{-2, -1, 0, 1, 2\} \cap \{0, 2, 4, 6\}$.

61. Solve. Then graph. $|2x - 1| \leqslant 5$.

62. Solve. Then graph. $|2y + 3| > 7$.

63. Graph: $2x - 3y < -6$.

Chapter 11

64. Find the sum of the even numbers from 2 to 200.

65. Find the sum of the infinite geometric sequence $8 + 2 + \dfrac{1}{2} + \cdots$.

66. Find the fifth term in the expansion of $(2 + y)^{10}$.

67. What is the probability of tossing 4 pennies and having them all turn up heads?

APPENDIX
AND TABLES

APPENDIX 1: GEOMETRIC FORMULAS

Plane Geometry:

Rectangle

Area: $A = lw$

Perimeter: $P = 2l + 2w$

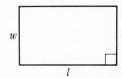

Square

Area: $A = s^2$

Perimeter: $P = 4s$

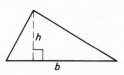

Triangle

Area: $A = \frac{1}{2}bh$

Sum of Angle Measures:

$A + B + C = 180°$

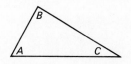

477

Right Triangle

Pythagorean Theorem:

$a^2 + b^2 = c^2$

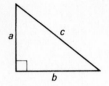

Parallelogram

Area: $A = bh$

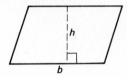

Trapezoid

Area: $A = \frac{1}{2}h(a + b)$

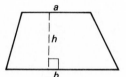

Circle

Area: $A = \pi r^2$

Circumference:

$C = \pi D = 2\pi r$

($\frac{22}{7}$ and 3.14 are different approximations for π)

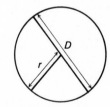

Solid Geometry:

Rectangular Solid

Volume: $V = lwh$

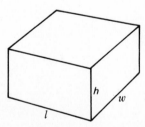

Cube

Volume: $V = s^3$

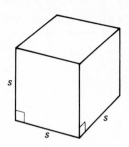

Right Circular Cylinder

Volume: $V = \pi r^2 h$

Lateral Surface Area: $L = 2\pi rh$

Total Surface Area:

$S = 2\pi rh + 2\pi r^2$

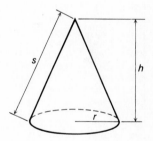

Right Circular Cone

Volume: $V = \frac{1}{3}\pi r^2 h$

Lateral Surface Area: $L = \pi rs$

Total Surface Area: $S = \pi r^2 + \pi rs$

Slant Height: $s = \sqrt{r^2 + h^2}$

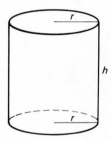

Sphere

Volume: $V = \frac{4}{3}\pi r^3$

Surface Area: $S = 4\pi r^2$

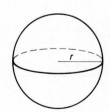

Table 1   Powers, Roots, and Reciprocals

n	n^2	n^3	$\sqrt{n}$	$\sqrt[3]{n}$	$\sqrt{10n}$	$\dfrac{1}{n}$	n	n^2	n^3	$\sqrt{n}$	$\sqrt[3]{n}$	$\sqrt{10n}$	$\dfrac{1}{n}$
1	1	1	1.000	1.000	3.162	1.0000	51	2,601	132,651	7.141	3.708	22.583	.0196
2	4	8	1.414	1.260	4.472	.5000	52	2,704	140,608	7.211	3.733	22.804	.0192
3	9	27	1.732	1.442	5.477	.3333	53	2,809	148,877	7.280	3.756	23.022	.0189
4	16	64	2.000	1.587	6.325	.2500	54	2,916	157,464	7.348	3.780	23.238	.0185
5	25	125	2.236	1.710	7.071	.2000	55	3,025	166,375	7.416	3.803	23.452	.0182
6	36	216	2.449	1.817	7.746	.1667	56	3,136	175,616	7.483	3.826	23.664	.0179
7	49	343	2.646	1.913	8.367	.1429	57	3,249	185,193	7.550	3.849	23.875	.0175
8	64	512	2.828	2.000	8.944	.1250	58	3,364	195,112	7.616	3.871	24.083	.0172
9	81	729	3.000	2.080	9.487	.1111	59	3,481	205,379	7.681	3.893	24.290	.0169
10	100	1,000	3.162	2.154	10.000	.1000	60	3,600	216,000	7.746	3.915	24.495	.0167
11	121	1,331	3.317	2.224	10.488	.0909	61	3,721	226,981	7.810	3.936	24.698	.0164
12	144	1,728	3.464	2.289	10.954	.0833	62	3,844	238,328	7.874	3.958	24.900	.0161
13	169	2,197	3.606	2.351	11.402	.0769	63	3,969	250,047	7.937	3.979	25.100	.0159
14	196	2,744	3.742	2.410	11.832	.0714	64	4,096	262,144	8.000	4.000	25.298	.0156
15	225	3,375	3.873	2.466	12.247	.0667	65	4,225	274,625	8.062	4.021	25.495	.0154
16	256	4,096	4.000	2.520	12.648	.0625	66	4,356	287,496	8.124	4.041	25.690	.0152
17	289	4,913	4.123	2.571	13.038	.0588	67	4,489	300,763	8.185	4.062	25.884	.0149
18	324	5,832	4.243	2.621	13.416	.0556	68	4,624	314,432	8.246	4.082	26.077	.0147
19	361	6,859	4.359	2.668	13.784	.0526	69	4,761	328,509	8.307	4.102	26.268	.0145
20	400	8,000	4.472	2.714	14.142	.0500	70	4,900	343,000	8.367	4.121	26.458	.0143
21	441	9,261	4.583	2.759	14.491	.0476	71	5,041	357,911	8.426	4.141	26.646	.0141
22	484	10,648	4.690	2.802	14.832	.0455	72	5,184	373,248	8.485	4.160	26.833	.0139
23	529	12,167	4.796	2.844	15.166	.0435	73	5,329	389,017	8.544	4.179	27.019	.0137
24	576	13,824	4.899	2.884	15.492	.0417	74	5,476	405,224	8.602	4.198	27.203	.0135
25	625	15,625	5.000	2.924	15.811	.0400	75	5,625	421,875	8.660	4.217	27.386	.0133
26	676	17,576	5.099	2.962	16.125	.0385	76	5,776	438,976	8.718	4.236	27.568	.0132
27	729	19,683	5.196	3.000	16.432	.0370	77	5,929	456,533	8.775	4.254	27.749	.0130
28	784	21,952	5.292	3.037	16.733	.0357	78	6,084	474,552	8.832	4.273	27.928	.0128
29	841	24,389	5.385	3.072	17.029	.0345	79	6,241	493,039	8.888	4.291	28.107	.0127
30	900	27,000	5.477	3.107	17.321	.0333	80	6,400	512,000	8.944	4.309	28.284	.0125
31	961	29,791	5.568	3.141	17.607	.0323	81	6,561	531,441	9.000	4.327	28.460	.0123
32	1,024	32,768	5.657	3.175	17.889	.0312	82	6,724	551,368	9.055	4.344	28.636	.0122
33	1,089	35,937	5.745	3.208	18.166	.0303	83	6,889	571,787	9.110	4.362	28.810	.0120
34	1,156	39,304	5.831	3.240	18.439	.0294	84	7.056	592,704	9.165	4.380	28.983	.0119
35	1,225	42,875	5.916	3.271	18.708	.0286	85	7,225	614,125	9.220	4.397	29.155	.0118
36	1,296	46,656	6.000	3.302	18.974	.0278	86	7,396	636,056	9.274	4.414	29.326	.0116
37	1,369	50,653	6.083	3.332	19.235	.0270	87	7,569	658,503	9.327	4.431	29.496	.0115
38	1,444	54,872	6.164	3.362	19.494	.0263	88	7,744	681,472	9.381	4.448	29.665	.0114
39	1,521	59,319	6.245	3.391	19.748	.0256	89	7,921	704,969	9.434	4.465	29.833	.0112
40	1,600	64,000	6.325	3.420	20.000	.0250	90	8,100	729,000	9.487	4.481	30.000	.0111
41	1,681	68,921	6.403	3.448	20.248	.0244	91	8,281	753,571	9.539	4.498	30.166	.0110
42	1,764	74,088	6.481	3.476	20.494	.0238	92	8,464	778,688	9.592	4.514	30.332	.0109
43	1,849	79,507	6.557	3.503	20.736	.0233	93	8,649	804,357	9.644	4.531	30.496	.0108
44	1,936	85,184	6.633	3.530	20.976	.0227	94	8,836	830,584	9.695	4.547	30.659	.0106
45	2,025	91,125	6.708	3.557	21.213	.0222	95	9,025	857,375	9.747	4.563	30.822	.0105
46	2,116	97,336	6.782	3.583	21.448	.0217	96	9,216	884,736	9.798	4.579	30.984	.0104
47	2,209	103,823	6.856	3.609	21.679	.0213	97	9,409	912,673	9.849	4.595	31.145	.0103
48	2,304	110,592	6.928	3.634	21.909	.0208	98	9,604	941,192	9.899	4.610	31.305	.0102
49	2,401	117,649	7.000	3.659	22.136	.0204	99	9,801	970,299	9.950	4.626	31.464	.0101
50	2,500	125,000	7.071	3.684	22.361	.0200	100	10,000	1,000,000	10.000	4.642	31.623	.0100

Table 2   Common Logarithms

x	0	1	2	3	4	5	6	7	8	9
1.0	.0000	.0043	.0086	.0128	.0170	.0212	.0253	.0294	.0334	.0374
1.1	.0414	.0453	.0492	.0531	.0569	.0607	.0645	.0682	.0719	.0755
1.2	.0792	.0828	.0864	.0899	.0934	.0969	.1004	.1038	.1072	.1106
1.3	.1139	.1173	.1206	.1239	.1271	.1303	.1335	.1367	.1399	.1430
1.4	.1461	.1492	.1523	.1553	.1584	.1614	.1644	.1673	.1703	.1732
1.5	.1761	.1790	.1818	.1847	.1875	.1903	.1931	.1959	.1987	.2014
1.6	.2041	.2068	.2095	.2122	.2148	.2175	.2201	.2227	.2253	.2279
1.7	.2304	.2330	.2355	.2380	.2405	.2430	.2455	.2480	.2504	.2529
1.8	.2553	.2577	.2601	.2625	.2648	.2672	.2695	.2718	.2742	.2765
1.9	.2788	.2810	.2833	.2856	.2878	.2900	.2923	.2945	.2967	.2989
2.0	.3010	.3032	.3054	.3075	.3096	.3118	.3139	.3160	.3181	.3201
2.1	.3222	.3243	.3263	.3284	.3304	.3324	.3345	.3365	.3385	.3404
2.2	.3424	.3444	.3464	.3483	.3502	.3522	.3541	.3560	.3579	.3598
2.3	.3617	.3636	.3655	.3674	.3692	.3711	.3729	.3747	.3766	.3784
2.4	.3802	.3820	.3838	.3856	.3874	.3892	.3909	.3927	.3945	.3962
2.5	.3979	.3997	.4014	.4031	.4048	.4065	.4082	.4099	.4116	.4133
2.6	.4150	.4166	.4183	.4200	.4216	.4232	.4249	.4265	.4281	.4298
2.7	.4314	.4330	.4346	.4362	.4378	.4393	.4409	.4425	.4440	.4456
2.8	.4472	.4487	.4502	.4518	.4533	.4548	.4564	.4579	.4594	.4609
2.9	.4624	.4639	.4654	.4669	.4683	.4698	.4713	.4728	.4742	.4757
3.0	.4771	.4786	.4800	.4814	.4829	.4843	.4857	.4871	.4886	.4900
3.1	.4914	.4928	.4942	.4955	.4969	.4983	.4997	.5011	.5024	.5038
3.2	.5051	.5065	.5079	.5092	.5105	.5119	.5132	.5145	.5159	.5172
3.3	.5185	.5198	.5211	.5224	.5237	.5250	.5263	.5276	.5289	.5307
3.4	.5315	.5328	.5340	.5353	.5366	.5378	.5391	.5403	.5416	.5428
3.5	.5441	.5453	.5465	.5478	.5490	.5502	.5514	.5527	.5539	.5551
3.6	.5563	.5575	.5587	.5599	.5611	.5623	.5635	.5647	.5658	.5670
3.7	.5682	.5694	.5705	.5717	.5729	.5740	.5752	.5763	.5775	.5786
3.8	.5798	.5809	.5821	.5832	.5843	.5855	.5866	.5877	.5888	.5899
3.9	.5911	.5922	.5933	.5944	.5955	.5966	.5977	.5988	.5999	.6010
4.0	.6021	.6031	.6042	.6053	.6064	.6075	.6085	.6096	.6107	.6117
4.1	.6128	.6138	.6149	.6160	.6170	.6180	.6191	.6201	.6212	.6222
4.2	.6232	.6243	.6253	.6263	.6274	.6284	.6294	.6304	.6314	.6325
4.3	.6335	.6345	.6355	.6365	.6375	.6385	.6395	.6405	.6415	.6425
4.4	.6435	.6444	.6454	.6464	.6474	.6484	.6493	.6503	.6513	.6522
4.5	.6532	.6542	.6551	.6561	.6571	.6580	.6590	.6599	.6609	.6618
4.6	.6628	.6637	.6646	.6656	.6665	.6675	.6684	.6693	.6702	.6712
4.7	.6721	.6730	.6739	.6749	.6758	.6767	.6776	.6785	.6794	.6803
4.8	.6812	.6821	.6830	.6839	.6848	.6857	.6866	.6875	.6884	.6893
4.9	.6902	.6911	.6920	.6928	.6937	.6946	.6955	.6964	.6972	.6981
5.0	.6990	.6998	.7007	.7016	.7024	.7033	.7042	.7050	.7059	.7067
5.1	.7076	.7084	.7093	.7101	.7110	.7118	.7126	.7135	.7143	.7152
5.2	.7160	.7168	.7177	.7185	.7193	.7202	.7210	.7218	.7226	.7235
5.3	.7243	.7251	.7259	.7267	.7275	.7284	.7292	.7300	.7308	.7316
5.4	.7324	.7332	.7340	.7348	.7356	.7364	.7372	.7380	.7388	.7396
x	0	1	2	3	4	5	6	7	8	9

continued

Table 2   (*continued*)

x	0	1	2	3	4	5	6	7	8	9
5.5	.7404	.7412	.7419	.7427	.7435	.7443	.7451	.7459	.7466	.7474
5.6	.7482	.7490	.7497	.7505	.7513	.7520	.7528	.7536	.7543	.7551
5.7	.7559	.7566	.7574	.7582	.7589	.7597	.7604	.7612	.7619	.7627
5.8	.7634	.7642	.7649	.7657	.7664	.7672	.7679	.7686	.7694	.7701
5.9	.7709	.7716	.7723	.7731	.7738	.7745	.7752	.7760	.7767	.7774
6.0	.7782	.7789	.7796	.7803	.7810	.7818	.7825	.7832	.7839	.7846
6.1	.7853	.7860	.7868	.7875	.7882	.7889	.7896	.7903	.7910	.7917
6.2	.7924	.7931	.7938	.7945	.7952	.7959	.7966	.7973	.7980	.7987
6.3	.7993	.8000	.8007	.8014	.8021	.8028	.8035	.8041	.8048	.8055
6.4	.8062	.8069	.8075	.8082	.8089	.8096	.8102	.8109	.8116	.8122
6.5	.8129	.8136	.8142	.8149	.8156	.8162	.8169	.8176	.8182	.8189
6.6	.8195	.8202	.8209	.8215	.8222	.8228	.8235	.8241	.8248	.8254
6.7	.8261	.8267	.8274	.8280	.8287	.8293	.8299	.8306	.8312	.8319
6.8	.8325	.8331	.8338	.8344	.8351	.8357	.8363	.8370	.8376	.8382
6.9	.8388	.8395	.8401	.8407	.8414	.8420	.8426	.8432	.8439	.8445
7.0	.8451	.8457	.8463	.8470	.8476	.8482	.8488	.8494	.8500	.8506
7.1	.8513	.8519	.8525	.8531	.8537	.8543	.8549	.8555	.8561	.8567
7.2	.8573	.8579	.8585	.8591	.8597	.8603	.8609	.8615	.8621	.8627
7.3	.8633	.8639	.8645	.8651	.8657	.8663	.8669	.8675	.8681	.8686
7.4	.8692	.8698	.8704	.8710	.8716	.8722	.8727	.8733	.8739	.8745
7.5	.8751	.8756	.8762	.8768	.8774	.8779	.8785	.8791	.8797	.8802
7.6	.8808	.8814	.8820	.8825	.8831	.8837	.8842	.8848	.8854	.8859
7.7	.8865	.8871	.8876	.8882	.8887	.8893	.8899	.8904	.8910	.8915
7.8	.8921	.8927	.8932	.8938	.8943	.8949	.8954	.8960	.8965	.8971
7.9	.8976	.8982	.8987	.8993	.8998	.9004	.9009	.9015	.9020	.9025
8.0	.9031	.9036	.9042	.9047	.9053	.9058	.9063	.9069	.9074	.9079
8.1	.9085	.9090	.9096	.9101	.9106	.9112	.9117	.9122	.9128	.9133
8.2	.9138	.9143	.9149	.9154	.9159	.9165	.9170	.9175	.9180	.9186
8.3	.9191	.9196	.9201	.9206	.9212	.9217	.9222	.9227	.9232	.9238
8.4	.9243	.9248	.9253	.9258	.9263	.9269	.9274	.9279	.9284	.9289
8.5	.9294	.9299	.9304	.9309	.9315	.9320	.9325	.9330	.9335	.9340
8.6	.9345	.9350	.9555	.9360	.9365	.9370	.9375	.9380	.9385	.9390
8.7	.9395	.9400	.9405	.9410	.9415	.9420	.9425	.9430	.9435	.9440
8.8	.9445	.9450	.9455	.9460	.9465	.9469	.9474	.9479	.9484	.9489
8.9	.9494	.9499	.9504	.9509	.9513	.9518	.9523	.9528	.9533	.9538
9.0	.9542	.9547	.9552	.9557	.9562	.9566	.9571	.9576	.9581	.9586
9.1	.9590	.9595	.9600	.9605	.9609	.9614	.9619	.9624	.9628	.9633
9.2	.9638	.9643	.9647	.9652	.9657	.9661	.9666	.9671	.9675	.9680
9.3	.9685	.9689	.9694	.9699	.9703	.9708	.9713	.9717	.9722	.9727
9.4	.9731	.9736	.9741	.9745	.9750	.9754	.9759	.9763	.9768	.9773
9.5	.9777	.9782	.9786	.9791	.9795	.9800	.9805	.9809	.9814	.9818
9.6	.9823	.9827	.9832	.9836	.9841	.9845	.9850	.9854	.9859	.9863
9.7	.9868	.9872	.9877	.9881	.9886	.9890	.9894	.9899	.9903	.9908
9.8	.9912	.9917	.9921	.9926	.9930	.9934	.9939	.9943	.9948	.9952
9.9	.9956	.9961	.9965	.9969	.9974	.9978	.9983	.9987	.9991	.9996
x	0	1	2	3	4	5	6	7	8	9

ANSWERS

CHAPTER 1

EXERCISE SET 1.1, pp. 7–8

1. 2, 14   **3.** All are rational.   **5.** 0.375   **7.** 1.666...   **9.** $\dfrac{27}{10}$ **11.** $\dfrac{145}{1000}$ **13.** 80%   **15.** 37.5%

17. 240%   **19.** 3.75%   **21.** 41.2%   **23.** 0.1%   **25.** 125%   **27.** 0.567, $\dfrac{567}{1000}$

29. 0.0685, $\dfrac{685}{10{,}000}$, or $\dfrac{137}{2000}$ **31.** 1.2, $\dfrac{12}{10}$, or $\dfrac{6}{5}$ **33.** 0.005, $\dfrac{5}{1000}$, or $\dfrac{1}{200}$ **35.** $55 > 40$ **37.** $2.3 < 3.2$

39. $-10 < -4$ **41.** $-2.3 > -3.2$ **43.** 16   **45.** 32   **47.** 0   **49.** $0.21\overline{42857}$ **51.** $0.42\overline{51}$

53. $0.\overline{213456789}$

55. The denominator, when factored into a product of primes, will contain a factor different from, or in addition to, 2 or 5.

57. $\dfrac{1}{8}\%$, 0.3%, 0.009, 1%, 1.1%, $\dfrac{9}{100}$, $\dfrac{1}{11}$, $\dfrac{99}{1000}$, 0.11, $\dfrac{1}{8}$, $\dfrac{2}{7}$, 0.286

EXERCISE SET 1.2, pp. 13–14

1. -28 **3.** -16 **5.** 5   **7.** 4   **9.** -7 **11.** -24 **13.** 1.2   **15.** -8.86 **17.** $\dfrac{1}{7}$ **19.** $-\dfrac{4}{3}$

21. $\dfrac{1}{10}$ **23.** -7 **25.** 2.8   **27.** -10 **29.** 0   **31.** -2 **33.** -12 **35.** 5   **37.** 15   **39.** -11.6

41. -29.25 **43.** $-\dfrac{7}{2}$ **45.** $-\dfrac{5}{12}$ **47.** Commutative law of addition   **49.** Associative law of addition

51. Commutative and associative laws of addition   **53.** 2

55. 19   **57.** -43 **59.** 0   **61.** -19.45

EXERCISE SET 1.3, pp. 20–21

1. -21 **3.** -8 **5.** 16   **7.** -126 **9.** 34.2   **11.** 26.46   **13.** 2   **15.** 60   **17.** $-\dfrac{12}{35}$ **19.** 1

21. $-\dfrac{8}{27}$ **23.** -2 **25.** -7 **27.** 7   **29.** 0.3   **31.** Not possible   **33.** 0   **35.** Not possible   **37.** $\dfrac{4}{3}$

39. $-\dfrac{8}{7}$ **41.** $\dfrac{1}{15}$, or $0.06\overline{6}$ **43.** $\dfrac{1}{4.4}$, or $0.227\overline{27}$ **45.** $-\dfrac{6}{77}$ **47.** 25   **49.** -6 **51.** 5   **53.** 46,871,451

55. $\dfrac{3}{10}$ **57.** 0.18   **59.** $3(3 + 10 - 8)$ **61.** $(20 \div 2)(5 - 2) + 14$

EXERCISE SET 1.4, pp. 25–26

1. -8 **3.** -4 **5.** -18 **7.** 127.20 **9.** $3a + 6$ **11.** $4x - 4y$ **13.** $-10a - 15b$

15. $2ab - 2ac + 2ad$ **17.** $2\pi rh + 2\pi r$ **19.** $\dfrac{1}{2}ha + \dfrac{1}{2}hb$ **21.** $8(x + y)$ **23.** $9(p - q)$

25. $7(x - 3)$ **27.** $x(y + z)$ **29.** $2(x - y + z)$ **31.** $3(x + 2y - 3z)$ **33.** $a(b + c - d)$

35. $\pi r(r + s)$ **37.** $9a$ **39.** $-3b$ **41.** $15y$ **43.** $11a$ **45.** $-8t$ **47.** $10x$ **49.** $8x - 8y$

51. $2c + 10d$ **53.** $22x + 18y$ **55.** $2x - 33y$ **57.** $\dfrac{1}{4}$

EXERCISE SET 1.5, pp. 30–31

1. $4b$ **3.** $-a - 2$ **5.** $-b + 3$ **7.** $-t + y$ or $y - t$ **9.** $-a - b - c$ **11.** $-8x + 6y - 13$

13. $2c - 5d + 3e - 4f$ **15.** $4a, -5b, 6$ **17.** $2x, -3y, -2z$ **19.** $3a + 5$ **21.** $m + 1$

23. $5d - 12$ **25.** $-7x + 14$ **27.** $-9x + 21$ **29.** $44a - 22$ **31.** -190 **33.** $-12y - 145$

35. $0.01583x + 0.00001y$ **37.** $4x + 12y$ **39.** $-31a$

EXERCISE SET 1.6, pp. 33–34

1. 4^5 **3.** 5^6 **5.** m^4 **7.** $(3a)^4$ **9.** $5^2c^3d^4$ **11.** $2 \cdot 2 \cdot 2 \cdot 2 \cdot 2$, or 32

13. $(-3) \cdot (-3) \cdot (-3) \cdot (-3)$, or 81 **15.** $x \cdot x \cdot x \cdot x$ **17.** $4b \cdot 4b \cdot 4b$ **19.** $ab \cdot ab \cdot ab \cdot ab$ **21.** 5

23. $\dfrac{7}{8}$ **25.** 1 **27.** 1 **29.** $\dfrac{1}{6^3}$ **31.** $\dfrac{1}{9^5}$ **33.** $-\dfrac{1}{11}$ **35.** 3^{-4} **37.** 10^{-3} **39.** $(-16)^{-2}$

41. $1 + 2 + 3 + 4 + 5 + 6 + 7 + 8 \times 9 = 100$ **43.** $\dfrac{5}{3}$ **45.** $\dfrac{328}{3}$ **47.** False

EXERCISE SET 1.7, p. 39

1. 5^9 **3.** 8^{-4} **5.** 8^{-6} **7.** b^{-3} **9.** a^3 **11.** $72x^5$ **13.** $-28m^5n^5$ **15.** $-14x^{-11}$, or $\dfrac{-14}{x^{11}}$ **17.** 6^5

19. 4^5 **21.** 10^{-9} **23.** 9^2 **25.** a^5 **27.** 1 **29.** $-\dfrac{4}{3}x^9y^{-2}$, or $\dfrac{-4x^9}{3y^2}$ **31.** $\dfrac{3}{2}x^3y^{-2}$, or $\dfrac{3x^3}{2y^2}$ **33.** 4^6

35. 8^{-12} **37.** 6^{12} **39.** $27x^6y^6$ **41.** $\dfrac{1}{4}x^{-6}y^8$, or $\dfrac{y^8}{4x^6}$ **43.** $\dfrac{1}{36}a^4b^{-6}c^{-2}$, or $\dfrac{a^4}{36b^6c^2}$ **45.** $\dfrac{4^{-9}}{3^{12}}$, or $\dfrac{1}{4^9 \cdot 3^{12}}$

47. $\dfrac{8}{27}x^9y^3$ **49.** 2^{21} **51.** $\dfrac{16}{729}x^{28}y^{-62}$

TEST OR REVIEW—CHAPTER 1, pp. 40–41

1. -2.6 **2.** $\dfrac{326}{100}$, or $\dfrac{163}{50}$ **3.** 3.6% **4.** 18.75% **5.** 0.0571 **6.** 0.006 **7.** $-4.5 > -8.7$ **8.** $\dfrac{7}{8}$

9. 13.4 **10.** 0 **11.** -2 **12.** -13.1 **13.** $-\dfrac{17}{4}$ **14.** -8 **15.** 13 **16.** 0 **17.** -1 **18.** -29.7

19. $\dfrac{25}{4}$ **20.** -33.62 **21.** $\dfrac{3}{4}$ **22.** -264 **23.** -5 **24.** 15 **25.** $-\dfrac{7}{5}$, or -1.4 **26.** $-\dfrac{28}{15}$ **27.** $\dfrac{22}{7}$

28. -1.5 **29.** $-6a + 8b$ **30.** $3\pi rs + 3\pi r$ **31.** $a(b - c + 2d)$ **32.** $h(2a + 1)$ **33.** $10y - 5x$

34. $21a + 14$ **35.** $9x - 7y + 22$ **36.** $-7x + 14$ **37.** $-11y + 30$ **38.** $10x - 21$ **39.** $-a + 68$

40. $-6a^9 b^{-5}$, or $\dfrac{-6a^9}{b^5}$ **41.** $-\dfrac{3}{2} x^{-4} y^2$, or $\dfrac{-3y^2}{2x^4}$ **42.** $\dfrac{1}{81} a^{12} b^{-8} c^{-4}$, or $\dfrac{a^{12}}{81 b^8 c^4}$

43. $-\dfrac{125}{27} x^{-15} y^{24}$, or $\dfrac{-125 y^{24}}{27 x^{15}}$

CHAPTER 2

EXERCISE SET 2.1, p. 49

1. 9 **3.** -4 **5.** 23 **7.** -39 **9.** 1135 **11.** 4 **13.** -22 **15.** $\dfrac{1}{6}$ **17.** 32 **19.** -6 **21.** 18

23. 11 **25.** -12 **27.** 8 **29.** 2 **31.** 21 **33.** -12 **35.** 2 **37.** -1 **39.** $\dfrac{18}{5}$ **41.** 0 **43.** 1

45. 0.007 **47.** $\dfrac{17}{6}$ **49.** $\dfrac{155}{8}$ **51.** 5 **53.** $\dfrac{1}{9}$

EXERCISE SET 2.2, p. 52

1. 2 **3.** 2 **5.** 7 **7.** 5 **9.** $-\dfrac{3}{2}$ **11.** 5 **13.** $-2, 5$ **15.** 8, 9 **17.** $\dfrac{3}{2}, \dfrac{2}{3}$ **19.** 0, 8 **21.** $-4, 3$

23. $0, 1, -2$ **25.** 8.414 **27.** -6 **29.** No solution **31.** $-\dfrac{16}{3}$

EXERCISE SET 2.3, pp. 57–59

1. 8 in.; 4 in. **3.** $1\dfrac{3}{5}$ m; $2\dfrac{2}{5}$ m **5.** $14.75 **7.** $59.50 **9.** 4 **11.** 5 **13.** $45 **15.** $650

17. 32°, 96°, 52° **19.** $l = 31$ m; $w = 17$ m **21.** 12, 13, 14 **23.** $8000 **25.** $1644

27. 98 % **29.** 84 **31.** 143 gallons

EXERCISE SET 2.4, p. 62

1. $l = \dfrac{A}{w}$ **3.** $I = \dfrac{W}{E}$ **5.** $m = \dfrac{F}{a}$ **7.** $t = \dfrac{I}{Pr}$ **9.** $m = \dfrac{E}{c^2}$ **11.** $l = \dfrac{P - 2w}{2}$ **13.** $a^2 = c^2 - b^2$

15. $r^2 = \dfrac{A}{\pi}$ **17.** $h = \dfrac{2}{11} W + 40$ **19.** $r^3 = \dfrac{3V}{4\pi}$ **21.** $h = \dfrac{2A}{a + b}$ **23.** $m = \dfrac{Fr}{v^2}$ **25.** $\dfrac{4}{5}$ yr

27. $a = \dfrac{2(s - v_1 t)}{t^2}$ **29.** $V_1 = \dfrac{P_2 V_2 T_1}{P_1 T_2}$

EXERCISE SET 2.5, p. 67

1.

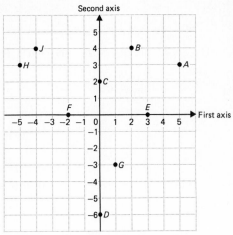

3.

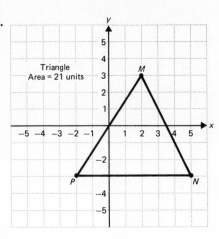

Triangle
Area = 21 units

5. Yes   **7.** No   **9.**

$y = 5x$

11.

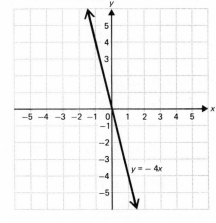

$y = -4x$

13.

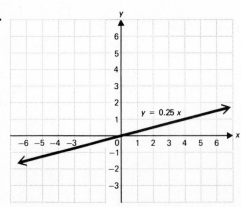

$y = 0.25\,x$

15. $(-1, -2)$, $(13, 10)$, or $(-19, -2)$ **17.** a, d

EXERCISE SET 2.6, p. 75

1. Slope is 4; y-intercept is $(0, 5)$.   **3.** Slope is -2; y-intercept is $(0, -6)$.

5. Slope is $-\dfrac{6}{16}$; y-intercept is $(0, -0.2)$.   **7.** Slope is 15.3; y-intercept is $(0, 4.04)$.

9. Slope is $\dfrac{5}{2}$; y-intercept is (0.1).   **11.** Slope is $-\dfrac{5}{2}$; y-intercept is $(0, -4)$.

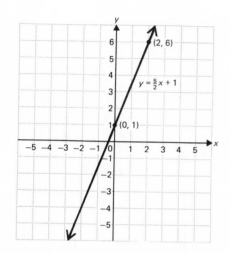

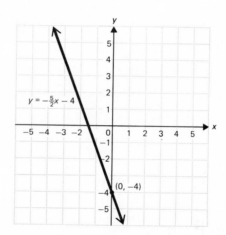

13.

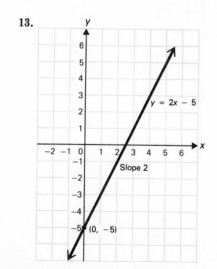

15.

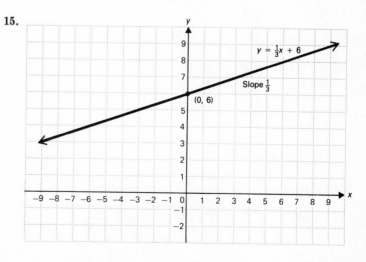

17. Slope is $-0.25x$; y-intercept is 2.

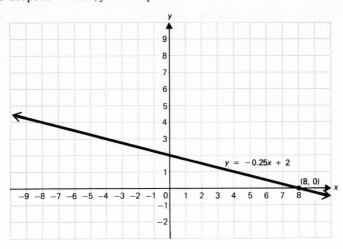

19.

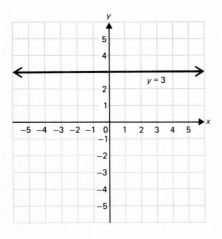

21.

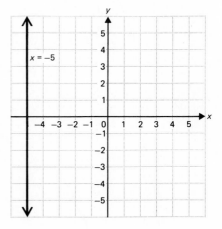

23. Linear　**25.** Linear　**27.** Not linear　**29.** Not linear　**31.** $m = \dfrac{1}{5}$, y-intercept $= -\dfrac{2}{3}$

33. $m = 0.01$, y-intercept $= -0.1$

EXERCISE SET 2.7, pp. 80–81

1.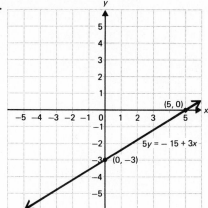

3.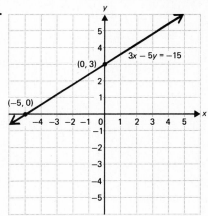

5.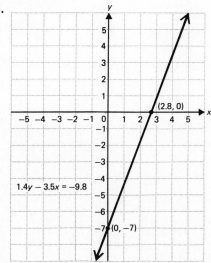

7. -2 **9.** 5 **11.** $-\dfrac{37}{111}$ **13.** Does not exist   **17.** Slope does not exist.   **19.** Zero slope

21. Zero slope   **23.** Slope does not exist.   **27.** Slope does not exist.   **29.** Zero slope

31. $a = \dfrac{5}{8}$ **33.** $(25, -1)$, $(-25, 1)$, $(50, -2)$, $(-50, 2)$. Answers may vary.

EXERCISE SET 2.8, pp. 85–86

1. $y - 2 = 4x - 12$, or $y = 4x - 10$ **3.** $y - 7 = -2x + 8$, or $y = -2x + 15$ **5.** $y + 4 = 3x + 6$.

or $y = 3x + 2$ **7.** $y = -2x + 16$ **9.** $y + 7 = 0$, or $y = -7$ **11.** $y = \dfrac{1}{2}x + \dfrac{7}{2}$ **13.** $y = x$

15. $y = \dfrac{5}{2}x + 5$ **17.** $y = \dfrac{3}{2}x$ **19.** $y = \dfrac{2}{5}x$ **21.** $y = -\dfrac{1}{2}x + \dfrac{17}{2}$ **23.** $y = \dfrac{5}{7}x - \dfrac{17}{7}$

25. $y = \dfrac{1}{2}x + 4$ **27.** $y = \dfrac{4}{3}x - 6$ **29.** $y = -\dfrac{7}{3}x + \dfrac{22}{3}$

31. The side that contains (3, 4) and (6, 9) has slope $\dfrac{5}{3}$. The side that contains $(-2, 7)$ and (3, 4)

has slope $-\dfrac{3}{5}$. Since $\dfrac{5}{3} \cdot \left(-\dfrac{3}{5}\right) = -1$, these two sides are perpendicular.

33. (1) and (4)   **35.** $k = 7$ **37.** $y = -0.5x - 0.6$

EXERCISE SET 2.9, pp. 89–90

1. (a) $E = 0.15t + 72$; (b) 76.5, 77.25   **3.** (a) $H = \dfrac{3}{20}W + \dfrac{905}{20}$; (b) $64\dfrac{3}{4}$ in.

5. (a) $R = -0.0075t + 3.85$; (b) 3.475, 3.445; (c) 2003   **7.** (a) $C = 0.25m + 1.25$; (b) \$3.00

TEST OR REVIEW—CHAPTER 2, pp. 90–91

1. 22   **2.** $\dfrac{2}{3}$ **3.** $\dfrac{11}{5}$ **4.** -2 **5.** $-1, 3$ **6.** $W = \dfrac{11}{2}(h - 40)$

7. $y = -\dfrac{2}{3}x + 2$; slope is $-\dfrac{2}{3}$; y-intercept is (0, 2).   **8.**

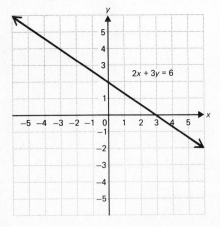

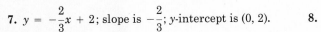

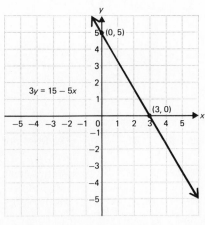

9. $-\dfrac{9}{7}$ **10.** $y = 2x + 1$ **11.** $y = -\dfrac{3}{2}x$ **12.** $y = -\dfrac{5}{7}x + \dfrac{8}{7}$ **13.** $y = \dfrac{1}{3}x + \dfrac{1}{3}$ **14.** 10

15. \$12.50

CHAPTER 3

EXERCISE SET 3.1, p. 96

1. Yes   **3.** No   **5.** Yes   **7.** No

9.

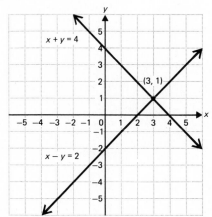

11.

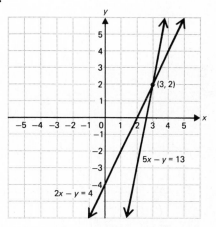

13.

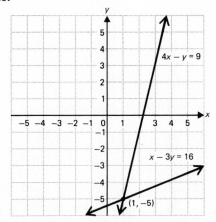

15.

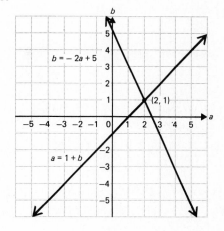

17.

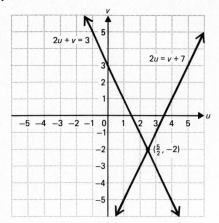

19.

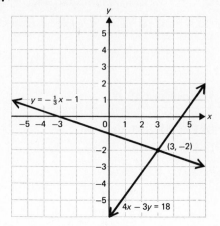

21.

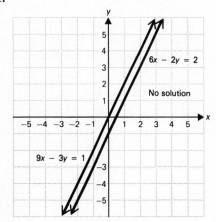

EXERCISE SET 3.2, pp. 101–102

1. $(-4, 3)$ **3.** $(-3, -15)$ **5.** $(2, -2)$ **7.** $(-2, 1)$ **9.** $(1, 2)$ **11.** $(3, 0)$ **13.** $(-1, 2)$

15. $\left(\dfrac{128}{31}, -\dfrac{17}{31}\right)$ **17.** $(6, 2)$ **19.** $\left(\dfrac{140}{13}, -\dfrac{50}{13}\right)$ **21.** $(4, 6)$ **23.** $(23.118878, -12.039964)$

25. $(-2, -9)$ **27.** $(10, 5)$ **29.** $(-20, 20)$ **31.** $\left(\dfrac{a + 2b}{7}, \dfrac{a - 5b}{7}\right)$ **33.** $p = 2,\ q = -\dfrac{1}{3}$

EXERCISE SET 3.3, pp. 107–109

1. 5 and -47 **3.** 24 and 8   **5.** $l = 160$ m, $w = 154$ m   **7.** 8 white, 22 yellow

9. 150 lb of soybean meal, 200 lb of corn meal   **11.** $4100 at 7%, $4700 at 8%

13. $30,000 at 5%, $40,000 at 6%   **15.** Ann 40, son 20   **17.** John 28, Sue 20

19. $l = 31$ in., $w = 12$ in.   **21.** 32 at $8.50, 13 at $9.75   **23.** $825 at 6%, $325 at 5%   **25.** $4\frac{4}{7}$ liters

27. 82   **29.** 180   **31.** $m = -\dfrac{1}{2}, b = \dfrac{5}{2}$

EXERCISE SET 3.4, pp. 114–115

1. Yes   **3.** $(1, 2, 3)$ **5.** $(-1, 5, -2)$ **7.** $(3, 1, 2)$ **9.** $(-3, -4, 2)$ **11.** $(2, 4, 1)$ **13.** $(-3, 0, 4)$

15. $(0, 2, -1, 1)$ **17.** $\left(-1, \dfrac{1}{2}, 1\right)$ **19.** $(w, x, y, z) = (-2, 0, 0, 1)$

EXERCISE SET 3.5, pp. 117–119

1. 17, 9, 79   **3.** 4, 2, -1 **5.** 34°, 104°, 42°   **7.** 25°, 50°, 105°   **9.** Thurs., $21; Fri., $18; Sat., $27

11. 82, 76, 97   **13.** A, 1500; B, 1900; C, 2300   **15.** A, 900 gph; B, 1300 gph; C, 1500 gph

17. 20   **19.** Student, 2050; adult, 845; children, 210

EXERCISE SET 3.6, p. 124

1. 3   **3.** 36   **5.** -10 **7.** -3 **9.** 5   **11.** $(2, 0)$ **13.** $\left(-\dfrac{25}{2}, -\dfrac{11}{2}\right)$ **15.** $\left(\dfrac{3}{2}, \dfrac{13}{14}, \dfrac{33}{14}\right)$

17. $(2, -1, 4)$ **19.** $(1, 2, 3)$

21. An equation of the line through (x_1, y_1) and (x_2, y_2) is

$$y - y_1 = \frac{y_2 - y_1}{x_2 - x_1}(x - x_1),$$

which is equivalent to

$$yx_2 - yx_1 - y_1x_2 + y_1x_1 = y_2x - y_2x_1 - y_1x + y_1x_1$$

and

$$y_2x_1 + y_1x - y_2x + yx_2 - yx_1 - y_1x_2 = 0;$$

$$\begin{vmatrix} x & y & 1 \\ x_1 & y_1 & 1 \\ x_2 & y_2 & 1 \end{vmatrix} = 0 \text{ is equivalent to}$$

$$x(y_1 - y_2) - x_1(y - y_2) + x_2(y - y_1) = 0$$

and

$$xy_1 - xy_2 - x_1y + x_1y_2 + x_2y - x_2y_1 = 0$$

23. 3   **25.** $\begin{vmatrix} 3 & -5 \\ b & a \end{vmatrix}$, answers may vary

EXERCISE SET 3.7, pp. 128–129

1. (a) $P = 20x - 600,000$; (b) 30,000 units   **3.** (a) $P = 50x - 120,000$; (b) 2400 units

5. (a) $P = 80x - 10,000$; (b) 125 units   **7.** ($10, 1400)   **9.** ($22, 474)   **11.** ($50, 6250)

13. (a) $C = 40x + 22,500$; (b) $R = 85x$; (c) $P = 45x - 22,500$; (d) profit of $112,500; (e) 500 units

EXERCISE SET 3.8, p. 132

1. Inconsistent   **3.** Consistent   **5.** Consistent   **7.** Inconsistent   **9.** Consistent

11. Dependent   **13.** Independent   **15.** Dependent   **17.** Independent   **19.** Dependent

TEST OR REVIEW—CHAPTER 3, pp. 132–133

1.

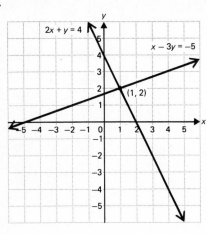

2. $(-1, 2)$ **3.** $l = 235$ m; $w = 120$ m   **4.** $(-3, 0, 3)$

5. 72, 85, 91   **6.** -2 **7.** $P = 35x - 90,000$; 2572

8. Inconsistent   **9.** Independent

CHAPTER 4

EXERCISE SET 4.1, pp. 140–141

1. $8x^2 + (-2x) + (-5)$; 5   **3.** $18xy + (-8x^3) + (-y) + 50$; 20   **5.** $5x^3, 7x^2, -3x, -9$; 5, 7, -3, -9

7. $-3xyz, 7x^2y^2, -5xy^2z, 4xyz^2$; $-3, 7, -5, 4$ **9.** 2, 5, 7, 0; 7   **11.** 5, 6, 2, 1, 0; 6   **13.** $2x^2$

15. $3x + y$ **17.** $a + 6$ **19.** $-6a^2b - 2b^2$ **21.** $9x^2 + 2xy + 15y^2$ **23.** $-x^2y + 4y + 9xy^2$

25. $x^3 - 3x^2 + x + 1$ **27.** $5a^5 - a^3 + 6a^2 + a - 9$ **29.** $-7 - y + 3y^2 + 2y^3 + y^4$

31. $-x + x^2 - 9x^3 + 18x^4 + 5x^5$ **33.** $-xy^3 + x^2y^2 + x^3y + 1$ **35.** $3xy^3 + x^2y^2 - 9x^3y + 2x^4$

37. $0.81125x^2y^2 - 0.07111x^3y$ **39.** $\dfrac{11}{4}$ **41.** 0.000000798

EXERCISE SET 4.2, pp. 143–144

1. $5x^2 + 2y^2 + 5$ **3.** $6a + b + c$ **5.** $-4a^2 - b^2 + 3c^2$ **7.** $-2x^2 + x - xy - 1$

9. $5x^2y - 4xy^2 + 5xy$ **11.** $9r^2 + 9r - 9$ **13.** $-5x^3 + 7x^2 - 3x + 6$ **15.** $13x - 6$

17. $-4x^2 - 3x + 13$ **19.** $2a - 4b + 3c$ **21.** $-2x^2 + 6x$ **23.** $-4a^2 + 8ab - 5b^2$

25. $27.94x^2 + 9.105y^2 + 8.15x$ **27.** $-3a^2 - 55ab - 40b^2$ **29.** $0.657u^2 + 5.003v^2$

EXERCISE SET 4.3, pp. 146–147

1. $10y^3$ **3.** $-20x^3y$ **5.** $-10x^5y^6$ **7.** $6x - 2x^2$ **9.** $3a^2b + 3ab^2$ **11.** $15c^3d^2 - 25c^2d^3$

13. $6x^2 + x - 12$ **15.** $s^2 - 9t^2$ **17.** $x^2 - 2xy + y^2$ **19.** $2y^2 + 9xy - 56x^2$

21. $a^4 - 5a^2b^2 + 6b^4$ **23.** $x^3 - 64$ **25.** $x^3 + y^3$ **27.** $a^4 + 5a^3 - 2a^2 - 9a + 5$

29. $4a^3b^2 + 4a^3b - 10a^2b^2 - 2a^2b + 3ab^3 + 7ab^2 - 6b^3$ **31.** $x^2 + 4xh$

EXERCISE SET 4.4, pp. 151–152

1. $a^2 + 5a + 6$ **3.** $y^2 + y - 6$ **5.** $b^2 - \dfrac{5}{6}b + \dfrac{1}{6}$ **7.** $2x^2 + 13x + 18$ **9.** $4x^2 - 4xy - 3y^2$

11. $x^2 + 6x + 9$ **13.** $4a^2 + \dfrac{4}{3}a + \dfrac{1}{9}$ **15.** $x^2 - 4xy + 4y^2$ **17.** $4x^4 - 12x^2y^2 + 9y^4$

19. $a^4b^4 + 2a^2b^2 + 1$ **21.** $c^2 - 4$ **23.** $4a^2 - 1$ **25.** $9m^2 - 4n^2$ **27.** $x^4 - y^2z^2$

29. $-\dfrac{4}{3}x^6y^{11}$ **31.** $-\dfrac{9}{4}r^{22}s^{14}$ **33.** $\dfrac{8}{21}x^8y^5$ **35.** $-16s^8t + 24s^6t^5 - 8s^4t^8 + 8s^3t^{11}$

37. $16x^4 - 32x^3 + 16x^2$ **39.** $-a^4 - 2a^3b + 25a^2 + 2ab^3 - 25b^2 + b^4$ **41.** $r^8 - 2r^4s^4 + s^8$

43. $a^2 + 2ac - b^2 - 2bd + c^2 - d^2$ **45.** $10y^2 - 38xy + 24x^2 - 212y + 306x + 930$

47. $16x^4 + 4x^2y^2 + y^4$ **49.** $x^{4a} - y^{4b}$ **51.** $a^3 - b^3 + 3b^2 - 3b + 1$ **53.** $\dfrac{1}{81}x^{12} - \dfrac{8}{81}x^6y^4 + \dfrac{16}{81}y^8$

55. $x^{a^2 - b^2}$

EXERCISE SET 4.4A, pp. 152–153

1. $8x^2 - 6xy + y^2$ **3.** $4c^2 + 28cd + 49d^2$ **5.** $9c^2 - 4d^2$ **7.** $2x^4 + 3x^2y - 2y^2$

9. $2m^4 + 5m^2n^2 - 3n^4$ **11.** $9x^2 - 24xy + 16y^2$ **13.** $a^4 - b^2$ **15.** $\dfrac{1}{25} - x^2$

17. $8x^4 + 2x^2y^2 - 15y^4$ **19.** $\dfrac{1}{4}x^2 - 6x + 36$ **21.** $x^4 - 1$ **23.** $a^4 - b^4$ **25.** $a^2 + 2ab + b^2 - 1$

27. $4x^2 + 12xy + 9y^2 - 16$ **29.** 1404.93

EXERCISE SET 4.5, pp. 155–156

1. $2a(2a + 1)$ **3.** $3(y^2 - y - 3)$ **5.** $3x^2(2 - x^2)$ **7.** $2a(2b - 3c + 6d)$

9. $5(2a^4 + 3a^2 - 5a - 6)$ **11.** $(a + c)(b - 2)$ **13.** $(x - 2)(2x + 13)$ **15.** $2a^2(x - y)$

17. $(a + b)(c + d)$ **19.** $(b^2 + 2)(b - 1)$ **21.** $(y - 1)(y - 8)$ **23.** $(x + 4)(x - 4)$

25. $6(x + y)(x - y)$ **27.** $4x(y^2 + z^2)(y + z)(y - z)$

29. $a^2(3a + 5b^2)(3a - 5b^2)$ **31.** $\left(\dfrac{1}{5} + x\right)\left(\dfrac{1}{5} - x\right)$ **33.** $(0.2x + 0.3y)(0.2x - 0.3y)$ **35.** 5

EXERCISE SET 4.6, pp. 160–161

1. $(y - 3)^2$ **3.** $(x + 7)^2$ **5.** $y(y - 9)^2$ **7.** $3(2a + 3)^2$ **9.** $2(x - 10)^2$ **11.** $(0.5x + 0.3)^2$

13. $(x + 5)(x + 4)$ **15.** $2(y - 4)^2$ **17.** $(x - 9)(x + 3)$ **19.** $y(8 - y)(4 + y)$ **21.** $(t + 5)(t + 3)$

23. $(x^2 + 16)(x^2 - 5)$ **25.** $(3x + 2)(x - 6)$ **27.** $x(3x - 5)(2x + 3)$ **29.** $(3a - 4)(a - 2)$

31. $(5y + 2)(7y + 4)$ **33.** $2(5t - 3)(t + 1)$ **35.** $4(2x + 1)(x - 4)$ **37.** 263.76 cm^2

EXERCISE SET 4.7, p. 164

1. $(a + b + 3)(a + b - 3)$ **3.** $(r - 1 + 2s)(r - 1 - 2s)$ **5.** $2(m + n + 5b)(m + n - 5b)$

7. $[3 + (a + b)][3 - (a + b)]$ **9.** $x^2 + 16x + 64$ **11.** $x^2 - 4.2x + 4.41$ **13.** $x^2 + \dfrac{2}{3}bx + \dfrac{1}{9}b^2$

15. $(x + 17)(x + 7)$ **17.** $(x - 5)(x - 21)$ **19.** $2(x + 19)(x + 9)$ **21.** $2(x - 12a)(x - 4a)$

23. $(x + 2.3)(x + 0.3)$ **25.** $\left(x + \dfrac{4}{10}\right)\left(x + \dfrac{2}{10}\right)$, or $\left(x + \dfrac{2}{5}\right)\left(x + \dfrac{1}{5}\right)$ **27.** $(x + 5.766)(x - 1.284)$

EXERCISE SET 4.8, pp. 166–167

1. $(x + 2)(x^2 - 2x + 4)$ **3.** $2(y - 4)(y^2 + 4y + 16)$ **5.** $3(2a + 1)(4a^2 - 2a + 1)$

7. $(2 - 3b)(4 + 6b + 9b^2)$ **9.** $(2x + 3)(4x^2 - 6x + 9)$ **11.** $\left(a + \dfrac{1}{2}\right)\left(a^2 - \dfrac{1}{2}a + \dfrac{1}{4}\right)$

13. $r(s + 4)(s^2 - 4s + 16)$ **15.** $5(x - 2z)(x^2 + 2xz + 4z^2)$ **17.** $(x + 0.1)(x^2 - 0.1x + 0.01)$

19. $8(2x^2 - t^2)(4x^4 + 2x^2t^2 + t^4)$ **21.** $x^2 - 16$

EXERCISE SET 4.9, pp. 170–171

1. $(x + 12)(x - 12)$ **3.** $(2x + 3)(x + 4)$ **5.** $3(x^2 + 2)(x^2 - 2)$ **7.** $(a + 5)^2$ **9.** $2(x - 11)(x + 6)$

11. $(3x + 5y)(3x - 5y)$ **13.** $(2c - d)^2$ **15.** $(x^2 + 2)(2x - 7)$ **17.** $2(5x - 4y)(25x^2 + 20xy + 16y^2)$

19. $(m^3 + 10)(m^3 - 2)$ **21.** $(c - b)(a + d)$ **23.** $(m + 1)(m^2 - m + 1)(m - 1)(m^2 + m + 1)$

25. $(x + y + 3)(x - y + 3)$ **27.** $(6y - 5)(6y + 7)$ **29.** $(a^4 + b^4)(a^2 + b^2)(a + b)(a - b)$

31. $ab(a + 4b)(a - 4b)$ **33.** $2(x + 2)(x - 2)(x + 3)$ **35.** $2(2x + 3y)(4x^2 - 6xy + 9y^2)$

37. $(2x - 3)(10x + 13)$ **39.** $4(2a - 1)(9a + 40)$ **41.** $7\left(x - \dfrac{1}{2}\right)\left(x^2 + \dfrac{1}{2}x + \dfrac{1}{4}\right)$

43. $(c - 2d)^2(c^2 - cd + d^2)^2$ **45.** $(x + 1)(x - 1)(x + 7)(x - 7)$

47. $(s - 3t)(s + 3t)(s^2 + 3st + 9t^2)(s^2 - 3st + 9t^2)$ **49.** $(3x^{2s} + 4y^t)(9x^{4s} - 12x^{2s}y^t + 16y^{2t})$

51. $(2x + y - r + 3s)(2x + y + r - 3s)$ **53.** $(c^2d^2 + a^8)(cd - a^4)(cd + a^4)$ **55.** $c(c^w + 1)^2$

57. $y(y^4 + 1)(y^2 + 1)(y + 1)(y - 1)$ **59.** $3(a + b - c - d)(a + b + c + d)$ **61.** $4(3a^2 + 4)$

63. 0

EXERCISE SET 4.10, pp. 173–175

1. $-7, 4$ **3.** 4   **5.** $-5, -4$ **7.** $0, -8$ **9.** $-6, 6$ **11.** $-5, -9$ **13.** $7, 4$ **15.** $\dfrac{1}{2}, 7$ **17.** $7, -2$

19. $-\dfrac{3}{2}, \dfrac{7}{2}$ **21.** $l = 12$ in.; $w = 7$ in.   **23.** 9 and 11   **25.** $-13, 12$ **27.** $l = 12$ cm; $w = 8$ cm

29. 2   **31.** h = 7 cm; $b = 16$ cm   **33.** $-10, -8$, and -6; 6, 8, and 10   **35.** $l = 28$ cm; $w = 14$ cm

37. Length = 40 inches, width = 30 inches, depth = 20 inches   **39.** 54 cm^2

EXERCISE SET 4.11, pp. 177–178

1. $y = \dfrac{15 - b}{3 + a}$ **3.** $z = a$ **5.** $x = -\dfrac{a}{9}$ **7.** $x = 11$ **9.** $x = b + a$ **11.** $r^2 = \dfrac{V}{\pi h}$ **13.** $t^2 = \dfrac{2s}{g}$

15. $h = \dfrac{A - \pi r^2}{2\pi r}$ **17.** $r = \dfrac{A - P}{Pt}$ **19.** $x = \dfrac{14 - 2b}{5}$

EXERCISE SET 4.12, pp. 182–183

1. Yes   **3.** 2   **5.** -7 **7.** 3   **9.** 12   **11.** 6   **13.** 4   **15.** $4\sqrt{3}$ **17.** 36π **19.** Yes   **21.** Yes

23. Yes   **25.** (a) $x^3 + 2x^2 - x + 3$; (b) $x^5 + x^4 - 2x^3 - 2x^2 + x + 1$; (c) x^6; (d) 1193; (e) 0;

(f) 64; (g) -41; (h) -4; (i) $\dfrac{1}{16}$

TEST OR REVIEW—CHAPTER 4, pp. 183–184

1. (a) $5x^5y^4 - 2x^4y - 4x^2y + 3xy^3$; (b) 9   **2.** $3xy + 3xy^2$

3. $3x^2y^4 - 2x^2y^2 + 2x^2y - 3xy^2 - 5x + 4$ **4.** $2y^2 + 5y + y^3$ **5.** $-8x^5 - 12x^4 + 24x^3 - 19x + 15$

6. $9x^2 - 24xy + 16y^2$ **7.** $9x^2 - 25$ **8.** $3a^2b(2 - 7ab + b^2)$ **9.** $(y - 4)(y + 5)$

10. $5(2a - b)(2a + b)$ **11.** $(3x - 5)^2$ **12.** $2(4x - 1)(3x - 5)$ **13.** $(y + 4 + 10t)(y + 4 - 10t)$

14. $x^2 - 5x + \dfrac{25}{4}$ **15.** $(x + 4.2)(x + 0.2)$ **16.** $2ab(2a^2 + 3b^2)(4a^4 - 6a^2b^2 + 9b^4)$ **17.** $\dfrac{1}{2}, -3$

18. $l = 8 \text{ cm}; w = 5 \text{ cm}$ **19.** $r = \dfrac{mv^2}{F}$ **20.** 15

CHAPTER 5

EXERCISE SET 5.1, pp. 191–192

1. $\dfrac{3x(x + 1)}{3x(x + 3)}$ **3.** $\dfrac{(t - 3)(t + 3)}{(t + 2)(t + 3)}$ **5.** $\dfrac{3y}{5}$ **7.** $a - 3$ **9.** $\dfrac{y - 3}{y + 3}$ **11.** $\dfrac{t + 4}{t - 4}$ **13.** $\dfrac{x^2 - 16}{x^2 + 3x}$

15. $\dfrac{y + 4}{2}$ **17.** $\dfrac{2x^2 + 13x + 15}{7x}$ **19.** $c - 2$ **21.** $\dfrac{1}{x + y}$ **23.** 3   **25.** $\dfrac{y^2 - y - 6}{y}$ **27.** $\dfrac{2a + 1}{a + 2}$

29. $\dfrac{x^2 + 6x + 8}{3x - 15}$ **31.** $\dfrac{y^3 + 3y}{y^2 + y - 6}$ **33.** $\dfrac{x^2 + 4x + 16}{x^2 + 8x + 16}$ **35.** $\dfrac{2s}{r + 2s}$ **37.** $\dfrac{21{,}934.2x^2}{y^2 - 182{,}499.84}$

39. $\dfrac{x - 3}{(x + 1)(x + 3)}$ **41.** $\dfrac{m - t}{m + t + 1}$ **43.** $\dfrac{x^2 + xy + y^2 + x + y}{x - y}$ **45.** $\dfrac{-2x}{x - 1}$

EXERCISE SET 5.2, p. 195

1. 36   **3.** 144   **5.** 72   **7.** 45   **9.** $\dfrac{11}{10}$ **11.** $\dfrac{43}{36}$ **13.** $\dfrac{79}{60}$ **15.** $12x^2y$ **17.** $3(y - 3)(y + 3)$

19. $30a^3b^2$ **21.** $5(y - 3)(y - 3)$ **23.** $(x + 2)(x - 2)$, or $(x + 2)(2 - x)$

25. $(2r + 3)(r - 4)(3r - 1)(r + 4)$ **27.** 516,600

29. $8a^4b^7, 8a^4b^6, 8a^4b^5, 8a^4b^4, 8a^4b^3, 8a^4b^2, 8a^4b, 8a^4$

EXERCISE SET 5.3, pp. 200–201

1. 2   **3.** $\dfrac{3y + 5}{y - 2}$ **5.** $a + b$ **7.** $\dfrac{11}{x}$ **9.** $\dfrac{1}{x + 5}$ **11.** $\dfrac{2y^2 + 22}{y^2 - y - 20}$ **13.** $\dfrac{x + y}{x - y}$ **15.** $\dfrac{3x - 4}{x^2 - 3x + 2}$

17. $\dfrac{8x + 1}{x^2 - 1}$ **19.** $\dfrac{2x - 14}{15(x + 5)}$ **21.** $\dfrac{-a^2 + 7ab - b^2}{a^2 - b^2}$ **23.** $\dfrac{y}{y^2 - 5y + 6}$ **25.** $\dfrac{3y - 10}{y^2 - y - 20}$

27. $\dfrac{3y^2 - 3y - 29}{(y + 8)(y - 3)(y - 4)}$ **29.** $\dfrac{2x^2 - 13x + 7}{(x + 3)(x - 1)(x - 3)}$ **31.** 0   **33.** $\dfrac{3}{x + 2}$ **35.** $\dfrac{-3x^2 - 3x - 4}{x^2 - 1}$

37. $\dfrac{2y^2 + 3 - 7x^3y}{x^2y^2}$ **39.** $\dfrac{5y + 23}{5 - 2y}$

EXERCISE SET 5.4, pp. 204–205

1. $\dfrac{1 + 4x}{1 - 3x}$ **3.** $\dfrac{x^2 - 1}{x^2 + 1}$ **5.** $\dfrac{3y + 4x}{4y - 3x}$ **7.** $\dfrac{x + y}{x}$ **9.** $\dfrac{a^2(b - 3)}{b^2(a - 1)}$ **11.** $\dfrac{1}{a - b}$ **13.** $\dfrac{-1}{x(x + h)}$

15. $\dfrac{y - 3}{y + 5}$ **17.** $\dfrac{5}{6}$ **19.** $\dfrac{11x + 8}{4x + 3}$ **21.** $\dfrac{b - a}{ab}$ **23.** a

EXERCISE SET 5.5, pp. 209–210

1. $6x^4 - 3x^2 + 8$ **3.** $y^3 - 2y^2 + 3y$ **5.** $x + 7$ **7.** $a - 12$, R 32, or $a - 12 + \dfrac{32}{a + 4}$

9. $x + 2$, R 4; or $x + 2 + \dfrac{4}{x + 5}$ **11.** $2y^2 - y + 2$, R 6; or $2y^2 - y + 2 + \dfrac{6}{2y + 4}$

13. $2y^2 + 2y - 1$, R 8; or $2y^2 + 2y - 1 + \dfrac{8}{5y - 2}$

15. $2x^2 - x - 9$, R $(3x + 12)$; or $2x^2 - x - 9 + \dfrac{3x + 12}{x^2 + 2}$ **17.** $2x^4 - 2x^3 + 5x^2 - 4x - 1$, R $5x + 2$

19. $x^2 + 2y$ **21.** $x^3 + x^2y + xy^2 + y^3$ **23.** $k = \dfrac{14}{3}$

EXERCISE SET 5.6, p. 212

1. $x^2 - x + 1$, R -4 **3.** $a + 7$, R -47 **5.** $x^2 - 5x - 23$, R -43 **7.** $3x^2 - 2x + 2$, R -3

9. $y^2 + 2y + 1$, R 12   **11.** $3x^3 + 9x^2 + 2x + 6$ **13.** $x^2 + 3x + 9$ **15.** $y^3 + y^2 + y + 1$

17. $3.41x^3 + 8.2181x^2 - 4.444379x - 10.710953$, R -39.283396

EXERCISE SET 5.7, pp. 218–219

1. $\dfrac{51}{2}$ **3.** -2 **5.** 144   **7.** $-5, -1$ **9.** 2   **11.** $\dfrac{17}{4}$ **13.** 11   **15.** No solution   **17.** 2

19. $\dfrac{3}{5}$ **21.** 5   **23.** -145 **25.** $-\dfrac{10}{3}$ **27.** -3 **29.** $-6, 5$ **31.** No solution

33. All reals except 1 and -1

EXERCISE SET 5.8, pp. 224–225

1. $3\dfrac{3}{14}$ hours   **3.** $8\dfrac{4}{7}$ hours   **5.** $3\dfrac{9}{52}$ hours   **7.** A, 10 da; B, 40 da   **9.** 7 mph   **11.** 375 km

13. Train A, 46 mph; train B, 58 mph   **15.** 9 km/h   **17.** $21\frac{9}{11}$ min after 4:00

19. Boat: 14 km/h; stream: 10 km/h   **21.** 3.75 km/h   **23.** 48 km/h   **25.** $t = \frac{2}{3}$ hr

EXERCISE SET 5.9, p. 228

1. $d_1 = \dfrac{d_2 W_1}{W_2}$ **3.** $t = \dfrac{2s}{v_1 + v_2}$ **5.** $r_2 = \dfrac{Rr_1}{r_1 - R}$ **7.** $s = \dfrac{Rg}{g - R}$ **9.** $p = \dfrac{qf}{q - f}$

11. $r = \dfrac{nE - IR}{In}$ **13.** $H = Sm(t_1 - t_2)$ **15.** $e = \dfrac{rE}{R + r}$

EXERCISE SET 5.10, pp. 233–234

1. $8; y = 8x$ **3.** $3.6; y = 3.6x$ **5.** $\dfrac{8}{5}; y = \dfrac{8}{5}x$ **7.** 6 amperes   **9.** 125,000   **11.** $60; y = \dfrac{60}{x}$

13. $36; y = \dfrac{36}{x}$ **15.** $0.32; y = \dfrac{0.32}{x}$ **17.** $\dfrac{2}{9}$ ampere   **19.** $685\frac{5}{7}$ kg   **21.** $I = 0.185P$

23. Steve: 56; Stephanie: 42

TEST OR REVIEW—CHAPTER 5, p. 235

1. $\dfrac{2x + 10}{x - 2}$ **2.** $\dfrac{y + 4}{2}$ **3.** $(x + 3)(x - 2)(x + 5)$ **4.** $x + y$ **5.** $\dfrac{3x}{(x - 1)(x + 1)}$ **6.** $\dfrac{x + 1}{x}$

7. $x - 7$, R -46 **8.** $6y^3 + 3y^2 - 6y - 16$, R 38   **9.** 9   **10.** $2\frac{2}{5}$ hours   **11.** 11 km/h

12. $t = \dfrac{mr}{2 - m}$ **13.** $495   **14.** $2\frac{5}{8}$ hours

CHAPTER 6

EXERCISE SET 6.1, pp. 242–243

1. 1   **3.** $x^2 + y$ **5.** $-15r^{-4}s^2$, or $\dfrac{-15s^2}{r^4}$ **7.** $-8m^{-6}n^9$, or $\dfrac{-8n^9}{m^6}$ **9.** $\dfrac{x^6}{5}$ **11.** $-12a^{-2}b^2$

13. $a^{-2}b^{-1}$, or $\dfrac{1}{a^2b}$ **15.** $-31,104y^6z^{-8}$, or $\dfrac{-31,104y^6}{z^8}$ **17.** $\dfrac{x^{12}y^6}{8}$ **19.** $\dfrac{27a^{-6}b^3}{c^6}$, or $\dfrac{27b^3}{a^6c^6}$

21. 4.7×10^{10} **23.** 8.63×10^{17} **25.** 1.6×10^{-8} **27.** 7×10^{-11} **29.** 673,000,000

31. 0.0000000008923 **33.** 9.66×10^{-5} **35.** 1.3338×10^{-11} **37.** 2.5×10^3 **39.** 5×10^{-4}

41. Answers may vary: 4.2×10^2 **43.** 1.4675×10^{12} **45.** $\dfrac{1}{3}$ **47.** 7 **49.** y^{-4} **51.** $\dfrac{3^{12s+4}}{625}$

EXERCISE SET 6.2, pp. 247–248

1. $4, -4$ **3.** $12, -12$ **5.** $20, -20$ **7.** $-\dfrac{7}{6}$ **9.** 14 **11.** $-\dfrac{4}{9}$ **13.** 0.3 **15.** -0.07 **17.** $p^2 + 4$

19. $\dfrac{x}{y + 4}$ **21.** $|4x|$ or $4|x|$ **23.** $|7c|$ or $7|c|$ **25.** $|a + 1|$ **27.** $|x - 2|$ **29.** $|2x + 7|$ **31.** 3

33. $-4x$ **35.** -6 **37.** $0.7(x + 1)$ **39.** 5 **41.** -1 **43.** $-\dfrac{2}{3}$ **45.** $|x|$ **47.** $|5a|$ or $5|a|$ **49.** 6

51. $|a + b|$ **53.** y **55.** $x - 2$ **57.** (a) 12.5; (b) 15; (c) 17.5; (d) 20

EXERCISE SET 6.3, pp. 252–253

1. $\sqrt{6}$ **3.** $\sqrt[3]{10}$ **5.** $\sqrt[4]{72}$ **7.** $\sqrt{30ab}$ **9.** $\sqrt[5]{18t^3}$ **11.** $\sqrt{x^2 - a^2}$ **13.** $2\sqrt{2}$ **15.** $2\sqrt{6}$

17. $2\sqrt{10}$ **19.** $6x^2\sqrt{5}$ **21.** $3x^2\sqrt[3]{2x^2}$ **23.** $3\sqrt{10}$ **25.** $3\sqrt[3]{2}$ **27.** $30\sqrt{3}$ **29.** $5bc^2\sqrt{2b}$

31. $2y^3\sqrt[3]{2}$ **33.** $(b + 3)^2$ **35.** $4a^3b\sqrt{6ab}$ **37.** 13.4 **39.** 7.5 **41.** Rational **43.** Rational

45. Rational **47.** Irrational **49.** (a) 20 mph; (b) 37.4 mph; (c) 42.4 mph **51.** $572\sqrt{3}$

EXERCISE SET 6.4, pp. 256–257

1. $\dfrac{4}{5}$ **3.** $\dfrac{4}{3}$ **5.** $\dfrac{7}{y}$ **7.** $\dfrac{5y\sqrt{y}}{x^2}$ **9.** $\dfrac{2x\sqrt[3]{x^2}}{3y}$ **11.** $\sqrt{7}$ **13.** 3 **15.** $y\sqrt{5y}$ **17.** $2\sqrt[3]{a^2b}$ **19.** $3\sqrt{xy}$

21. $\sqrt{x^2 + xy + y^2}$ **23.** $6a\sqrt{6a}$ **25.** $4b\sqrt[3]{4b}$ **27.** $54a^3b\sqrt{2b}$ **29.** $2c\sqrt[3]{18cd^2}$

31. (a) 1.62 sec; (b) 1.99 sec; (c) 2.20 sec **33.** $9\sqrt[3]{9n^2}$

EXERCISE SET 6.5, pp. 258–259

1. $8\sqrt{3}$ **3.** $3\sqrt[3]{5}$ **5.** $13\sqrt[3]{y}$ **7.** $7\sqrt{2}$ **9.** $6\sqrt[3]{3}$ **11.** $21\sqrt{3}$ **13.** $38\sqrt{5}$ **15.** $122\sqrt{2}$

17. $9\sqrt[3]{2}$ **19.** $29\sqrt{2}$ **21.** $(1 + 6a)\sqrt{5a}$ **23.** $(2 - x)\sqrt[3]{3x}$ **25.** $3\sqrt{2y - 2}$ **27.** $(x + 3)\sqrt{x - 1}$

31. $(25x - 30y - 9xy)\sqrt{2xy}$ **33.** $(7x^2 - 2y^2)\sqrt{x + y}$

EXERCISE SET 6.6, pp. 260–261

1. $2\sqrt{6} - 18$ **3.** $\sqrt{6} - \sqrt{10}$ **5.** $2\sqrt{15} - 6\sqrt{3}$ **7.** -6 **9.** $3a\sqrt[3]{2}$ **11.** 1 **13.** -12

15. $a - b$ **17.** $1 + \sqrt{5}$ **19.** $7 + 3\sqrt{3}$ **21.** -6 **23.** $a + \sqrt{3a} + \sqrt{2a} + \sqrt{6}$

25. $2\sqrt[3]{9} - 3\sqrt[3]{6} - 2\sqrt[3]{4}$ **27.** $7 + 4\sqrt{3}$ **29.** (a) 19.485281; (b) 19.485281 **31.** $2x - 2\sqrt{x^2 - 4}$

33. $\sqrt[3]{y} - y$ **35.** $12 + 6\sqrt{3}$ **37.** $16(a + 3) + 38a\sqrt{3(a + 3)} - 15a^2$ **39.** $2\sqrt{34}$

EXERCISE SET 6.7, pp. 265–267

1. $\dfrac{\sqrt{30}}{5}$ 3. $\dfrac{\sqrt{70}}{7}$ 5. $\dfrac{2\sqrt{15}}{5}$ 7. $\dfrac{2\sqrt[3]{6}}{3}$ 9. $\dfrac{\sqrt[3]{75ac^2}}{5c}$ 11. $\dfrac{y\sqrt[3]{9yx^2}}{3x^2}$ 13. $\dfrac{\sqrt[3]{x^2y^2}}{xy}$ 15. $\dfrac{7}{\sqrt{21x}}$ 17. $\dfrac{2}{\sqrt{6}}$

19. $\dfrac{52}{3\sqrt{91}}$ 21. $\dfrac{7}{\sqrt[3]{98}}$ 23. $\dfrac{7x}{\sqrt{21xy}}$ 25. $\dfrac{5y^2}{x\sqrt[3]{150x^2y^2}}$ 27. $\dfrac{ab}{3\sqrt{ab}}$ 29. $\dfrac{5(8+\sqrt{6})}{58}$

31. $-2\sqrt{7}(\sqrt{5}+\sqrt{3})$ 33. $-\dfrac{\sqrt{15}+20-6\sqrt{2}-8\sqrt{30}}{77}$ 35. $\dfrac{x-2\sqrt{xy}+y}{x-y}$ 37. $\dfrac{3\sqrt{6}+4}{2}$

39. $\dfrac{-11}{4(\sqrt{3}-5)}$ 41. $\dfrac{-22}{\sqrt{6}+5\sqrt{2}+5\sqrt{3}+25}$ 43. $\dfrac{x-y}{x+2\sqrt{xy}+y}$ 45. $\dfrac{7}{43\sqrt{2}+66}$ 47. 6.708

49. $\dfrac{\sqrt{10}+2\sqrt{5}-2\sqrt{3}}{10}$ 51. $\dfrac{b^2+\sqrt{b}}{b^2+b+1}$ 53. $6a\sqrt[3]{36ab^2}$ 55. $\dfrac{x-19}{x+31+10\sqrt{x+6}}$

57. $\left(\dfrac{5x+4y-3}{xy}\right)\sqrt{xy}$ 59. $\dfrac{7\sqrt{3}}{39}$

EXERCISE SET 6.8, pp. 272–273

1. $\sqrt[4]{x}$ 3. 2   5. $\sqrt[5]{a^2b^2}$ 7. 8   9. $20^{1/3}$ 11. $(xy^2z)^{1/5}$ 13. $(3mn)^{3/2}$ 15. $(8x^2y)^{5/7}$ 17. $\dfrac{1}{x^{1/3}}$

19. $x^{2/3}$ 21. $5^{7/8}$ 23. $7^{1/4}$ 25. $8.3^{7/20}$ 27. $10^{3/20}$ 29. $\sqrt[3]{a^2}$ 31. $2y^2$ 33. $2\sqrt[4]{2}$ 35. $\sqrt[3]{2x}$

37. $2c^2d^3$ 39. $\dfrac{m^2n^4}{2}$ 41. $\sqrt[4]{r^2s}$ 43. $3ab^3$ 45. $\sqrt[6]{200}$ 47. $\sqrt[6]{4x^5}$ 49. $\sqrt[6]{x^5-4x^4+4x^3}$

51. $\sqrt[6]{a+b}$ 53. $\sqrt[12]{a^8b^9}$ 55. $\sqrt[4]{st^4}=t\sqrt[4]{s}$ 57. $\sqrt[10]{x^9y^7}$ 59. $(a+b)\sqrt[12]{(a+b)^{11}}$ 61. $343w^4$

63. $\sqrt[3]{9}+\sqrt[3]{6}+\sqrt[3]{4}$ 65. $x^5y^7z\sqrt[8]{yz^3}$

EXERCISE SET 6.9, p. 277

1. 168   3. $\dfrac{80}{3}$ 5. -27 7. No solution   9. -6 11. 5   13. 9   15. 7   17. $\dfrac{80}{9}$ 19. 6, 2

21. $-\dfrac{8}{9}$ 23. 6912   25. 0   27. $-1, 6$ 29. 2   31. $0, \dfrac{125}{4}$ 33. 2   35. 3

EXERCISE SET 6.10, pp. 282–283

1. $\sqrt{15}\,i$ 3. $4i$ 5. $-2\sqrt{3}\,i$ 7. $8+i$ 9. $9-5i$ 11. $7+4i$ 13. $-2-3i$ 15. $-1+i$

17. $11+6i$ 19. $1+5i$ 21. $18+14i$ 23. $38+9i$ 25. $2-46i$ 27. -1 29. $-125i$

31. $5-12i$ 33. $-5+12i$ 35. $-5-12i$ 37. $\dfrac{8}{5}+\dfrac{1}{5}i$ 39. $-i$ 41. $-\dfrac{21}{49}-\dfrac{56}{49}i$, or $-\dfrac{3}{7}-\dfrac{8}{7}i$

43. Yes **45.** No **47.** $-4 - 8i$ **49.** $-3 - 4i$ **51.** $-88i$ **53.** 8 **55.** $\frac{3}{5} + \frac{9}{5}i$ **57.** 1

59. $-\frac{7}{13} + \frac{2\sqrt{30}}{13}i$

TEST OR REVIEW—CHAPTER 6, pp. 284–285

1. $\frac{x^{-1}y^{-1}}{108}$, or $\frac{1}{108xy}$ **2.** 1.344×10^{-4} **3.** $|6y|$ or $6|y|$ **4.** $|x + 5|$ **5.** -2 **6.** 4 **7.** $2x^3$ **8.** 12.2

9. $x\sqrt{5x}$ **10.** $4a\sqrt[3]{4ab^2}$ **11.** $38\sqrt{2}$ **12.** -20 **13.** $\frac{a + 2\sqrt{ab} + b}{a - b}$ **14.** $(5xy^2)^{5/2}$

15. $\sqrt[4]{2x^2y - 12xy + 18y}$ **16.** 7 **17.** $3\sqrt{2}i$ **18.** $7 + 5i$ **19.** No **20.** $-\frac{77}{50} + \frac{14}{50}i$

CHAPTER 7

EXERCISE SET 7.1, pp. 293–294

1. $\pm\sqrt{5}$ **3.** $\pm\frac{2}{5}i$ **5.** $\pm\frac{\sqrt{6}}{2}$ **7.** $0, -\frac{9}{14}$ **9.** $0, -2$ **11.** $5, 1$ **13.** $\frac{2}{3}, -\frac{1}{2}$ **15.** $-5, -3$

17. $-2, -1$ **19.** $10, 5$ **21.** 2 in. **23.** Length is 6 ft; width is 2 ft **25.** Length is 24 yd;

width is 12 yd **27.** 10 ft, 24 ft **29.** 4 ft, 3 ft **31.** ± 8 **33.** $-8, -\frac{10}{3}, 0, \frac{7}{2}$

35. A: 15 km/h, B: 8 km/h

EXERCISE SET 7.2, p. 298

1. $-3 \pm \sqrt{5}$ **3.** $-1, -\frac{5}{3}$ **5.** $\frac{1 \pm i\sqrt{3}}{2}$ **7.** $2 \pm 3i$ **9.** $\frac{-3 \pm \sqrt{41}}{2}$ **11.** $-1 \pm 2i$ **13.** $0, -1$

15. $\frac{3}{4}, -2$ **17.** $\frac{1 \pm 3i}{2}$ **19.** $1.3, -5.3$ **21.** $2.8, -1.3$ **23.** $0.5700731, -0.7973458$

25. $\frac{-y \pm \sqrt{-47y^2 + 108}}{6}$ **27.** $\frac{1 \pm \sqrt{1 + 8\sqrt{5}}}{4}$ **29.** $\frac{-i \pm i\sqrt{1 + 4i}}{2}$

31. $x = \frac{-y \pm \sqrt{-47y^2 + 108}}{6}$

EXERCISE SET 7.3, pp. 303–305

1. $6, -4$ **3.** $\pm\frac{3}{2}$ **5.** $25, -20$ **7.** $4, -1$ **9.** $\pm\sqrt{37}$ **11.** First part, 40 mph; second part, 35 mph

13. 35 mph   **15.** A, 350 km/h; B, 400 km/h   **17.** 6 hours   **19.** 3.24 km/h

21. A, 24.0 hours; B, 51.0 hours   **23.** $\dfrac{-3 \pm \sqrt{57}}{2}$ **25.** $\dfrac{-i \pm \sqrt{23}}{4}$ **27.** \$2.50

EXERCISE SET 7.4, pp. 307–308

1. One real   **3.** Two nonreal   **5.** Two real   **7.** One real   **9.** Two nonreal   **11.** Two real

13. Two real   **15.** One real   **17.** $x^2 + 2x - 99 = 0$ **19.** $x^2 - 14x + 49 = 0$

21. $25x^2 - 20x - 12 = 0$ **23.** $4x^2 - 2(c + d)x + cd = 0$ **25.** $x^2 - 4\sqrt{2}x + 6 = 0$

27. (a) $k = 2$; (b) $\dfrac{11}{2}$

29. The solutions of $ax^2 + bx + c = 0$ are given by $x = \dfrac{-b \pm \sqrt{b^2 - 4ac}}{2a}$.

(a) $\dfrac{-b + \sqrt{b^2 - 4ac}}{2a} + \dfrac{-b - \sqrt{b^2 - 4ac}}{2a} = \dfrac{-2b}{2a} = -\dfrac{b}{a}$;

(b) $\dfrac{-b + \sqrt{b^2 - 4ac}}{2a} \cdot \dfrac{-b - \sqrt{b^2 - 4ac}}{2a} = \dfrac{(-b)^2 - (\sqrt{b^2 - 4ac})^2}{4a^2} = \dfrac{b^2 - b^2 + 4ac}{4a^2} = \dfrac{4ac}{4a^2} = \dfrac{c}{a}$.

31. (a) $k = -\dfrac{3}{5}$; (b) $-\dfrac{1}{3}$ **33.** $x^2 - \sqrt{3}x + 8 = 0$ **35.** $h = -36, k = 15$ **37.** $|k| = 2$ **39.** $k = 6$

EXERCISE SET 7.5, pp. 311–312

1. $s = \sqrt{\dfrac{A}{6}}$ **3.** $r = \sqrt{\dfrac{Gm_1 m_2}{F}}$ **5.** $c = \sqrt{\dfrac{E}{m}}$ **7.** $b = \sqrt{c^2 - a^2}$ **9.** $k = \dfrac{3 + \sqrt{9 + 8N}}{2}$

11. $r = \dfrac{-\pi h + \sqrt{\pi^2 h^2 + 2\pi A}}{2\pi}$

13. $g = \dfrac{4\pi^2 l}{T^2}$ **15.** $D = \sqrt{\dfrac{32LV}{g(P_1 - P_2)}}$ **17.** $v = \dfrac{c\sqrt{m^2 - m_0^2}}{m}$ **19.** $t = \dfrac{10 \pm \sqrt{102 - s}}{2}$, 1.4 sec

EXERCISE SET 7.6, pp. 314–315

1. 81, 1   **3.** $\pm\sqrt{5}$ **5.** 7, $-1, 5, 1$ **7.** 4, 1, 6, -1 **9.** $\pm\sqrt{2 + \sqrt{6}}, \pm\sqrt{2 - \sqrt{6}}$ **11.** $\dfrac{1}{3}, -\dfrac{1}{2}$

13. 2, -1 **15.** 28.5   **17.** $3 \pm \sqrt{10}, -1 \pm \sqrt{2}$ **19.** 259

EXERCISE SET 7.7, pp. 318–320

1. $y = 15x^2$ **3.** $y = \dfrac{0.0015}{x^2}$ **5.** $y = xz$ **7.** $y = \dfrac{3}{10}xz^2$ **9.** $y = \dfrac{xz}{5wp}$ **11.** $355\dfrac{5}{9}$ ft   **13.** 94.03 kg

15. 97   **17.** y is tripled   **19.** (a) $N = \dfrac{0.001 P_1 P_2}{d^2}$; (b) 1173 km

EXERCISE SET 7.8, p. 328

1.

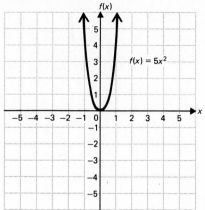

$f(x) = 5x^2$

3.

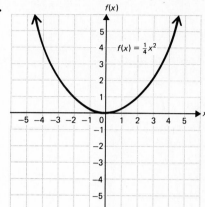

$f(x) = \frac{1}{4}x^2$

5.

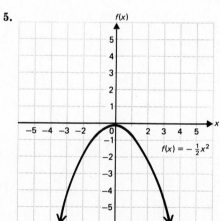

$f(x) = -\frac{1}{2}x^2$

7.

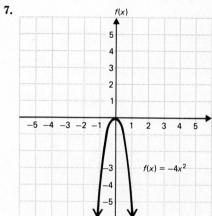

$f(x) = -4x^2$

9.

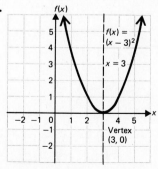

$f(x) = (x - 3)^2$

$x = 3$

Vertex
(3, 0)

11.

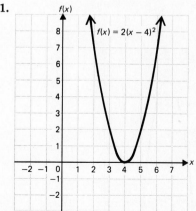

$f(x) = 2(x - 4)^2$

13.

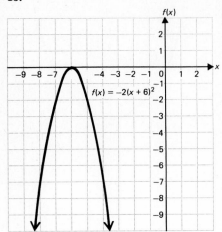

15.

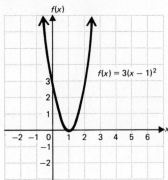

17.

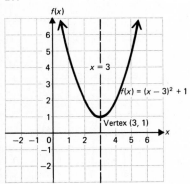

19.

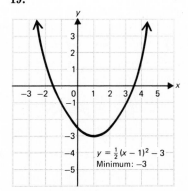

21.

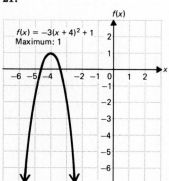

23.

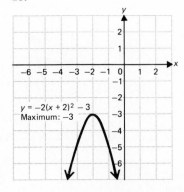

25.

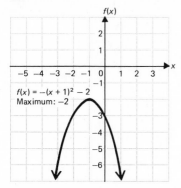

$f(x) = -(x + 1)^2 - 2$
Maximum: -2

EXERCISE SET 7.9, p. 332

1.

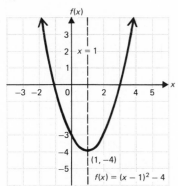

$x = 1$

$(1, -4)$

$f(x) = (x - 1)^2 - 4$

3.

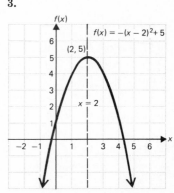

$f(x) = -(x - 2)^2 + 5$

$(2, 5)$

$x = 2$

5.

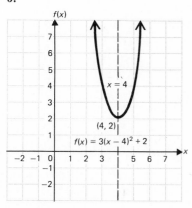

$x = 4$

$(4, 2)$

$f(x) = 3(x - 4)^2 + 2$

7.

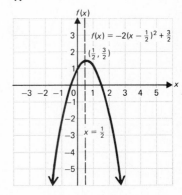

$f(x) = -2(x - \frac{1}{2})^2 + \frac{3}{2}$

$(\frac{1}{2}, \frac{3}{2})$

$x = \frac{1}{2}$

9. $(2 - \sqrt{3}, 0), (2 + \sqrt{3}, 0)$ **11.** $(-1, 0), (3, 0)$ **13.** None   **15.** $\left(-\dfrac{3}{2}, 0\right)$

17. $(3.1585884, 0), (-1.0103267, 0)$ **19.** $x = \dfrac{p + q}{2}$

EXERCISE SET 7.10, pp. 335–336

1. 19 ft by 19 ft; 361 ft^2 **3.** 121; 11 and 11   **5.** -4; 2 and -2 **7.** $-\dfrac{25}{4}; \dfrac{5}{2}$ and $-\dfrac{5}{2}$

9. $f(x) = 2x^2 + 3x - 1$ **11.** $f(x) = -3x^2 + 13x - 5$ **13.** (a) $f(x) = -4x^2 + 40x + 2$; (b) \$98

15. (a) $f(x) = 0.05x^2 - 10.5x + 650$; (b) 250   **17.** Minimum: -7.0234008

19. $b = 19$ cm, $h = 19$ cm, $A = 180.5$ cm^2 **21.** 4050 ft^2

TEST OR REVIEW—CHAPTER 7, p. 337

1. 9, 2   **2.** Length is 8.5 in.; width is 4 in.   **3.** $\dfrac{-1 \pm i\sqrt{3}}{2}$ **4.** $0.4, -4.4$ **5.** $0, 2$ **6.** 2 hours

7. Two nonreal   **8.** $x^2 - 4\sqrt{3}x + 9 = 0$ **9.** $r = \sqrt{\dfrac{3V}{\pi} - R^2}$ **10.** $\pm\sqrt{\dfrac{5 + \sqrt{5}}{2}}, \pm\sqrt{\dfrac{5 - \sqrt{5}}{2}}$

11. 6.664 cm^2

12.

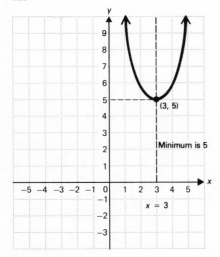

13. $2(x^2 + 2x + 1) - 8$, or $2(x + 1)^2 - 8$

14. $(3, 0)$ and $(-2, 0)$

15. -16; 4 and -4

16. $\dfrac{x^2}{5} - \dfrac{3x}{5} = f(x)$

CHAPTER 8

EXERCISE SET 8.1, pp. 343–344

1. 5   **3.** 12   **5.** 10   **7.** $\sqrt{74} \approx 8.602$ **9.** $\sqrt{2} \approx 1.414$ **11.** $2\sqrt{3} \approx 3.464$ **13.** 5

15. $3\sqrt{2} \approx 4.242$ **17.** $4\sqrt{2} \approx 5.656$ **19.** Yes

21. Let (a, c) and (b, c) be two points on a horizontal line. Then $d = \sqrt{(a - b)^2 + (c - c)^2} = \sqrt{(a - b)^2} = |a - b|$. Let (c, a) and (c, b) be two points on a vertical line. Then $d = \sqrt{(c - c)^2 + (a - b)^2} = \sqrt{(a - b)^2} = |a - b|$.

23. Let $P_1(x_1 y_1)$ and $P_2(x_2 y_2)$ be the endpoints of a segment, and let M be the midpoint $\left(\dfrac{x_1 + x_2}{2}, \dfrac{y_1 + y_2}{2}\right)$. If the $d(P_1 M) = d(M P_2) = \dfrac{1}{2} d(P_1 P_2)$, then M is the midpoint of $P_1 P_2$.

$$d(P_1 M) = \sqrt{\left(\frac{x_1 + x_2}{2} - x_1\right)^2 + \left(\frac{y_1 + y_2}{2} - y_1\right)^2} = \frac{1}{2}\sqrt{(x_2 - x_1)^2 + (y_2 - y_1)^2}.$$

$$d(M P_2) = \sqrt{\left(x_2 - \frac{x_1 + x_2}{2}\right)^2 + \left(y_2 - \frac{y_1 + y_2}{2}\right)^2} = \frac{1}{2}\sqrt{(x_2 - x_1)^2 + (y_2 - y_1)^2}.$$

$$d(P_1 P_2) = \sqrt{(x_2 - x_1)^2 + (y_2 - y_1)^2}.$$

$\therefore$ M is the midpoint of $P_1 P_2$.

25. Yes

EXERCISE SET 8.2, pp. 348–349

1.

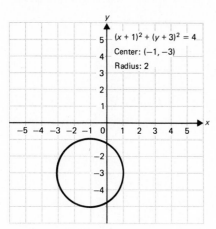

$(x + 1)^2 + (y + 3)^2 = 4$

Center: $(-1, -3)$

Radius: 2

3.

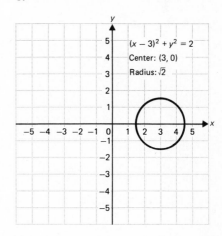

$(x - 3)^2 + y^2 = 2$

Center: $(3, 0)$

Radius: $\sqrt{2}$

5.

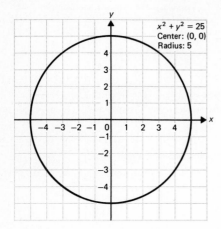

7. $x^2 + y^2 = 49$ **9.** $(x + 2)^2 + (y - 7)^2 = 5$ **11.** $(-4, 3), r = 2\sqrt{10}$ **13.** $(4, -1), r = 2$

15. $(2, 0), r = 2$

17.

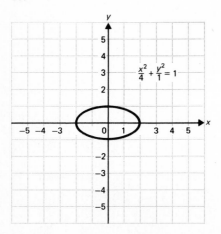

19.

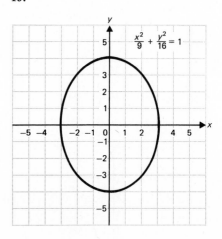

21. (a) $x^2 + y^2 = 2$; (b) $(x + 4)^2 + (y - 1)^2 = 72$, (c) $(x + 3)^2 + (y + 2)^2 = 9$;
(d) $(x - 3)^2 + y^2 = 25$

23. (a) $\dfrac{x^2}{64} + \dfrac{y^2}{4} = 1$; (b) $\dfrac{(x - 5)^2}{25} + \dfrac{(y - 2)^2}{9} = 1$

EXERCISE SET 8.3, pp. 357-358

1.

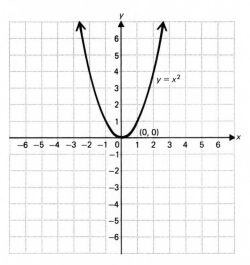

3.

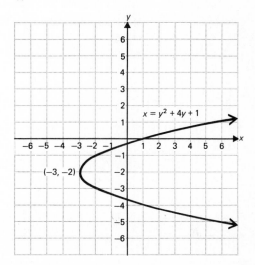

5.

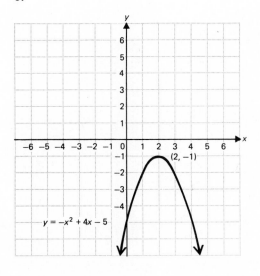

7.

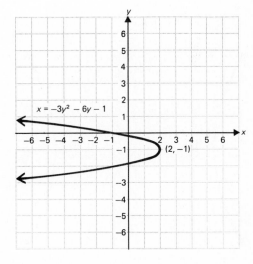

9.

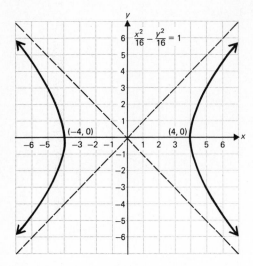

$$\frac{x^2}{16} - \frac{y^2}{16} = 1$$

$(-4, 0)$ $(4, 0)$

11.

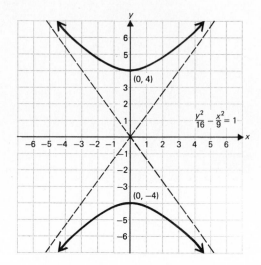

$(0, 4)$

$$\frac{y^2}{16} - \frac{x^2}{9} = 1$$

$(0, -4)$

13.

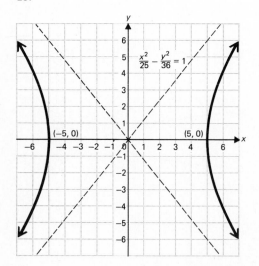

$$\frac{x^2}{25} - \frac{y^2}{36} = 1$$

$(-5, 0)$ $(5, 0)$

15.

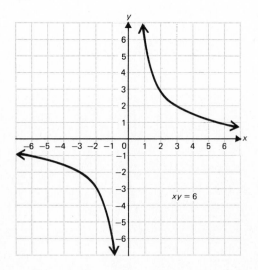

$xy = 6$

17.

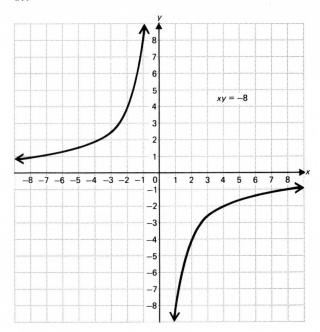

$xy = -8$

19.

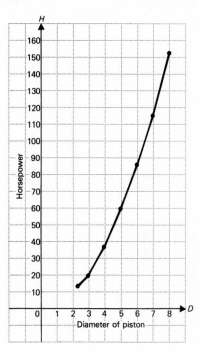

21. (a) Circle; (b) parabola; (c) hyperbola; (d) parabola; (e) ellipse; (f) circle

EXERCISE SET 8.4, p. 362

1. $(-4, -3), (3, 4)$ **3.** $(0, 2), (3, 0)$ **5.** $(-2, 1)$ **7.** $\left(\dfrac{5 + \sqrt{70}}{3}, \dfrac{-1 + \sqrt{70}}{3}\right), \left(\dfrac{5 - \sqrt{70}}{3}, \dfrac{-1 - \sqrt{70}}{3}\right)$

9. 10 and 3   **11.** Length is 8 ft; width is 6 ft   **13.** $\left(x + \dfrac{5}{13}\right)^2 + \left(y - \dfrac{32}{13}\right)^2 = \dfrac{5365}{169}$

15. $(x + 2)^2 + (y - 1)^2 = 4$

EXERCISE SET 8.5, p. 365

1. $(-5, 0), (4, 3), (4, -3)$ **3.** $(3, 0), (-3, 0)$ **5.** $(1, 2), (-1, -2), (2, 1), (-2, -1)$

7. $(3, 2), (-3, -2)$ **9.** 13 and 12, -13 and -12 **11.** Length is $\sqrt{3}$ m; width is 1 m

13. $\dfrac{1}{2}, \dfrac{1}{4}$ **15.** 10 in. by 7 in. by 5 in.

TEST OR REVIEW—CHAPTER 8, pp. 365–366

1. $\sqrt{14} \approx 3.742$ **2.** $6\sqrt{5} \approx 13.416$ **3.** $(x - 2)^2 + (y + 7)^2 = 18$ **4.** Center: $(-4, 5)$; Radius: 7

5.

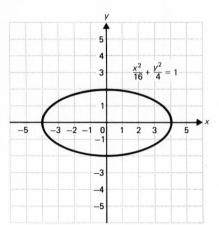

$$\frac{x^2}{16} + \frac{y^2}{4} = 1$$

6.

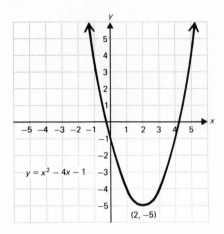

$y = x^2 - 4x - 1$

$(2, -5)$

7.

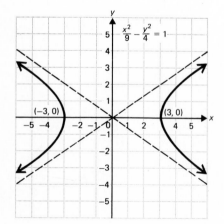

$$\frac{x^2}{9} - \frac{y^2}{4} = 1$$

$(-3, 0)$ $(3, 0)$

8.

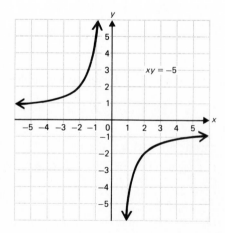

$xy = -5$

CHAPTER 9

EXERCISE SET 9.1, pp. 372–373

1.

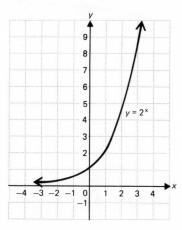

3.

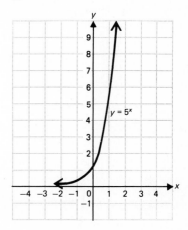

5.

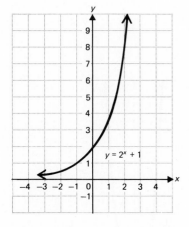

7.

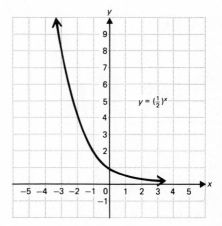

9.

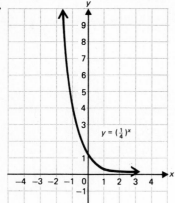

$y = (\frac{1}{4})^x$

11.

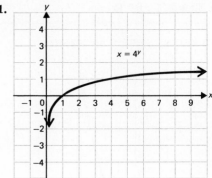

$x = 4^y$

13.

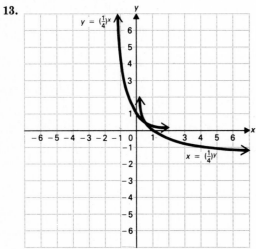

$y = (\frac{1}{4})^x$

$x = (\frac{1}{4})^y$

15.

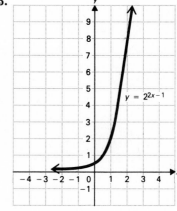

$y = 2^{2x-1}$

17.

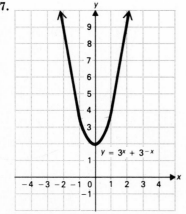

$y = 3^x + 3^{-x}$

19.

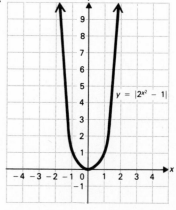

$y = |2^{x^2} - 1|$

EXERCISE SET 9.2, pp. 376–377

1.

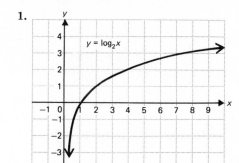

3. $3 = \log_{10} 1000$ **5.** $-3 = \log_5 \dfrac{1}{125}$ **7.** $\dfrac{1}{3} = \log_8 2$

9. $0.3010 = \log_{10} 2$ **11.** $3^t = 8$ **13.** $5^2 = 25$

15. $10^{-1} = 0.1$ **17.** $10^{0.845} = 7$ **19.** 9 **21.** 4 **23.** $\dfrac{1}{2}$

25. 2 **27.** 2 **29.** -1 **31.** 0 **33.** 4 **35.** -2 **37.** 0

39.

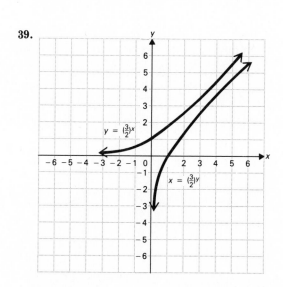

41.

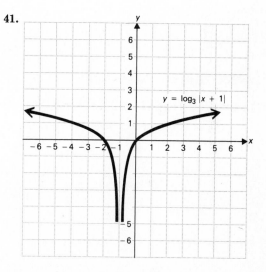

43. 25 **45.** $\dfrac{\sqrt{5}}{25}$ **47.** 6 **49.** 1296 **51.** $\dfrac{2}{3}$ **53.** x^2 **55.** 0

57. 1 **59.** -2

EXERCISE SET 9.3, pp. 381–382

1. $\log_2 32 + \log_2 8$ **3.** $\log_4 64 + \log_4 16$ **5.** $\log_c B + \log_c x$ **7.** $\log_a (6 \cdot 70)$ **9.** $\log_c (K \cdot y)$

11. $3 \log_a x$ **13.** $\log_a 67 - \log_a 5$ **15.** $\log_a \dfrac{15}{7}$ **17.** $2 \log_a x + 3 \log_a y + \log_a z$

19. $\log_b x + 2 \log_b y - 3 \log_b z$ **21.** $\log_a x^{2/3} y^{-1/2}$ **23.** $\log_a \dfrac{2x^4}{y^3}$ **25.** $\log_a (x + 2)$ **27.** True

29. False   **31.** $\log_a (x^6 - x^4y^2 + x^2y^4 - y^6)$ **33.** $\frac{1}{2} \log_a (1 - s) + \frac{1}{2} \log_a (1 + s)$ **35.** -0.544

37. 1.322

EXERCISE SET 9.4, pp. 389–390

1. $\log (7 \times 2) = \log 7 + \log 2 \approx 0.8451 + 0.3010 \approx 1.1461$, antilog $1.1461 \approx 14$

3. $\log \frac{12}{3} = \log 12 - \log 3 \approx 1.0792 - 0.4771 \approx 0.6021$, antilog $0.6021 \approx 4$ **5.** 0.3909 **7.** 0.7251

9. 0.0294 **11.** 2.5403 **13.** 1.7202 **15.** 3.5877 **17.** $0.8062 + (-1)$ **19.** $0.6064 + (-5)$ **21.** 2330

23. 18 **25.** 0.613 **27.** 0.00346 **29.** 25.2 **31.** 0.00973 **33.** 0.000613

EXERCISE SET 9.5, p. 393

1. 189.6 **3.** 272.4 **5.** 8.770 **7.** 3.846 **9.** 3.637 **11.** 0.2560

EXERCISE SET 9.6, pp. 396–397

1. 1.6194 **3.** 0.4689 **5.** 2.8130 **7.** $9.1538 - 10$ **9.** $7.6291 - 10$ **11.** $9.2494 - 10$ **13.** 2.7786

15. 43.15 **17.** 4.444 **19.** 1.082 **21.** 0.7185 **23.** 0.002587 **25.** 0.01589 **27.** 4.834

EXERCISE SET 9.7, pp. 399–400

1. 3.170 **3.** 3.32193 **5.** $\frac{5}{2}$ **7.** $0.1538 + (-1)$ **9.** $0.6291 + (-3)$ **11.** $0.2494 + (-1)$ **13.** $\frac{1}{3}$

15. 3 **17.** 7.27 yr   **19.** 6.7 **21.** -4 **23.** 4 **25.** 2 **27.** $\pm\sqrt{34}$ **29.** $10^{100,000}$ **31.** $\frac{9}{8}$ **33.** 11

35. $\frac{1}{100,000}, 100,000$ **37.** $-7, 3$ **39.** $\frac{1}{10}, \frac{1}{1000}$ **41.** $-\frac{1}{3}$ **43.** $\frac{3}{2}$ **45.** 38

47. $(21, 100, 10)$ or $(-25, 169, -13)$

TEST OR REVIEW—CHAPTER 9, p. 400

1.

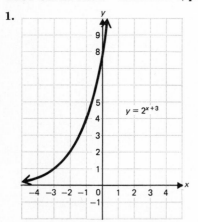

2.

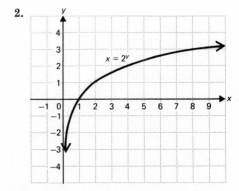

3.

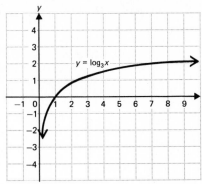

$y = \log_3 x$

4. $-3 = \log_4 x$ **5.** $4^2 = 16$

6. $3 \log a + \dfrac{1}{2} \log b - 2 \log c$ **7.** $\log_a \dfrac{z^2 \sqrt[3]{x}}{y^3}$

8. 0.008358 **9.** $0.0899 + (-2)$ **10.** $445{,}000$

11. 0.0000001825 **12.** $0.0913 + (-2)$ **13.** 0.0937

14. 9

CHAPTER 10

EXERCISE SET 10.1, pp. 406–407

1.

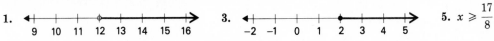

3.   **5.** $x \geqslant \dfrac{17}{8}$

7. $x \leqslant 3$ **9.**

11.   **13.** $-1 < x \leqslant 6$ **15.** $x \leqslant -9 \text{ or } x > -5$

17. $x \leqslant -8.92 \text{ or } x > -4.8003$ **19.** $-\dfrac{1}{6} < x \leqslant \dfrac{37}{6}$ **21.** $x \leqslant \dfrac{1}{2}$ **23.** $10 < x \leqslant 18$

25. No solution

EXERCISE SET 10.2, pp. 411–413

1. $\{0, 2, 4, 6, 8\}$ **3.** $\{11, 13, 15, \cdots\}$ **5.** $\{x \mid x > \sqrt{3}\}$ **7.** $\{-5, 5\}$ **9.** $\{3, -2\}$ **11.** $\{x \mid x > -5\}$

13. True   **15.** False   **17.** False   **19.** True   **21.** True   **23.** True   **25.** $\{-4, 7\}, \{-4\}, \{7\}, \emptyset$

27. $\{6, 8\}$ **29.** $\{1, 4, 5, 6, 7, 8, 11\}$ **31.** $\emptyset$ **33.** $\{1, 2, 3, 4\}$ **35.** $\{4, 8, 11\}$

37.

39.   **41.** 2^n

43. True   **45.** True   **47.** True   **49.** True   **51.** False   **53.** False   **55.** Ø   **57.** Ø

59. $\{-2, 0, 2\}$ **61.** $\{x|x \leqslant -9 \text{ or } x \geqslant 5\}$ **63.** $\{\ldots -14, -13, -12, -11, 5, 6, 7, 8, \ldots\}$

65. $\left\{x \left| -\dfrac{10}{3} < x \leqslant 3\right.\right\}$

EXERCISE SET 10.3, pp. 417–418

1. $\{x \mid x \leqslant 6\}$ **3.** $\left\{y \mid y < -\dfrac{1}{8}\right\}$ **5.** $\left\{x \mid x < -\dfrac{1}{56}\right\}$ **7.** $\left\{x \mid x \leqslant -\dfrac{2}{3}\right\}$ **9.** $\{x \mid 2 < x \leqslant 4\}$

11. 620.5 or less   **13.** \$20,000   **15.** Gross sales greater than \$15,000

17. $\{m|m \leqslant 7\}$ **19.** $\{x|-4 < x \leqslant 1\}$ **21.** Ø   **23.** $\left\{x \left| \dfrac{2}{5} \leqslant x \leqslant \dfrac{4}{5}\right.\right\}$ **25.** $\left\{x \left| -\dfrac{1}{8} < x < \dfrac{1}{2}\right.\right\}$

27. True   **29.** False   **31.** False

EXERCISE SET 10.4, p. 422

1. 34   **3.** 11   **5.** 33   **7.** 2   **9.** $3, -3$,

11. $-3 < x < 3$

13. $x \leqslant -2 \text{ or } x \geqslant 2$

15. $-9 < x < 15$

17. $-\dfrac{7}{2} \leqslant x \leqslant \dfrac{1}{2}$

19. $y < 1 \text{ or } y > 6$

21. $x \leqslant -\dfrac{5}{4} \text{ or } x \geqslant \dfrac{23}{4}$ **23.** $-1 \leqslant x \leqslant 2$

25. $\left\{-\dfrac{7}{3}, \dfrac{7}{3}\right\}$ **27.** All reals   **29.** $\left\{x \left| -\dfrac{13}{54} < x < -\dfrac{7}{54}\right.\right\}$ **31.** All reals   **33.** $\left\{-\dfrac{1}{4}, 1\right\}$

EXERCISE SET 10.5, pp. 427–428

1. Yes   **3.** Yes

5.

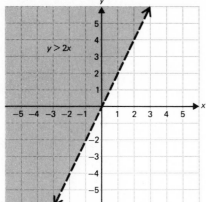

$y > 2x$

7.

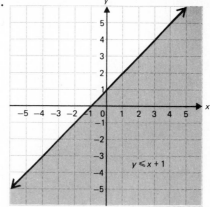

$y \leqslant x + 1$

9.

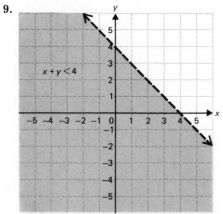

$x + y < 4$

11.

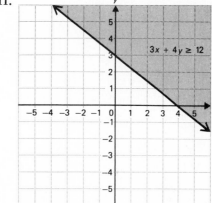

$3x + 4y \geq 12$

13.

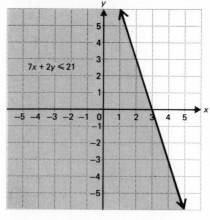

$7x + 2y \leqslant 21$

15.

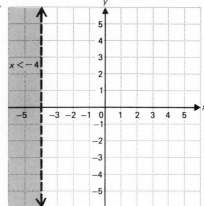

$x < -4$

17.

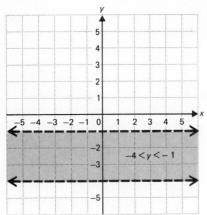

$-4 < y < -1$

19.

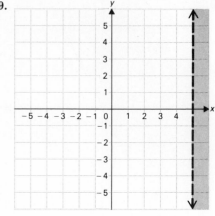

21.

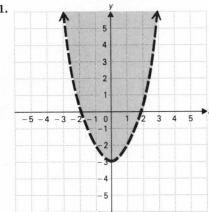

23. $\left\{ x \mid -2 < x < \dfrac{3}{2} \right\}$

25.

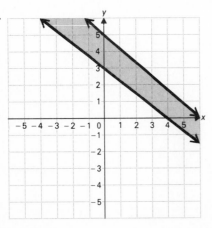

27.

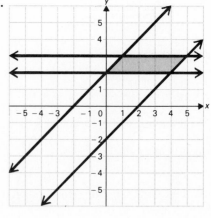

TEST OR REVIEW—CHAPTER 10, pp. 428–429

1.

$$\begin{array}{c} \end{array}$$

     -1 0 1 $\frac{4}{3}$ 2 3$\frac{10}{3}$ 4 5

2. $\{-2, 3\}$ **3.** $\{x \mid x > -5\}$ **4.** $\{x \mid x > -3\}$ **5.** a, b, c

6. $\{3, 5\}$ **7.** $\{1, 3, 5, 7, 9, 11, 13\}$ **8.**

     -6 -5 -4 -3 -2 -1 0 1 2 3 4 5 6

9.

     -6 -5 -4 -3 -2 -1 0 1 2 3 4 5 6

10. $-\dfrac{5}{3} < x < 3$

     -4 -3 -2 -1 0 1 2 3 4 5

11. $x \leqslant -\dfrac{3}{2}$ or $x \geqslant \dfrac{9}{2}$

     -7 -6 -5 -4 -3 -2 -1 0 1 2 3 4 5 6 7 8

12.

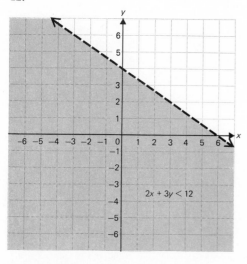

$2x + 3y < 12$

CHAPTER 11

EXERCISE SET 11.1, pp. 437–438

1. (a) 0; (b) 1; (c) 57; (d) $5t^2 + 4t$; (e) $5t^2 - 6t + 1$; (f) $10a + 5h + 4$ **3.** (a) 5; (b) -2; (c) -4;

(d) $4|y| + 6y$; (e) $2|a + h| + 3a + 3h$; (f) $\dfrac{2|a + h| + 3h - 2|a|}{h}$ **5.** (a) 3.14977; (b) 55.73147; (c) 3178.20675;

(d) 1166.70323   **7.** (a) $\dfrac{2}{3}$; (b) $\dfrac{10}{9}$; (c) 0; (d) Undefined   **9.** $f^{-1}(x) = \dfrac{x - 2}{3}$ **11.** $g^{-1}(x) = \dfrac{3x - 1}{2}$

13. $g^{-1}(x) = \sqrt{\dfrac{x-3}{2}}$ **15.** $f^{-1}(x) = \log_{10} x$

17.

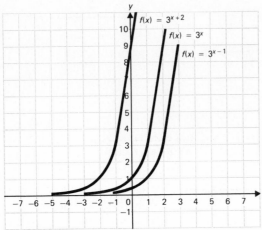

19.

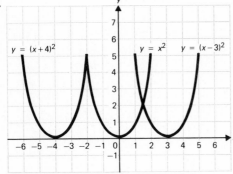

21.

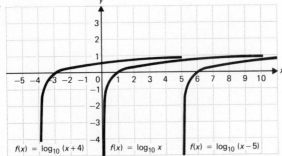

23.

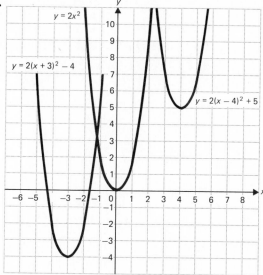

$y = 2x^2$

$y = 2(x + 3)^2 - 4$

$y = 2(x - 4)^2 + 5$

25. $\dfrac{-1}{x(x + h)}$ **27.** $\dfrac{1}{\sqrt{x + h} + \sqrt{x}}$

EXERCISE SET 11.2, pp. 440–441

1. 4, 7, 10, 13; 31; 46   **3.** $1, \dfrac{1}{2}, \dfrac{1}{3}, \dfrac{1}{4}; \dfrac{1}{10}; \dfrac{1}{15}$ **5.** $\dfrac{1}{2}, \dfrac{2}{3}, \dfrac{3}{4}, \dfrac{4}{5}; \dfrac{10}{11}; \dfrac{15}{16}$ **7.** $2n - 1$ **9.** $\dfrac{n + 1}{n + 2}$

11. $(\sqrt{3})^n$ **13.** $n - 1$ **15.** 28   **17.** 30   **19.** $\dfrac{3}{2}, \dfrac{3}{2}, \dfrac{3}{2}, \dfrac{3}{2}, \dfrac{3}{2}$

21. 2, 2.25, 2.37037, 2.441406, 2.488320, 2.521626

EXERCISE SET 11.3, pp. 444–445

1. 46   **3.** 27th   **5.** 101   **7.** 5   **9.** 28   **11.** $a_1 = 8; d = -3; 8, 5, 2, -1, -4$ **13.** 670   **15.** 2500
17. 990   **19.** 465   **21.** n^2

EXERCISE SET 11.4, pp. 449–450

1. $\dfrac{81}{64}$ **3.** 1275   **5.** $\dfrac{547}{18}$ **7.** \$933.12   **9.** $\dfrac{1}{256}$ ft   **11.** \$5866.60   **13.** 8   **15.** 125   **17.** 10,485.76 in.

EXERCISE SET 11.5, p. 453

1. $15a^4b^2$ **3.** $-745,472a^3$ **5.** $1120x^{12}y^2$ **7.** $m^5 + 5m^4n + 10m^3n^2 + 10m^2n^3 + 5mn^4 + n^5$

9. $x^{10} - 15x^8y + 90x^6y^2 - 270x^4y^3 + 405x^2y^4 - 243y^5$ **11.** $1 - n + \dfrac{n(n-1)}{1 \cdot 2} - \dfrac{n(n-1)(n-2)}{1 \cdot 2 \cdot 3}$

$+ \cdots + \dfrac{n(n-1)(n-2)\cdots(1)}{1 \cdot 2 \cdot 3 \cdots n}(-1)^n$ **13.** $9 - 12\sqrt{3}t + 18t^2 - 4\sqrt{3}t^3 + t^4$ **15.** $-1,959,552u^5v^{10}$

17. $x^{-8} + 4x^{-4} + 6 + 4x^4 + x^8$ **19.** $140\sqrt{2}$

EXERCISE SET 11.6, pp. 457–458

1. $\dfrac{2}{7}$ **3.** 0 **5.** $\dfrac{11}{4165}$ **7.** $\dfrac{60}{143}$ **9.** $\dfrac{5}{36}$ **11.** $\dfrac{1}{36}$ **13.** $\dfrac{245}{1938}$ **15.** $\dfrac{1}{221}$ **17.** $\dfrac{9}{19}$ **19.** $\dfrac{18}{19}$ **21.** $\dfrac{1}{38}$

23. $\dfrac{9}{19}$

EXERCISE SET 11.7, pp. 461–462

1. $\dfrac{3}{2}, \dfrac{5}{2}$ **3.** $(-1, 2, -2)$ **5.** $\left(\dfrac{1}{2}, \dfrac{3}{2}\right)$ **7.** $\left(\dfrac{3}{2}, -4, 3\right)$ **9.** $(r - 2, 3 - 2r, r)$ **11.** $(-3, -2, -1, 1)$

13. 4 dimes, 30 nickels   **15.** 10 nickels, 4 dimes, 8 quarters   **17.** 5 lb of $4.05, 10 lb of $2.70

EXERCISE SET 11.8, pp. 468–469

1. $\begin{bmatrix} -2 & 7 \\ 6 & 2 \end{bmatrix}$ **3.** $\begin{bmatrix} 1 & 3 \\ 2 & 6 \end{bmatrix}$ **5.** $\begin{bmatrix} 9 & 9 \\ -3 & -3 \end{bmatrix}$ **7.** $\begin{bmatrix} 11 & 13 \\ 5 & 3 \end{bmatrix}$ **9.** $\begin{bmatrix} -4 & 3 \\ -2 & -4 \end{bmatrix}$ **11.** $\begin{bmatrix} 17 & 9 \\ -2 & 1 \end{bmatrix}$

13. $\begin{bmatrix} 0 & 0 \\ 0 & 0 \end{bmatrix}$ **15.** $\begin{bmatrix} 1 & 2 \\ 4 & 3 \end{bmatrix}$ **17.** $\begin{bmatrix} -5 & 4 & 3 \\ 5 & -9 & 4 \\ 7 & -18 & 17 \end{bmatrix}$ **19.** $\begin{bmatrix} -2 & 9 & 6 \\ -3 & 3 & 4 \\ 2 & -2 & 1 \end{bmatrix}$ **21.** $\begin{bmatrix} -16 \end{bmatrix}$

23. $\begin{bmatrix} 2 & -19 \end{bmatrix}$ **25.** $\begin{bmatrix} 3 & -2 & 4 \\ 2 & 1 & -5 \end{bmatrix} \begin{bmatrix} x \\ y \\ z \end{bmatrix} = \begin{bmatrix} 17 \\ 13 \end{bmatrix}$ **27.** $\begin{bmatrix} 1 & -1 & 2 & -4 \\ 2 & -1 & -1 & 1 \\ 1 & 4 & -3 & -1 \\ 3 & 5 & -7 & 2 \end{bmatrix} \begin{bmatrix} x \\ y \\ z \\ w \end{bmatrix} = \begin{bmatrix} 12 \\ 0 \\ 1 \\ 9 \end{bmatrix}$

29. $\begin{bmatrix} -40.19 & 37.94 & 142.24 \\ -36.78 & 16.63 & 119.62 \\ -1.659 & 14.97 & 12.65 \end{bmatrix}$

TEXT OR REVIEW—CHAPTER 11, pp. 469–470

1. (a) $f(3a) = 27a^2 - 6a$; (b) $f(a + h) = 3a^2 + 6ah + 3h^2 - 2a - 2h$; (c) $6a + 3h - 2$

2. $g^{-1}(x) = \dfrac{3x + 2}{5}$ **3.** $\dfrac{15}{4}$ **4.** 531   **5.** 11   **6.** 4   **7.** 12,144; 13,824   **8.** $\dfrac{1}{12}$ **9.** $\dfrac{3}{28}$

10. $\left(\dfrac{3}{2}, \dfrac{13}{14}, \dfrac{33}{14}\right)$ **11.** $220a^9x^3$ **12.** $x^{10} - 15x^8y + 90x^6y^2 - 270x^4y^3 + 405x^2y^4 - 243y^5$

13. $\begin{bmatrix} 0 & -1 & 6 \\ 3 & 1 & -2 \\ -2 & 1 & -2 \end{bmatrix}$ **14.** $\begin{bmatrix} -3 & 3 & 0 \\ -6 & -9 & 6 \\ 6 & 0 & -3 \end{bmatrix}$ **15.** $\begin{bmatrix} -1 & 1 & 0 \\ -2 & -3 & 2 \\ 2 & 0 & -1 \end{bmatrix}$ **16.** $\begin{bmatrix} -2 & 2 & 6 \\ 1 & -8 & 18 \\ 2 & 1 & -15 \end{bmatrix}$

17. $\begin{bmatrix} 2 & -1 & -6 \\ 1 & 5 & -2 \\ -2 & -1 & 4 \end{bmatrix}$

FINAL EXAMINATION, pp. 471–475

1. 18   **2.** $\dfrac{7}{6}$ **3.** $-3y + 17$ **4.** $-45x^6y^{-4}$ **5.** $\dfrac{2}{3}$ **6.** 64, 66, 68   **7.** $r = \dfrac{A - P}{Pt}$ **8.** $y = -x + 3$

9.

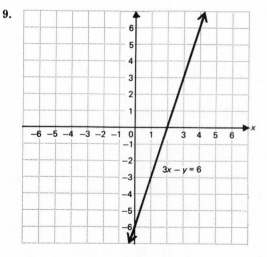

$3x - y = 6$

10. -2 **11.** $(1, -1)$ **12.** $(2, -1, 1)$ **13.** 25 and 19   **14.** 56 L of beer A and 24 L of beer B

15. $13x^3 - 7x^2 - 6x + 6$ **16.** $3a^2 - 8ab - 15b^2$ **17.** $6a^2 + 7a - 5$ **18.** $9x^4 - 30x^2y + 25y^2$

19. $(2x - 3)^2$ **20.** $(3a - 2)(9a^2 + 6a + 4)$ **21.** $(a + b + 5)(a + b - 5)$ **22.** $8, -6$

23. $x = \dfrac{c^2}{a - b}$ **24.** 20   **25.** $-\dfrac{6}{5}, 4$ **26.** $R = \dfrac{E - IR}{I}$ **27.** $\dfrac{4}{x + 2}$ **28.** $\dfrac{x - 4}{3(x + 2)}$ **29.** $\dfrac{(x + y)^2}{x^2 + y^2}$

30. $x - a$ **31.** $7x^3 + 9x^2 + 19x + 38$, R 72   **32.** 350 mph   **33.** $8\dfrac{2}{5}$ min   **34.** 6.764×10^{-12}

35. $8x^2\sqrt{y}$ **36.** $\dfrac{\sqrt[3]{5xy}}{y}$ **37.** $\dfrac{1 - 2\sqrt{x} + x}{1 - x}$ **38.** $\sqrt[6]{x + 1}$ **39.** 9   **40.** $26 - 13i$ **41.** $\pm i\sqrt{2}$

42. $-2 \pm \sqrt{7}$ **43.** $x^2 - 50 = 0$ **44.** $5, -5, 2, -2$ **45.**

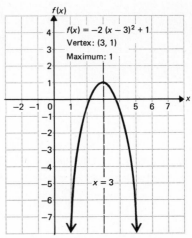

$f(x) = -2(x - 3)^2 + 1$
Vertex: (3, 1)
Maximum: 1
$x = 3$

46. 20   **47.** 1250 ft^2 **48.** $(\sqrt{5}, \sqrt{3}), (\sqrt{5}, -\sqrt{3}), (-\sqrt{5}, \sqrt{3}), (-\sqrt{5}, -\sqrt{3})$ **49.** 5 ft by 12 ft

50. Center, $(2, -3)$; radius, 6

51.

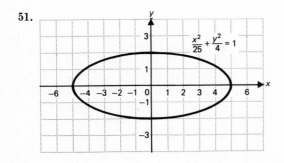

$\dfrac{x^2}{25} + \dfrac{y^2}{4} = 1$

52.

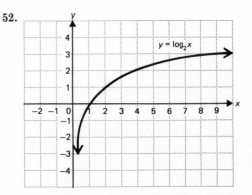

$y = \log_2 x$

53. $\log_a \dfrac{\sqrt[3]{x^2 \cdot z^5}}{\sqrt{y}}$ **54.** $a^5 = c$ **55.** 101   **56.** 1.1492   **57.** 0.006037   **58.** 1.2920   **59.** 1005

60. $\{0, 2\}$ **61.** $-2 \leqslant x \leqslant 3$

62. $y < -5$ or $y > 2$

63.

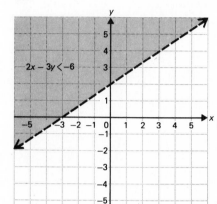

64. 10,000   **65.** $\dfrac{32}{3}$ **66.** $13,440y^4$ **67.** $\dfrac{1}{16}$

INDEX

Applied Problems
53–57 integer, percent, perimeter
87–89

good examples and exercises